世界超级工程
——中国三峡工程建设开发的实践与经验

A Mega Project in the World
—Practice and Experiences in the Construction and Development
of the Three Gorges Project in China

本书编委会 编著

中国三峡出版传媒
中国三峡出版社

2020年11月，三峡工程建设任务全面完成，工程质量满足规程规范和设计要求、总体优良，运行持续保持良好状态。至此，从1918年孙中山提出开发三峡设想，到2020年三峡工程完成整体竣工验收，中华民族实现百年三峡梦想。

三峡工程是中国开发、治理长江的关键性骨干工程，是当今世界上最大的水利枢纽工程，兼具防洪、发电、航运、补水、生态等综合效益。它对于保障中国江汉平原和洞庭湖地区1500多万人民群众生命财产安全具有十分重要的作用，对保障国家能源安全、改善电力结构、推动清洁能源建设具有十分重要的地位，对促进长江经济带经济社会生态有序发展、增强中国综合国力具有十分重要的意义。三峡工程，是唯一经全国人民代表大会审议批准，举全国之力兴建的国家重大公共工程。它规模巨大、技术复杂，功能目标和效益多元，战略地位重要，对推动经济和社会发展具有重大而深远的影响。

三峡工程建设取得的伟大成就、实现的重大创新、积累的重要经验，是三峡工程建设宝贵的精神财富。2018年4月24日，习近平总书记视察三峡工程时指出，三峡工程是国之重器，是靠劳动者的辛勤劳动自力更生创造出来的，看了以后非常振奋。三峡工程的成功建成和运转，使多少代中国人开发和利用三峡资源的梦想变为现实，成为改革开放以来我国发展的重要标志。这是我国社会主义制度能够集中力量办大事优越性的典范，是中国人民富于智慧和创造性的典范，是中华民族日益走向繁荣强盛的典范。

为了让社会大众更全面、深刻地认识三峡工程，本书从三峡工程的缘起、论证决策、建设历程、运行情况、综合效益、贡献成就、环境生态保护等方面进行了介绍，全面展示了三峡工程建设伟大成就和三峡工程产生的巨大综合效益。

In November 2020, the Three Gorges Project fully completed its construction. The project quality meets the specifications and design requirements. Thanks to its overall excellent quality, the project has been good in operation. This marked that the Chinese nation has realized the centenary dream of the Three Gorges, spanning from 1918, when Sun Yat-sen proposed the idea of developing resources at the Three Gorges, to 2020, when the Three Gorges Project finished all the required procedures and construction works for overall completion acceptance.

The Three Gorges Project is a core part for China's development and management of the Yangtze River. It is the world's largest water conservancy project, with comprehensive benefits for flood control, power generation, shipping, water replenishment and ecology. It plays an essential role in guaranteeing the safety of life and property of more than 15 million people in China's Jianghan Plain and Dongting Lake area, and is of great importance in ensuring national energy security, improving power structure and promoting clean energy development. It also means a lot for the orderly economic, social and ecological development of the Yangtze River Economic Zone and for enhancing China's overall national strength. The Three Gorges Project is the only major national public project that has been considered and approved by the National People's Congress and built with efforts from the entire country. It is huge in scale, complex in technology, diversified in functional objectives and benefits, and important in strategic position. Accordingly, it has a significant and far-reaching impact on promoting economic and social development.

The great achievements, major innovations and important experiences accumulated during the construction of the Three Gorges Project are an invaluable source of inspiration. On April 24, 2018, CPC General Secretary Xi Jinping pointed out during an inspection to the Three Gorges Dam that it is a pillar project of the country, a result by the diligence of self-reliant laborers. And he felt inspired by the visit. The successful completion and operation of the Three Gorges Project has made the dream of generations of Chinese people to develop and utilize the resources of the Three Gorges a reality. It has become an important symbol of China's development since the reform and opening up. It exemplifies the strength of China's socialist system to pool resources in a major mission, the wisdom and creativity of the Chinese people, and the growing prosperity and strength of the Chinese nation.

To give the public a more comprehensive and profound understanding of the Three Gorges Project, this book introduces the origins of the Three Gorges Project, its research and decision making, construction process, operation, comprehensive benefits, contribution and achievements, and environmental and ecological protection. The book presents a big picture of the project's great achievements and comprehensive benefits.

第一章　三峡工程的缘起
Chapter 1　Origin of the Three Gorges Project

第一节　长江洪水是中华民族的心腹大患 002
Section 1　Yangtze Floods: A Big Threat to the Chinese Nation 003

第二节　三峡工程的初步构想 020
Section 2　Preliminary Conception of the Three Gorges Project 021

第二章　三峡工程的研究论证
Chapter 2　Research and Feasibility of the Three Gorges Project

第一节　中华人民共和国成立后三峡工程的研究论证 038
Section 1　Research and Feasibility of the Three Gorges Project After the
　　　　　 Founding of the PRC 039

第二节　研究论证的主要内容 052
Section 2　Main Content of Research and Feasibility 053

第三节　科学论证形成决策 060
Section 3　Decision-making Based on Scientific Feasibility Work 061

第三章　三峡枢纽工程的建设
Chapter 3　Construction of the Three Gorges Project

第一节　三峡枢纽工程的构成 076
Section 1　Composition of the Three Gorges Project 077

第二节　三峡枢纽工程的建设过程 092
Section 2　Construction Process of the Three Gorges Project 093

第四章　三峡输变电工程的建设
Chapter 4　Construction of Three Gorges Power Transmission and Transformation Works

第一节　三峡输变电工程的综述 106
Section 1　Overview of the TGPTTW 107

第二节　三峡输变电工程的建设过程 116
Section 2　Construction Process of the TGPTTW 117

第五章　三峡移民工程的实施
Chapter 5　Implementation of the Three Gorges Resettlement

第一节　三峡移民工程的难点 130
Section 1　Difficulties of Three Gorges Resettlement 131

第二节　三峡移民工程的政策体系 136
Section 2　Policy System of the Three Gorges Resettlement 137

第三节　三峡移民工程的实施和成效 152
Section 3　Implementation and Results of the Three Gorges Resettlement 153

第六章　三峡工程的创新
Chapter 6　Innovations of the Three Gorges Project

第一节　三峡工程的建设管理创新 172
Section 1　Construction and Management Innovations of the TGP 173

第二节　三峡工程的科技创新 200
Section 2　Technological Innovations of the TGP 201

第七章 三峡工程的运行效益
Chapter 7 Operational Benefits of the TGP

第一节 三峡工程的防洪效益 226
Section 1　Benefits of Flood Control of the TGP 227

第二节 三峡工程的发电效益 238
Section 2　Power Generation Benefits of the TGP 239

第三节 三峡工程的通航效益 246
Section 3　Shipping Benefits of the TGP 247

第四节 三峡工程的水资源利用（配置）效益 254
Section 4　Water Resources Utilization (Allocation) Benefits of the TGP 255

第五节 三峡工程的运行管理与效益发挥 260
Section 5　Operation, Management and Benefits of the TGP 261

第八章 三峡工程的生态保护
Chapter 8 Ecological Conservation in the Three Gorges Project

第一节 三峡工程生态保护的论证和规划 274
Section 1　Appraisal and Planning of Ecological Conservation for the
　　　　　 Three Gorges Project 275

第二节 三峡工程的生态保护机制、措施和实施 284
Section 2　Ecological Conservation Mechanism and Measures for the
　　　　　 Three Gorges Project and Their Implementation 285

第三节 三峡工程的生态保护成效 306
Section 3　Effectiveness of Ecological Conservation in the Three Gorges Project ... 307

第九章 三峡工程的文物和自然人文景观保护
Chapter 9 Cultural Relics and Natural and Cultural Landscapes of the Three Gorges Project

第一节 三峡工程的文物保护规划 322
Section 1　Preservation Plan for Cultural Relics in the Three Gorges Project 323

第二节 三峡工程文物保护的成效 330
Section 2　Effectiveness of Cultural Heritage Protection in the Three
　　　　　 Gorges Project 331

第三节　三峡工程的文化保护规划及成效 340
Section 3　Planning for and Effectiveness of Cultural Preservation in the Three Gorges Project 341

第十章　三峡工程的国际合作和国际影响
Chapter 10　International Cooperation and the Global Influence of the Three Gorges Project

第一节　三峡工程中的国际合作 354
Section 1　International Cooperation in the Three Gorges Project 355

第二节　三峡工程的国际影响 366
Section 2　The International Influence of the Three Gorges Project 367

阅读提示：

长江是中华民族的母亲河。长江流域自然资源丰富，人口稠密，经济发达，自11世纪以来，一直是中国经济的中心地带。长江流域幅员辽阔，地理气候条件复杂，千百年来，长江尤其是中下游地区频繁发生严重洪水灾害，一直是中华民族的心腹大患。

"水利者，兴利除弊也。"中华民族5 000年的文明史，也是一部治水史。兴水利，除水害，历来被视为治国安邦的大事。兴建三峡工程是中华民族治水兴邦的必然选择。正确认识和深刻理解三峡工程，需要从更深远的历史坐标和更宽广的时代背景，以及自然环境的演变和人类社会的进步中，进行深入思考和系统回顾。

The Yangtze River is known as the mother river by all the Chinese people. Endowed with rich natural resources, the Yangtze River basin is densely populated and economically developed and has been the center of China's economy since the 11th century. As a large basin with complex geology and climate, it has suffered frequent severe floods, especially in its middle and lower reaches, long serving as a significant threat to the Chinese nation.

Water conservancy projects are aimed to maximize the benefits of water resources and minimize their detriments. Five thousand years of Chinese civilization is a history of water control. Water conservation and flood control are issues impacting national governance. Building the Three Gorges Project was an inevitable choice for the Chinese nation to harness water and allow the country to prosper. For a deeper understanding of the Three Gorges Project, in-depth thinking and a systematic review are required. Furthermore, an understanding of the natural environment and the progress of human society is needed.

Chapter 1 >>>>

三峡工程的缘起
Origin of the Three Gorges Project

第一节 长江洪水是中华民族的心腹大患

长江是孕育中华文明的母亲河。千百年来，以她生生不息的律动，带给两岸无尽的福泽与蓬勃生机。但同时，桀骜不驯的江水也给她滋养的生灵带来一次次灾难深重的洪患梦魇。驾驭洪魔、治水兴邦，成为沿江人民的千年企盼和不懈追求。

一、长江流域地形地貌特点

1. 长江概貌

长江是中国第一大河，就河长而言论，是世界第三大河。它发源于青藏高原唐古拉山脉主峰各拉丹冬雪山西南侧，全长6 300余千米，总落差5 400米左右，流域形状呈东西长南北短的狭长形。

长江流域面积约为180万平方千米，约占中国陆地总面积的18.75%。流经中国西南、华中、华东三大经济区，涉及19个省（自治区、直辖市）。其中，干流流经青海、西藏、云南、四川、重庆、湖北、湖南、江西、安徽、江苏、上海等11个省（自治区、直辖市），在上海汇入东海。支流布及甘肃、陕西、河南、贵州、广西、广东、福建、浙江等8个省（区）。

长江流域拥有丰富的水资源。多年平均径流量约9 600亿立方米，约占中国总水量的36%，相当于20条黄河。

长江流域的地势西高东低。流域内地势的最高峰位于四川西部贡嘎山，高程7 556米，最低为上海的吴淞零点。全流域平均高程约为1 650米。

Section 1 Yangtze Floods: A Big Threat to the Chinese Nation

The Yangtze River is the mother river that feeds Chinese civilization. For thousands of years, it has nourished the survival and development of the nation with its continuous supply of rich resources. However, the Chinese relationship with the river has been uneasy. Devastating floods are a recurring nightmare for people living there, and locals have been working relentlessly to prevent and control floods for thousands of years.

I. Topography of the Yangtze River Basin

1. A general picture of the Yangtze River

The Yangtze River is the longest river in China and the third longest in the world. It originates from the southwest side of Geladaindong Peak in the Tanggula Mountains of the Qinghai-Tibet Plateau, with a total length of more than 6,300 km and a gross head of appropriately 5,400 m, draining an area extending approximately from west to east China.

The Yangtze River has a drainage area of 1.8 million km^2, occupying 18.75% of China's land area. It flows through China's three major economic regions, i.e., Southwest China, Central China and East China, stretching over 19 provinces (including autonomous regions and municipalities directly under the Central Government). Its mainstem flows through 11 provinces (including autonomous regions and municipalities directly under the Central Government) – Qinghai, Tibet, Yunnan, Sichuan, Chongqing, Hubei, Hunan, Jiangxi, Anhui, Jiangsu and Shanghai – before finally pouring into the East China Sea here. Its tributaries stretch over eight provinces (regions): Gansu, Shaanxi, Henan, Guizhou, Guangxi, Guangdong, Fujian and Zhejiang.

The Yangtze River basin is rich in water resources. The mean annual runoff is about 960 billion m^3, accounting for 36% of the total runoff in China, 20 times that of the Yellow River.

The terrain of the Yangtze River basin is high in the west and low in the east. The high-

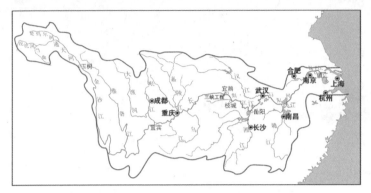

图 1.1　长江流域示意图

流域内的地貌类型众多，有山地、丘陵、盆地、高原和平原，山地、高原和丘陵约占 84.7%。其中，高山和高原主要分布于西部地区，中部地区以中山为主，低山多见于淮阳山地及江南丘陵地区，丘陵主要分布于川中、陕南及湘西、湘东、赣东、皖南等地。

长江流域森林主要分布在上游西部高原山地，其次在中下游的湘西、鄂西、皖南和江西等山地。全流域森林面积约为 7.16 亿亩，如加上灌木等，覆盖率达 27.4%。长江流域的地下矿产资源也比较丰富，各类矿种齐全。

2. 长江的上、中、下游

长江与黄河在名称上有着显著不同。黄河从源头到入海口通称为黄河；而长江在这一总名称下，干流有些江段又有它自己的名称：沱沱河、通天河、金沙江、川江、荆江、浔阳江、扬子江等，都是长江干流的一部分。

长江源头至湖北省宜昌市三峡出口的南津关为长江上游，长 4 512 千米，占总长度的 70.53%；流域面积约 100 万平方千米。自北向南流淌的金沙江，到云南省丽江市石鼓镇后，拐了一个小于 90 度的大弯，掉头流向东方；石鼓镇以上的金沙江，与澜沧江、怒江"三江并流"，形成独特的令人神往的自然风光。石鼓镇以下的金沙江，有闻名于世的虎跳峡，有绵延千里的金沙江大峡谷。

宜宾以上干流大多属峡谷河段，长 3 479 千米，落差约 5 100 米，约占干流总落差的 94%，蕴藏着巨大的水力资源；汇入的主要支流有北岸的雅砻江。宜宾至宜昌段长 1 033 千米，沿江山地、丘陵与阶地互间，汇入的主要支流，

est peak in its drainage area is Mount Gongga in western Sichuan, which reaches 7,556 m above sea level, and the lowest point is the Wusong Zero Datum in Shanghai. The average elevation of the whole basin is about 1,650 m.

Figure 1.1 Sketch Map of Yangtze River Basin

The Yangtze River basin enjoys a multi-terraced terrain since the river flows through mountains, hills, basins, plateaus and plains. The mountains, plateaus and hills account for approximately 84.7%, of which the high mountains and plateaus are mainly distributed in the western region, while mountains of moderate size and height dominate the central region. The low mountains are primarily found in the mountain areas in Huaiyang District and the hilly regions south of the Yangtze River. The hills are mainly distributed in central Sichuan, southern Shanxi, western and eastern Hunan, eastern Jiangxi and southern Anhui.

The forests in the Yangtze River basin are mainly distributed in the western plateaus and mountains in the upper reaches, followed by the mountains in western Hunan, western Hubei, southern Anhui and Jiangxi in the middle and lower reaches. The forest area of the whole basin is about 716 million *mu* (1 hm^2=15 *mu*), with a forest coverage rate of 27.4% if shrubs are included. There are also rich underground mineral resources, with a complete range of minerals.

2. The upper, middle and lower Yangtze River

The Yangtze River is quite different from the Yellow River's naming. The whole course of the Yellow River is called so from its source to the estuary, while the Yangtze River is known by different names over its course. For example, the Tuotuo River, the Tongtian River, the Jinsha River, the Chuanjiang River, the Jingjiang River, the Xunyang River and the Yangtze River are parts of its mainstem.

The upper Yangtze is the section from its source to the Nanjin pass, the natural exit of the Three Gorges in Hubei Province's Yichang City. This section is 4,512 km long, occupying 70.53% of the river's total length and drains an area of approximately 1 million km^2. The Jinsha River, which flows from north to south, does a U-turn at Shigu Town of Lijiang City in Yunnan Province and then runs eastward; the Jinsha River section above Shigu Town, the Lancang River and the Nujiang River form a unique and fascinating natural scenery of "Three Parallel Rivers". The section of Jinsha River below Shigu Town is world-renowned for the Tiger Leaping Gorge, a magnificent gorge that extends for thousands of *li*.

The mainstem of the Yangtze River above Yibin mostly runs through gorges, with a length of 3,479 km and a head of 5,100 m, taking approximately 94% of the gross head of the mainstem and containing enormous hydraulic resources. The main tributary flowing into this

图1.2 航拍穿梭于崇山峻岭之间的长江（摄影：王连生）

北岸有岷江、嘉陵江，南岸有乌江，奉节以下为雄伟秀丽的三峡河段，两岸悬崖峭壁，江面狭窄。

南津关至江西省鄱阳湖湖口为长江中游，长955千米，占总长度的14.93%；流域面积约68万平方千米。南津关至枝城为丘陵区，枝城以下进入平原区，干流河道坡降变小、水流平缓，河道与众多大小湖泊相连。汇入的主要支流，北岸有汉江；南岸有清江，洞庭湖水系的湘、资、沅、澧四水，鄱阳湖水系的赣、抚、信、饶、修五河。自枝城至城陵矶河段为著名的荆江，"万里长江，险在荆江"，两岸平原广阔，地势低洼，泥沙淤积使荆江成为地上悬河，全靠沿江两岸堤防保护江汉平原和洞庭湖区1500万人口、2300万亩耕地的安全。南岸有松滋、太平、藕池、调弦四口分流入洞庭湖，由洞庭湖汇集湘、资、沅、澧四水后，在城陵矶注入长江，江湖关系最为复杂。城陵矶以下至湖口，主要为宽窄相间的藕节状分汊河道，总体河势比较稳定。

鄱阳湖湖口至长江入海口为长江下游，长930千米，占总长度的14.54%；流域面积约12万平方千米。湖口以下干流沿岸有堤防保护，汇入的主要支流有南岸的青弋江、水阳江水系、太湖水系和北岸的巢湖水系，淮河部分水量通过

section is the Yalong River from the north bank. The Yibin to Yichang section runs 1,033 km through mountains, hills and terraces. The main tributaries flowing into this section are the Minjiang River and Jialing River from the north bank and the Wujiang River from the south bank. Below Fengjie are the majestic and narrow Three Gorges, with steep cliffs facing each other along the banks.

⊂ Figure 1.2　Aerial Photo of the Yangtze River Flowing Through Mountains (Photographed by Wang Liansheng)

The middle Yangtze is the section from the Nanjin Pass to the outlet of Poyang Lake in Jiangxi Province. This section is 955 km long, occupying 14.93% of the river's total length, and drains an area of approximately 680,000 km^2. It flows through a hilly area from the Nanjin Pass to Zhicheng and then to the plains below Zhicheng, where it slopes gently, and numerous lakes join it. Main tributaries flowing into it are the Hanjiang River from the north bank; Qingjiang River from the south bank; Xiangjiang River, Zi River, Yuan River and Li River that feed Dongting Lake; and the Gan River, Fu River, Xin River, Rao River and Xiu River that feed Poyang Lake. The section from Zhicheng to Chenglingji is the famous Jingjiang River. As a proverb states, "The Yangtze River runs for thousands of miles, and the Jingjiang section is the most dangerous." Located on a low-lying plain, the Jingjiang River is an earth-suspended river due to sediment accumulation. Levees were built to protect the safety of the 15 million people and 23 million *mu* (1 hm^2 = 15 *mu*) of farmland in the Jianghan Plain and Dongting Lake area. Tributaries Songzi, Taiping, Ouchi and Tiaoxian run into Dongting Lake from the south bank. After reaching Dongting Lake, the four rivers of Xiangjiang, Zi, Yuan and Feng flow into the Yangtze in Chenglingji. This section of the Yangtze features the most complicated river-lake relationship. The section from Chenglingji to the lake outlet features braided channels like lotus rhizomes that vary in width, and the channel bars are relatively stable.

The lower Yangtze is the section from the outlet of Poyang Lake to the river's estuary. This section is 930 km long, occupying 14.54% of the river's total length, and drains an area of approximately 120,000 km^2. Levees protect banks below the lake outlet. Main tributaries flowing into this section are the Qingyi River, Shuiyang River and Taihu Lake systems from the south bank and the Chaohu Lake system from the north bank, and some runoff of the Huaihe River also flows into the Yangtze River through the Huaihe River channel. Lower reaches are deep and wide, with a small water level range and high navigation capacity. Affected by the tides, the 600-km section below Datong often suffers severe bank failure; the Yangtze River Delta plain at the estuary is connected with a wide continental shelf in the east, which was formed by the sedimentation brought by the Yangtze River over time.

淮河入江水道汇入长江。下游河段水深江阔，水位变幅小，通航能力大，大通以下约 600 千米河段受潮汐影响，是坍岸较为严重江段；入海口的长江三角洲平原与东面宽广的大陆架相连，是长期以来长江带来的泥沙淤积而形成的。

3. 长江的三峡

长江三峡是瞿塘峡、巫峡、西陵峡的总称，上起重庆市奉节县白帝城，下至湖北省宜昌市南津关，全长 192 千米。三峡雄奇壮美，幽深秀丽，奇峰竞秀，千姿百态，令人目不暇接，是中国十大自然风景区之一，也是世界上著名的山水画廊。三峡大坝坝址位于西陵峡中段三斗坪，三峡工程因此而得名。

（1）瞿塘峡

瞿塘峡又叫夔峡，上起奉节县白帝城，下至巫山县大溪镇，长约 8 千米。在三个峡中，它虽然最短，却最为雄伟险峻。夔门右侧的高山叫"白盐山"，左侧的高山叫"赤甲山"，高耸入云，拔地而起，临江岩壁如刀削斧砍，恰似一座顶天立地的大门。这里江面宽仅有百余米，把滔滔大江紧束得像一条玉带。

（2）巫峡

巫峡上起重庆市巫山县大宁河口，下至湖北省巴东县官渡口，长 45 千米，是三峡中峡谷连绵最长、最整齐的一个峡，以幽深秀丽而闻名。巫峡的"巫山 12 峰"的峰顶海拔高程都在 1 000～1 500 米，形成了美丽的峰林地貌，乘

图 1.3　雄伟的瞿塘峡（摄影：黄正平）

3. The Three Gorges

The Three Gorges comprise the Qutang, Wu and Xiling gorges from west to east and span 192 km beginning at Baidi City of Fengjie, Chongqing Municipality, in the west and ending at Nanjin Pass, at Yichang City, Hubei Province, in the east. The Three Gorges are deep and magnificent, flanked by towering peaks. They are one of China's top ten natural scenic spots and a world-famous natural landscape gallery. The various names of the Three Gorges are due to their locations in Sandouping, the middle section of Xiling Gorge.

(1) Qutang Gorge

The Qutang Gorge, also known as the Kui Gorge, spans about 8 km from Baidi City of Fengjie to Daxi Town of Wushan County. It is the shortest of the Three Gorges but is also the narrowest and most dangerous. The Yangtze River passes between the Baiyan Mountain on the right and the Chijia Mountain on the north, and the point where the river passes between the mountains is called the Kuimen Gate. The cliffs on both sides appear to have been cut by knives or axes, towering like a gate. Though only about 100 m wide, it is like a jade belt confining the surging river.

(2) Wu Gorge

The Wu Gorge stretches 45 km from the Daning River mouth in Wushan County of Chongqing Municipality to Guandukou in Badong County of Hubei Province. Flanked by strange continuous peaks, the Wu Gorge is the longest and neatest among the Three Gorges and presents a panorama of lovely scenery. All of the Twelve Peaks along the banks of the Wu Gorge have a sea-level elevation of between 1,000 and 1,500 km. Traveling between the karst cliffs in a boat, one may feel as if sailing through a cave as the peaks seemingly connect with the sky. The world-famous Goddess Peak stands out in the clouds by the river. Beside the peak is a stone column like a graceful young girl, hence its name. The Goddess Peak is 922 m above sea level and 175 m above river level. So high and steep is the peak that people need to raise their necks upward to see it. As recorded in the *Shui Jing Zhu* (*Commentary on the Water Classic*) by Li Daoyuan, folk poetic lines in the Northern Wei Dynasty read, "Among Three Gorges of Badong County, Wu Gorge is the longest. One cannot help but shed tears once they hear the ape eek."

Figure 1.3　Imposing Qutang Gorge (Photographed by Huang Zhengping)

图 1.4 巫山神女峰

船进入峡谷中,给人以"峰与天关接,舟从地窟行"的感觉。举世闻名的神女峰,上入云端,下临大江,山峰旁有一形如古代少女的石柱,亭亭玉立,秀丽婀娜,那就是神女。神女的海拔高程为 922 米,而江面最高高程为 175 米,游客都要引颈仰视,才能一睹神女的风采。"巴东三峡巫峡长,猿鸣三声泪沾裳。"早在中国北魏时期民间就流传着这样的歌谣,被郦道元记录在《水经注》中。

(3) 西陵峡

西陵峡上起秭归县香溪河口,下至宜昌市南津关,长 66 千米。民间素有"瞿塘雄、巫峡幽、西陵险"的传说。葛洲坝工程蓄水前,西陵峡以风景秀丽、滩多水急而著称;葛洲坝工程蓄水后,民间改称为"瞿塘雄、巫峡幽、西陵秀"。西陵峡临江峡谷的山峰峰顶海拔高程较低,约为 500 米,使得西陵峡的气势没有瞿塘峡那样雄伟、巫峡那样幽深,但由于莲沱以下全是石灰岩,其风景在三个峡中是最为秀美的。两岸峭壁上镶珠嵌玉般的 174 个溶洞,以及仙人桥、笔架山,灯影峡峰顶上惟妙惟肖的唐僧师徒 4 人的天然石像,都给游人留下深刻印象。

全长 192 千米的长江三峡,只有西陵峡中段庙河至莲沱的 31 千米是花岗岩地质,致密、坚硬、完整且不透水;三峡工程的三斗坪坝址就位于这一江段,是适于建设混凝土高坝的好坝址。国际著名工程地质学权威、隧洞及地下洞室开挖"新奥法"创始人——奥地利的穆勒(Leopold Müller)教授,在 1986 年 5 月勘察三斗坪坝址时,赞叹地说:"这真是一个难得的好坝址,三峡坝址是上帝赐给中国人的一个好坝址。"

⊃ 图 1.5 傍晚的西陵峡

Figure 1.4 Goddess Peak of Wushan Mountain

(3) Xiling Gorge

The 66-km-long Xiling Gorge begins from the Xiangxi River mouth in Zigui County and ends at Nanjin Pass in Yichang City. Local people have long said, "Among the Three Gorges, Qutang is the most imposing, Wu the most secluded, and Xiling the most dangerous." Before the impoundment of Gezhouba Dam, the Xiling Gorge was famous for its stunning scenery, numerous shoals and turbulent rapids. After its impoundment, the Three Gorges began to be described as "Qutang is the most imposing, Wu the most secluded, and Xiling the most beautiful." The peak elevation of the valleys in the Xiling Gorge is relatively low, about 500 m high. Therefore, Xiling is not as imposing as Qutang and not as secluded as Wu. However, the limestone below the Liantuo Formation makes it the most beautiful among the Three Gorges. The gorge features some unusual and impressive sights, such as the 174 karst caves as if pearls and jade embedded in the cliffs, as well as the Xianren Bridge, the Bijia Mountain, and the four strange rocks that finely resemble the shapes of the four well-known figures - Monk Xuanzang (San Zang) and his three disciples, the Monkey King (Sun Wukong), Pig Zhu Bajie and Friar Sand (Sha Wujing) - in *Journey to the West*.

In the 192-km-long Three Gorges, only the 31-km section from Miaohe to Liantuo in the middle part of Xiling Gorge is granite, which is solid and dense, with complete rock mass, and almost impervious. The Sandouping dam is located here, an ideal site for building high concrete gravity dams. Internationally renowned Austrian geologist Leopold Müller, founder of the "New Austrian tunneling method" (NATM) for excavating tunnels and underground cavities, once praised that "The Three Gorges Dam site is a good dam site given by God to Chinese."

Figure 1.5 Xiling Gorge Before Sunset

二、长江流域的洪水灾害

1. 洪水成因及几次较大规模的洪灾

长江是中华民族的母亲河,但它同样是一条给两岸人民带来灾难的河流。长江流域幅员辽阔,地理条件复杂,自然灾害多种多样,而其中最为严重的是洪水灾害。

长江是条雨洪河流,造成洪水的主要原因是暴雨形成的洪峰超过河床的安全泄量。因此,流域可能发生洪灾的地区分布很广,在上游及支流的山区,有山洪及暴雨触发的滑坡、山崩、泥石流等洪水灾害;在上游主要支流的中下游沿岸及长江中下游冲积平原区,则有洪水泛滥或堤防溃决造成的洪水灾害;在河口地区,还有因台风、风暴潮造成的洪水灾害。

受长江洪灾危害最大的地区在中下游,尤其是荆江河段。但其根源却在上游。据测算,宜昌地区以上的长江平均流量,占荆江洪水流量的95%以上,占城陵矶洪水流量的70%,占汉口洪水流量的66%。也就是说,只要宜昌发生特大洪水,就会对中下游地区造成严重影响。

有历史文献记载的长江流域的洪水灾害,最早见于汉代高后三年(公元前185年),至清朝末年(1911年)的2 096年中,有记载的较大洪水灾害多达214余次,平均约10年发生1次。清末到民国时期发生的较大洪水灾害,约4年发生1次。

每次洪灾造成的损失都是非常严重的,特别是长江中下游平原地区受灾最为严重。如1788年洪水,中游的荆江大堤沙市以上溃决达22处之多,使荆州城被淹,大量人口死亡。在19世纪中叶的1860年(清咸丰十年)和1870年(清同治九年)发生了两次特大洪水,这两次洪水先后冲开了荆江南岸堤防,形成了藕池河、松滋河,造成两湖平原一片汪洋,受淹面积4 500万亩,损失惨重。

20世纪以来,长江流域又发生了1931年、1935年、1954年及1998年等年份的大洪水。其中,1931年洪灾受灾面积达13万平方公里,淹没农田339.33万公顷,被淹房屋180万间,受灾民众2 855万人,被淹死亡者达14.5万人;1935年洪灾,湖北、湖南、江西、安徽、江苏、浙江六省份均受灾,

II. Yangtze River Floods

1. Causes and several major floods

While feeding the Chinese nation, the Yangtze River is a source of hardship for those living along it. The Yangtze River drains a vast area of complex geographical conditions and diverse natural disasters, among which the most devastating are floods.

The Yangtze River is prone to floods during the rainy season due to the flood peak formed by rainstorms exceeding the safe discharge of the riverbed. Therefore, the areas of potential flooding in the river basin are widely spread. The mountainous regions of the upper reaches and tributaries suffer landslides and mudslides caused by mountain torrents and rainstorms; the middle and lower reaches of the main tributaries of the upper reaches and the alluvial plains area of the middle and lower reaches of the Yangtze River suffer flood damage caused by flooding or levee failure. The estuary area suffers floods caused by typhoons and storm surges.

Floods hit the middle and lower reaches the hardest, especially the Jingjiang River section. But the source comes from the upper reaches. It is estimated that the average flow of the Yangtze River above Yichang accounts for more than 95% of the flood discharge of the Jingjiang River, 70% of Chenglingji and 66% of Hankou. That is to say, the occurrence of any significant flood in Yichang is bound to have a severe impact on the middle and lower reaches of the Yangtze River.

Earliest historical records of flood disasters in the Yangtze River basin date back to the third year of Empress Gao's reign in the Han Dynasty (185 BC). Since then, during the subsequent 2,096 years until the last year of the Qing Dynasty (1911), there were more than 214 major flood disasters on record, an average of one every ten years. Major flood disasters from the end of the Qing Dynasty to the Republic of China period occurred about once every four years.

Each flood brought immense loss, especially in the plains in the middle and lower reaches of the Yangtze River. Taking the 1788 Yangtze flood, for example, 22 levee failures above Shashi at the Jingjiang Dyke in the middle Yangtze submerged Jingzhou City and took numerous lives. The year 1860 (the tenth year of Xianfeng Emperor's reign in the Qing Dynasty) and 1870 (the ninth year of Tongzhi Emperor's reign in the Qing Dynasty) witnessed two major floods, which broke the levees on the south bank of the Jingjiang River to form the Ouchi River and the Songzi River, and put the Jianghan and Dongting Lake Plain under a broad expanse of water and flooded an area of 45 million *mu* (1 hm^2 = 15 *mu*).

Since the 20th century, the Yangtze River basin has suffered several catastrophic floods:

图1.6 基座上1870年洪水题刻，高程150.35米是水利部长江水利委员会近年实测数据

淹没农田150.87万公顷，受灾人口1 000万人，被淹死亡者14.2万人；1954年洪灾，虽然荆江修建了分洪工程，采取了临时分洪措施和防洪抢险，保住了重点堤防和重要城市安全，但中下游地区仍遭受严重损失，淹没农田317万公顷，受灾人口1 800多万人，受灾县市123个，京广铁路不能通车达100多天；1998年全流域性洪水，中国政府动用大量人力、物力，进行了近3个月的抗洪抢险，各地调用130多亿元的抢险物资，高峰期有670万群众和数十万军队参加抗洪抢险，但仍有重大的损失，倒塌房屋212.85万间。长江流域洪水灾害的发生给整个国民经济和社会发展带来十分严重的影响。

管子说："五害之属，水最为大。"长江洪水就像悬在头上的一把利剑，年年岁岁威胁困扰着两岸人民的安居乐业。要减轻和消除长江洪水危害，就必须加强水利工程的运用，通过修建坝、堤、溢洪道、水闸、渠道、渡漕等不同类型的水工建筑物，应对缓解水资源时空分布不均带来的一系列问题，趋利避害、治水兴利，变洪水之害为水力之利。

2. 历史上的长江洪水治理

中国古代社会以农业耕作为经济基础，农业、商业及水运都离不开水。兴水利，除水害，历来被视为治国安邦的大事。历代君王都非常重视治水，素有"治国必先治水"之说。从某种意义上说，中华民族5 000多年的文明史和发展史也是一部治水史。

⊂ Figure 1.6   Hydrological inscription of the 1870 Yangtze flood on the footstone, the elevation 150.35 m was actually measured by the then Changjiang Water Resources Commission of the Ministry of Water Resources of the People's Republic of China

- 1931: 130,000 km² were flooded, impacting 3.3933 million hectares of farmland, 1.8 million homes and 28.55 million people, with 145,000 deaths.
- 1935: 1.5087 million hectares of farmland were flooded, impacting 10 million people in Hubei, Hunan, Jiangxi, Anhui, Jiangsu and Zhejiang provinces, with 142,000 deaths.
- 1954: Although the Jingjiang Flood Diversion Project was built, and temporary flood diversion and control and rescue measures were taken to maintain the safety of critical levees and major cities, the middle and lower reaches suffered severe losses. It flooded 3.17 million hectares of farmland, affecting more than 18 million people from 123 counties and cities, and the Beijing-Guangzhou Railway was not open to traffic for more than 100 days.
- 1998: The whole river basin was hit. All of China participated in the fight against the flood for nearly three months. Emergency rescue supplies of more than 13 billion yuan in various localities were deployed, and 6.7 million people and hundreds of thousands of soldiers actively participated. A total of 2.1285 million homes were damaged.

Floods in the Yangtze River basin have seriously impacted China's national economy and social development.

As ancient Chinese philosopher Guan Zhong (723-645 BC) once said, "Floods are the most devastating among the five major natural disasters." The Yangtze floods are like a sharp sword hanging above the head, posing a threat to the safety of local people year after year. Reducing and eliminating flood damage from the Yangtze River requires strengthening water conservation projects. Different hydraulic structures such as dams, levees, spillways, sluices, channels and aqueducts are needed to cope with a series of problems caused by the uneven spatial and temporal distribution of water resources and turn flood hazards into hydraulic benefits.

2. Historical practices of flood control in the Yangtze River

Ancient China's economy was based on agriculture, commerce and water transportation. The development of water conservancy, in addition to water damage, has always been regarded as a significant issue in the governance of the country and the state. Successive dynasties had attached great importance to water control. It has been said that "Before governing the country, China must control the floods first." In a sense, China's 5,000-year history is also a history of water management.

中国古代水问题主要是洪水与干旱，从殷周到民国，有史料记载的5 258次自然灾害，一大半是水旱灾害。古代治水围绕兴利和除害展开，经历了一个由"避"到"治"、由"堵"到"疏"、由"疏"到"导"的过程，同时也是一个由防水害到既防水害又兴水利、由被动治水到主动治水、由单一治水到全面治水的过程。

万里长江自古以来就是中国经济发达、人民富庶的理想栖息之地，拥有丰富的自然资源与优美的自然风光。然而，长江尤其是中下游地区频繁发生严重洪水灾害，一直是中华民族的心腹大患，也是中国历朝历代治水兴利、治水安邦的重点。

早在新石器时期，长江下游的马家浜和良渚文化层中，就发现了一批4 700年至7 000年前的水稻遗存。在湖州市邱城遗址下还发现了距今4 700年的9条排水沟和两条宽1.5米至2米的引水渠。这说明，早在新石器时代，长江流域的先民就开始了筑堤、开沟、引灌的历史。

到春秋战国时期，随着铁质工具的出现及战争对运输的需要，农田水利和航运都有了较快发展。尤其是吴国和楚国，相继开凿运河。公元前256年，

图1.7　现代仍在发挥作用的都江堰

Water-related natural disasters in ancient China were mainly floods and droughts. From the Yin and Zhou dynasties (c.1300-256 BC) to the Republic of China period (1912-1949), floods and droughts took the better part of the 5,258 natural disasters on record. Ancient water control practices mainly focused on water conservancy and flood control, which shifted from "avoiding" to "controlling", from "blocking" to "dredging", and from "dredging" to "channeling". It was also a process of preventing water hazards to both preventing water hazards and promoting water conservancy, from passive to active water control and from single to comprehensive water control.

Since ancient times, the Yangtze River basin has been an ideal habitat for China with a developed economy, abundant natural resources and beautiful natural scenery. However, the frequent occurrence of devastating floods in the Yangtze River, especially in its middle and lower reaches, has always been a major concern of the Chinese nation and the focus of water control and national governance in all its dynasties.

A batch of rice remains dating back 4,700 to 7,000 years in the Neolithic period were found in the Majiabang and Liangzhu cultures in the lower reaches of the Yangtze River. Under the Qiucheng Site in Huzhou City, nine drainage ditches and two diversion canals with a width of 1.5-2 m dating back 4,700 years were also found. All these indicate that the ancients living in the Yangtze River basin began building levees and opening ditches for irrigation early in the Neolithic age.

In the Spring and Autumn period and the Warring States period (770-221 BC), with the emergence of iron tools and the need for transportation in war, irrigation, water conservancy and shipping saw rapid development. In particular, the states of Wu and Chu successively dug canals. In 256 BC, Li Bing, governor of Shu during the Qin state, directed the construction of Dujiangyan, which turned the Chengdu Plain into a land of abundance where "People were able to control drought and flood disasters, and hunger has disappeared." Today, it still plays a significant role in irrigation.

⊂ Figure 1.7 Dujiangyan Irrigation System Still in Service Today

秦国的蜀郡守李冰主持修建的都江堰，更使成都平原成为"水旱从人，不知饥馑"的天府之国，至今仍发挥着巨大的灌溉效益。

秦汉时期，长江流域纳入了统一的国家体系，水利也有了长足的进步。如秦始皇时期史禄开凿的灵渠，汉武帝时期张汤父子开凿的褒斜道，召信臣和杜预对南阳地区的先后开发，促进了该地区当时的农业发展。

三国两晋南北朝时期，国家分裂，战乱频繁，人口大量南移，中国的经济重心亦随之南移，江南水利遂渐次开发，尤其以南京附近的太湖流域为重。东晋时期，桓温令陈遵修筑的万城堤，成为荆江大堤的雏形。

隋唐时期，中国重新统一，逐步形成政治中心在北而经济重心于南的局面，沟通南北的大运河成为维系这两个中心的纽带。太湖流域成为全国的粮仓，但也导致了开垦过量和人水矛盾的加剧。

宋元明清时期，经济中心完全移到长江以南，流域水利日趋发达，堤防体系形成，以及太湖流域持续开发是这一阶段的显著特点。在明清时期，荆江大堤、武汉市堤、黄广大堤、无为大堤已经联为整体，江浙海塘已形成完整体系。除太湖外，洞庭湖、鄱阳湖和江汉湖群也得到了开发，成为全国粮税收入的重要地区。湖区的圩堤也越建越多，导致洪水宣泄不畅，洪灾损失愈来愈重。

"为政之要，其枢在水。"纵观中国历史，举凡善治国者均以治水为重，善为国者必先除水旱之害。从上古时代的大禹，到秦皇汉武、唐宗宋祖，每一位试图有所作为的统治者都把治水作为富民安邦的重要手段。可以说，水利兴而天下定，水利兴而百业旺，水利兴而人心稳，水利兴而百姓富。

During the Qin and Han dynasties, the Yangtze River basin was incorporated into the unified national system, and water conservancy also saw considerable progress. For example, Shi Lu, in the reign of the First Emperor of Qin (221-206 BC), dug the Lingqu Canal, Zhang Tang and his son in the reign of Emperor Wu of Han (156-87 BC) dug the Baoxie Road and the subsequent development of Nanyang area by Zhao Xinchen and Du Yu all promoted local agricultural development at that time.

The Three Kingdoms, the Western and Eastern Jin Dynasties and the Southern and Northern Dynasties were a time of division and war. People moved southward in large numbers, and so did China's economic center. Water conservancy projects south of the Yangtze River were gradually developed, especially in the Taihu Lake basin near Nanjing. In the Eastern Jin Dynasty, Huan Wen ordered Chen Zun to build the Wancheng Dyke, the embryo of the later Jingjiang Dyke.

With the reunification of China during the Sui and Tang dynasties, such a situation gradually formed with the political center in the north while the economic center in the south and the Grand Canal became the link connecting the two centers. The Taihu Lake basin was the country's primary grain producer, resulting in excessive reclamation and intensified contradiction between humans and water.

During the Song, Yuan, Ming and Qing dynasties, the country's economic center ultimately moved to the south of the Yangtze River, where water conservancy projects were increasingly developed, the embankment system was formed, and the Taihu Lake basin was continuously exploited. In the Ming and Qing dynasties, the Jingjiang Dyke, Wuhan City Dyke, Huangguang Dyke and Wuwei Dyke were connected, and the Jiangsu-Zhejiang seawalls formed a complete system. In addition to Taihu Lake, the Dongting Lake, Poyang Lake and Jianghan Lake Group had also been developed, becoming an important area of the country's grain tax revenues. With the construction of more and more embankments in the lake areas, poor flood discharge had caused severe losses.

Water control is at the center of governance. Throughout Chinese history, those who excelled at governing the country prioritized controlling flood and drought disasters. Rulers who aimed to make a difference, such as Yu the Great in ancient China, the First Emperor of Qin, Emperor Wu of Han, Emperor Taizong of Tang and Emperor Taizu of Song, took water management as an essential means of national governance. Robust water conservancy meant a prosperous country, flourishing industries and a stable and affluent society.

第二节 三峡工程的初步构想

鸦片战争后，中国沦为半封建、半殖民地社会。在1840年至1949年的110年间，江河变迁，水害严重。尤其是1860年、1870年的两次长江特大洪水，冲击松滋口和藕池口，形成四口入洞庭湖的格局，江湖矛盾也日益突出。1931年、1935年和1954年的三次大洪水，对长江中下游造成极其严重的破坏，两岸百姓生命财产遭受巨大的损失。长江中下游防洪形势的变化推动了有识之士对长江治理的思索。

一、孙中山开发三峡的构想

1.《实业计划》

1919年2月，孙中山撰写的《国际共同发展中国实业计划》（以下简称"《实业计划》"）正式向外界公布，详细论述使中国步入现代强国的方案。后来，《实业计划》被译成中文，与《孙文学说》《民权初步》两部著作共同组成《建国方略》一书。

《实业计划》共包括六个计划，在第二个计划的第四部分"改良扬子江现有水路及运河"一节，提出改善三峡航

图 1.8 《实业计划》

Chapter 1　Origin of the Three Gorges Project

Section 2　Preliminary Conception of the Three Gorges Project

In the wake of the Opium War, China gradually became a semi-colonial and semi-feudal society. During the 110 years from 1840 to 1949, the Yangtze River basin suffered severe water damage. In particular, the two major floods of the Yangtze in 1860 and 1870 hit the Songzikou and Ouchikou rivers, resulting in four tributaries flowing into Dongting Lake, posing an increasingly prominent river-lake contradiction. The three devastating floods in 1931, 1935 and 1954 caused severe damage to the middle and lower reaches of the Yangtze and considerable losses in human life and property along the banks. The change in the dynamics of flood management in the middle and lower Yangtze prompted men of insight to ponder the harnessing of the Yangtze River.

I. Sun Yat-sen's Conception of Developing the Three Gorges

1. *International Development Plan of China*

In February 1919, Dr. Sun Yat-sen's *International Development of China* was officially released to the outside world, detailing the plan to make China a modern power; it was later translated into Chinese. The *International Development of China*, *Sun Wen Theory* and *Parliamentary Law* together constituted the *Plans for National Reconstruction*.

The *International Development of China* contains six programs. In Part IV of Program Two, "The Improvement of the Existing Waterways and Canals", Sun Yat-sen proposed improving the waterways and developing hydroelectric power in the Three Gorges, the earliest record of the concept of building the Three Gorges project ever published in China. He wrote,

↻ Figure 1.8　The *International Development of China*

道、开发三峡水力发电的设想,这是我国公开发表的兴建三峡工程设想的最早记载。他在文中写道:

> 自宜昌而上,入峡行,约一百英里①而达四川之低地,即地学家所谓红盆地也。此宜昌以上迄于江源一部分河流,两岸岩石束江,使窄且深,平均深有六寻②,最深有至三十寻者。急流与滩石,沿流皆是。
>
> 改良此上游一段,当以水闸堰其水,使舟得溯流以行,而又可资其水力。其滩石应行爆开除去,于是水深十尺③之航路,下起汉口,上达重庆,可得而致。

1921年,北洋政府成立长江水道讨论会,此后这一机构先后演变为扬子江水道讨论委员会、扬子江水道整理委员会、扬子江水利委员会和长江水利工程总局,成为治理开发长江的专门机构。与此同时,长江流域的水利也走上了近代化的道路。

2.《民生主义》演讲

1924年8月17日,孙中山先生在广州国立高等师范学校礼堂发表题为《民生主义》的演讲,再次谈到三峡水力资源的开发。他在演讲中指出:

> 像扬子江上游夔峡④的水力,更是很大。有人考察由宜昌到万县一带的水力,可以发生三千余万匹马力⑤的电力,像这样大的电力,比现在各国所发生的电力都要大得多。不但是可以供给全国火车、电车和各种工厂之用,并且可以用来制造大宗的肥料。

在当时的历史条件下,孙中山先生敏感地把握住水力发电这一新技术的出现,提出开发三峡水力资源的伟大构想,并把它作为振兴中国经济、富国兴邦的重要举措,这是十分难能可贵的。他在一系列的文章和演讲中,把开

---

① 1英里=1.6093千米。
② 英寻,1英寻=1.8288米。
③ 英尺,1英尺=0.3048米。
④ 即瞿塘峡。
⑤ 1马力=0.7355千瓦。

"The Yangtze River above Ichang (Yichang) enters the Gorges which run about a hundred miles[1] up to the Szechuen (Sichuan) depression, known as the Red Basin. This part of the river from Ichang (Yichang) right along to its source is confined by rocky banks, very narrow and deep, having an average depth of six fathoms[2] and at some particular points even thirty fathoms. Many rapids and obstructions occur along its course."

"To improve the Upper Yangtze, the rapids should be dammed up to form locks to enable crafts to ascend the river as well as to generate water power. Obstructions should be blasted and boulders removed. Thus, a ten-foot[3] channel right along from Hankow (Hankou) to Chungking (Chongqing) could be obtained."

In 1921, the Beiyang Government established the Changjiang River Waterway Discussion Commission, which later evolved into the Yangtze River Waterway Discussion Commission, the Yangtze River Waterway Preservation Commission, the Yangtze River Water Conservancy Commission and the General Administration for the Yangtze River Water Conservancy Engineering, becoming a specialized agency for the harnessing and development of the Yangtze River. At the same time, China's water conservancy practices in the Yangtze River basin also embarked on the road to modernization.

2. Lecture of *The Principle of Minsheng* (People's Livelihood)

On August 17, 1924, Dr. Sun Yat-sen delivered a lecture entitled "*The Principle of Minsheng*" in the auditorium of Guangzhou National Higher Normal School, in which he mentioned the development of hydropower resources in the Three Gorges once again. In his lecture, he pointed out that,

"Consider the tremendous water power in the Kui Gorge[4] of the upper Yangtze. Some who have studied the stretch of river between Ichang (Yichang) and Wanshien (Wanxian) estimate that the water power there could generate over 30 million electrical horsepower[5]. Such an immense power is much greater than that produced at present in any country. It would not only supply all the railways, electric lines, and factories in the country with power, but it could be used to manufacture staple fertilizers."

---

[1] 1 mi = 1.609,3 km.
[2] Fathom, 1 fm = 1.828,8 m.
[3] Foot, 1 ft = 0.304,8 m.
[4] More commonly known as the Qutang Xia.
[5] 1 ps = 0.735,5 kW.

发三峡水力资源的"三峡梦"和他的富国梦联系起来，这一构想对中国社会产生了积极而深远的影响。然而，遗憾的是，《实业计划》和《民生主义》均未涉及关系到上千万老百姓生命财产安全的长江中下游防洪问题。

二、萨凡奇的三峡计划

1. 潘绥报告

从1932年开始，围绕三峡水利资源开发，国民政府时断时续地做了一些工作。1932年10月，国民政府组织开展了三峡水力开发的第一次科学考察，推荐在葛洲坝修建一座装机容量30万千瓦的水电站。然而，由于受国力所限，国民政府对所上报计划暂时搁置。

1944年，国民政府成立了战时生产局，在该局担任专家的美国经济学家、美国驻华使馆经济专员潘绥（G. R. Paschal），于1944年4月提出《利用美贷修建中国水力发电厂与清偿贷款方法》的报告（以下简称《潘绥报告》）。该报告建议在长江三峡修建一座坝高120米、总装机容量1 050万千瓦的水电厂；同时建造一座年产量500万吨的化肥厂，利用三峡廉价的电力，生产化肥向美国出口。所需建筑费用由美国提供贷款，大约为9亿美元，化肥厂投产后，大约15年内还清全部债务。

由于当时中国经济窘迫、对能源十分渴求，于1939年4月成立了水力发电审议委员会，并于第二年成立了水力发电勘测总队，勘测了近200个水力发电坝址，其中包括长江三峡坝址。《潘绥报告》受到国民政府资源委员会的大力赞赏。同时，这份报告也成为美国内政部垦务局设计总工程师、享誉世界的大坝专家萨凡奇（J. L. Savage）亲赴三峡考察的缘由。

Thanks to his far sight under those historical conditions, Dr. Sun Yat-sen stayed sensitive to the technology of hydropower generation, put forward the great idea of developing hydropower resources in the Three Gorges and regarded it as an important measure to revitalize China's economy and enrich the country. In a series of articles and lectures, he linked the "Three Gorges Dream" of developing the hydropower resources in the Three Gorges with his dream of enriching the country, which had a positive and far-reaching influence on Chinese society. Regrettably, both the *International Development of China* and *The Principle of Minsheng* failed to consider flood control issues that concerned the safety of tens of millions of people in the middle and lower reaches of the Yangtze River.

II. J. L. Savage's Plan on the Three Gorges

1. The *Paschal Report*

Since 1932, the Nationalist Government of the Republic of China did some work intermittently on developing hydraulic resources in the Three Gorges. In October 1932, the Nationalist Government organized the first scientific investigation on the development of hydraulic resources in the Three Gorges. It recommended building a hydropower station with an installed capacity of 300 MW at Gezhouba. However, the plan was shelved due to limited national strength.

In 1944, the Nationalist Government set up the War Production Board. In April of the year, G. R. Paschal, a renowned American economist serving as an adviser to the Board and economic commissioner of the American Embassy in China, presented the *Report on Building a Dam-type Power Plant in China with the U.S. Loan and the Way to Repay the Loan* (hereinafter referred to as "the *Paschal Report*"). The report proposed building a hydropower plant in the Three Gorges area with a dam height of 120 m and an installed capacity of 10,500 MW. He also resurrected Dr. Sun Yan-sen's idea of using the power generated to produce electricity to build a fertilizer plant with an annual output of 5 million tons with U.S. investment, to be repaid to the United States with fertilizer over 15 years.

Due to China's economic distress and urgent demand for energy, the Nationalist Government established the Hydropower Deliberation Committee in April 1939 and the General Hydropower Survey Crew the following year, which surveyed nearly 200 hydropower dam sites, including the Three Gorges dam site. The *Paschal Report* was praised and strongly recommended by the Resources Commission of the Nationalist Government. After reading the report, Dr. J. L. Savage, the chief design engineer and a high dam expert with the U.S. Bureau of Reclamation, decided to visit the Three Gorges in person.

2. 萨凡奇计划

由于《潘绥报告》对国民政府资源委员会的触动很大，1944年5月，资源委员会决定邀请萨凡奇来华考察西南地区的水力资源。萨凡奇是20世纪世界著名的大坝专家，一生钟情于大坝的设计和建造，参与设计了60多座大坝，其中很多是世界著名大坝。

1944年8月下旬，萨凡奇的考察任务即将完成时，国民政府资源委员会主任翁文灏派人给他送来了《潘绥报告》，请他从专业的角度做出评价。他看后激动不已，不顾65岁高龄和三峡处于抗日前线，坚持一定要亲自前往三峡考察。

1944年9月上旬，萨凡奇前往三峡进行了为期10天的考察。10月，他完成了《扬子江三峡计划初步报告》（以下简称"《报告》"），即"萨凡奇计划"，并提交给翁文灏。当时，恰好美国罗斯福总统的特别代表纳尔逊（Donald Nelson）也在中国，为寻求美国援助，国民政府请他将《报告》转交给罗斯福总统。随即，美国白宫以中国将修建世界上最大水电工程为题向新闻界透露了萨凡奇的考察成果，一时间媒体竞相报道，在国际社会引起了巨大轰动。

《报告》中译本共16节，约3万字，主要内容有：

坝址选择：拦河坝坝址选择在湖北省宜昌上游5～15千米的范围内，即南津关坝址。

图1.9　萨凡奇设计的三峡大坝设计图

2. The *Savage Plan*

Greatly encouraged by the *Paschal Report*, the Resources Commission of the Nationalist Government decided to invite Dr. Savage to China to investigate the hydraulic resources in Southwest China in May 1944. Dr. Savage was a world-famous dam expert in the 20th century. He devoted his life to designing and constructing dams and participated in designing more than 60 dams, including many world-famous ones.

In late August 1944, when Dr. Savage's investigation was about to be completed, Weng Wenhao, then director of the Resources Commission of the Nationalist Government, sent him the *Paschal Report* and asked for his professional input. So excited he was that he insisted on visiting the Three Gorges personally even though he was 65 years old and that the Three Gorges were on the front line of the Chinese People's War of Resistance against Japanese Aggression.

In early September 1944, Dr. Savage paid a ten-day visit to the Three Gorges. In October, he completed the *Preliminary Report on Yangtze Gorges Project* (hereinafter referred to as "the *Report*"), or the stunning *Savage Plan*, and presented it to Weng Wenhao. At that time, Donald Nelson, who was described as President Roosevelt's personal representative, was also in China. To seek U.S. help, the Nationalist Government asked him to transfer the *Report* to President Roosevelt. Soon, The White House disclosed Dr. Savage's investigation results to the press circles, saying that China was to build the world's largest hydropower plant, which garnered global media coverage.

The Chinese version of the *Report* consists of 16 sections, with about 30,000 characters. The main content is as follows:

Dam site selection: The dam site will be located 5-15 km from the upper Yichang, Hubei Province, i.e., the Nanjin Pass dam site.

Development program: The key projects include the dam, hydropower houses and ship locks. The dam would be a concrete gravity dam with a maximum height of 225 m and an installed generating capacity of 10,560 MW. Navigation structures will also be built, and both the ship locks and equipment capacity would accommodate a fleet of 10,000 tons. The construction will be implemented in five phases.

Expected benefits: The project will be built mainly for power generation, with additional benefits like irrigation, flood control, shipping, water supply and tourism.

It should be said that the *Report* was the first concrete development plan for the Three Gorges with the benefit of creating conditions to fend off natural disasters. It was also ex-

⊂ Figure 1.9 Design Drawing of the Three Gorges by Dr. Savage

开发方案：主要枢纽工程有大坝、水电站厂房、船闸等。大坝采用混凝土重力坝，最大坝高 225 米。水电站总装机容量 1 056 万千瓦。另设通航建筑物，船闸容量及设备均以能航运万吨船队为目标。工程建设拟分 5 期进行。

工程效益：工程以发电为主，兼有灌溉、防洪、航运、供水及游览等效益。

应当说，《报告》是第一个比较具体的、具有兴利除害综合效益的三峡工程计划，对于当时中国抗日战争后的经济复兴应大有裨益，正如萨凡奇在《报告》临时结论的第八条中写道：

> 扬子江三峡计划为一杰作，关系中国前途至为重大，将鼓舞华中、华西一带工业之长足进步，将有广泛之就业机会，将提高人民之生活标准，将使中国转弱为强。为中国计、为全球计，建造扬子江三峡计划实属必要之图也。

为此，1944 年 11 月 12 日，国民政府经济部向萨凡奇颁发金色民生勋章，以表彰他对中国经济发展的贡献。

1946 年 3 月，受国民政府资源委员会邀请，萨凡奇再次来华考察三峡。4 月上旬考察结束后，萨凡奇与资源委员会商讨三峡水力发电工程设计事项，并计划中美合作设计三峡工程，即由美国内政部垦务局主持设计，国民政府资源委员会及各合作机关派员参加。

3. 中美联合勘测设计三峡工程

"萨凡奇计划"提出后，国民政府启动了实施三峡工程计划的筹备工作，先后成立了全国水力发电工程总处和三峡勘测队，三峡工程被明确为全国水电的第一优先项目。

1945 年 7 月 26 日，国民政府外交部部长宋子文与美国驻华大使赫尔利（Patrick Jay Hurley）等人，在重庆商讨长江三峡工程设计事宜，包括设计经费、工程初步预算等问题。10 月 1 日，美国内政部垦务局与国民政府资源委员会驻华盛顿办事处签署《美国内政部垦务局与中国资源委员会关于编制长江流域工程报告的合约》。该合约中的"工程"，除三峡工程外，还包括长江上游 5 条支流的水电、灌溉、防洪、航运等综合性开发工程。同年 12 月，萨

pected to be of great help to China's postwar economic recovery, just as Dr. Savage wrote in Article 8 of the provisional conclusion of the *Report*:

> "The Yangtze Gorge Project is a 'CLASSIC'. It will be of utmost importance to China. It will bring great industrial developments in Central and Western China. It will bring widespread employment. It will bring high standards of living. It will change China from a weak to a strong nation. The Yangtze Gorge Project should be constructed for the benefit of China and the world at large."

For this reason, on November 12, 1944, the Ministry of Economic Affairs of the Nationalist Government awarded Dr. Savage the "Golden *Minsheng* Medal" in recognition of his contribution to China's economic development.

In March 1946, Dr. Savage revisited the Three Gorges area at the invitation of the Resources Commission of the Nationalist Government. After an investigation in early April, Dr. Savage discussed with the Resources Commission the design of the Three Gorges hydropower project with the cooperation of the United States. The U.S. Bureau of Reclamation would direct the design, and the Resources Commission of the Nationalist Government and various cooperation offices would send representatives to participate.

3. China-US cooperation on survey and design of the Three Gorges project

After the *Savage Plan* was proposed, the Nationalist Government started the preparatory work for implementing the Three Gorges project and established the National General Office of Hydropower Projects and the Three Gorges Survey Team, making the Three Gorges project the top priority among hydropower projects across the country.

On July 26, 1945, Soong Tse-ven, minister of foreign affairs of the Nationalist Government, and Patrick Jay Hurley, US ambassador to China, discussed the design of the Three Gorges project in Chongqing, including the design costs and preliminary budget of the project. On October 1, the U.S. Bureau of Reclamation and the Washington office of the Resources Commission of the Nationalist Government signed the *Agreement between the Bureau of Reclamation of the U.S. Department of the Interior and the Resources Commission of China on Preparing the Report on the Yangtze River Basin Projects*. In addition to the Three Gorges project, the "projects" in the *Agreement* also included comprehensive development projects such as hydropower, irrigation, flood control and shipping of the five tributaries of the upper Yangtze. In December of the same year, Dr. Savage was employed as a senior advisory engineer of the Resources Commission of the Nationalist Government.

凡奇被聘为国民政府资源委员会高级顾问工程师。

围绕三峡工程计划，国民政府资源委员会也成立了一系列三峡水力发电计划工作机构。1945年5月，三峡水力发电计划技术研究委员会正式成立，由资源委员会、全国水利委员会、扬子江水利委员会、中央水利实验处、中央农业试验所、交通部航政司、中央地质调查所等部门组成。1945年5月，资源委员会在龙溪河水力发电工程处的基础上，在四川长寿成立了全国水力发电工程总处。1946年7月，全国水电工程总处下设立了扬子江三峡勘测处，主要做了筹备坝基钻探、坝址及水库测量，以及水库淹没调查等工作。

1946年春，国民政府资源委员会与美国内政部垦务局签订了三峡工程设计合同，并先后选派54名中国工程技术人员赴美国内政部垦务局参加了三峡工程的设计工作。随后，全国水力发电工程总处和扬子江三峡勘测处先后进行了一些地质勘探和地形测绘工作。

正当三峡工程的设计工作顺利进行时，国民党发动全面内战，国民党统治地区的经济形势日趋恶化。1947年4月，国民政府批准暂停三峡工程国内外的一切工作。5月16日，国民政府资源委员会正式通知美国内政部垦务局，中止共同设计三峡工程合约，停止三峡工程设计工作，召回中方工程师。

近代中国开发三峡水利的梦想，在旧中国风雨如晦、内忧外患的国情下，如同星空中的彗星一现，不久就被严峻的现实打破。那时的中国长江流域的水利开发十分缓慢，导致水利失修，水患频繁，航运不兴。河道长期失治，堤防残破不堪，水利设施寥寥无几，残缺不全。据水利部统计，截至1949年，偌大的中华大地上只有22座大中型水库和一些塘坝、小型水库，江河堤防仅4.2万千米。

Specific to the Three Gorges project plan, the Resources Commission set up a series of working bodies for the Three Gorges hydropower plan. In May 1945, the Three Gorges Hydro Power Plan and Technical Research Committee was jointly formed by the Resources Commission, the National Water Conservancy Commission, the Yangtze River Water Conservancy Commission, the Central Water Conservancy Experimental Office, the Central Agricultural Laboratory, the Navigation Administration Department of the Ministry of Transportation and Communications, and the Central Geological Survey Institute. In May 1945, based on the Longxi River Hydropower Engineering Office, the Resources Commission established the National General Office of Hydropower Projects in Changshou, Sichuan Province. In July 1946, the Yangtze River Three Gorges Survey Office was set up under the National General Office of Hydropower Projects, mainly to prepare for dam foundation drilling, the dam site and reservoir survey, and a reservoir inundation survey.

In the spring of 1946, the Resources Commission signed an agreement with the U.S. Bureau of Reclamation to entrust engineering to the latter and sent some 54 Chinese technical personnel to the United States to participate in the project. Later, the National General Office of Hydropower Projects and the Yangtze River Three Gorges Survey Office carried out some geological exploration and topographic mapping work.

While the design work of the Three Gorges project was going on wheels, the Kuomintang launched a full-scale civil war, and the economic situation in the Kuomintang-ruled areas was deteriorating. In April 1947, the Nationalist Government approved the suspension of all work related to the Three Gorges project at home and abroad. On May 16, the Resources Commission officially notified the U.S. Bureau of Reclamation of suspending the contract for the joint design of the Three Gorges project, terminated the design work and recalled the Chinese engineers.

China's Three Gorges dream in modern times, like a flash in the pan, was beaten down by the grim situation of domestic turmoil and foreign aggression in old China. China's water conservancy development in the Yangtze River basin moved slowly then. Chronic bad management of the river led to the disrepair of water conservancy facilities, frequent floods and poor shipping, leaving the levees in shreds and patches. According to Ministry of Water Resources statistics, by 1949, China had only 22 large and medium-scale dams and reservoirs and some ponds and small reservoirs, with only 42,000 km of embankments along the rivers, insufficient to meet the needs of the vast territory.

---本章小结：---

长江是中华文明的重要发祥地，也是中国经济社会可持续发展的重要命脉。长江流域地理和气候条件复杂，洪水灾害深重，治水兴利、根除洪水灾害是治国理政的必然要求。古代中国数千年的治水史实说明，要根本性解决长江洪水问题，必须兴建控制性主体工程。中华人民共和国成立后长江治水的系列实践和科学技术的持续进步进一步证实，作为长江中下游防洪体系的关键性骨干工程，三峡工程的兴建具有历史必然性和现实可行性。历史实践证明，兴建三峡工程是中华民族治水兴邦的必然选择。

参考文献：

［1］本书编委会. 百问三峡［M］. 北京：科学普及出版社，2012.

［2］水利部长江水利委员会. 长江流域综合规划（2012—2030）［EB］. 2012.

［3］长江水利委员会. 三峡大观［M］. 北京：水利水电出版社，1986.

［4］孙中山. 孙中山全集［M］. 北京：中华书局，1985.

［5］《中国三峡建设年鉴》编纂委员会. 中国三峡建设年鉴（2000）［J］. 宜昌：中国三峡建设年鉴社，2000.

［6］中国第二历史档案馆. 民国时期筹备三峡工程档案文献图片集［M］. 北京：中国三峡出版社，2016.

［7］黄山佐. 民国时期开发长江三峡水力资源筹划始末［J］. 中国科技史料，1984，5（3）.

［8］钱昌照. 钱昌照回忆录［M］. 北京：东方出版社，2011.

［9］《中国三峡建设年鉴》编纂委员会. 中国三峡建设年鉴（1997）［J］. 宜昌：中国三峡建设年鉴社，1997.

第一章 | 三峡工程的缘起
Chapter 1 Origin of the Three Gorges Project

Chapter Summary:

The Yangtze River is an important birthplace of the Chinese civilization and concerns the lifeline of China's sustainable economic and social development. The complex geology and climate makes it prone to floods. To harness flood resources while avoiding the damages is an inevitable requirement for the governance of the country. Thousands of years of water harnessing practices in ancient China tells us that main control constructions must be built in order to fundamentally solve the flood problem of the Yangtze River. After the founding of the People's Republic of China, the series of practices of water harnessing in the Yangtze River and the continuous progress of science and technology have further verified that the building of the Three Gorges Project, as the key backbone project of the flood control system in the middle and lower reaches of the Yangtze River, is historically necessary and practically feasible. Historical practices have proved that building the Three Gorges Project is an inevitable choice for the Chinese nation to harness water and prosper the country.

References:

[1] Editorial Board of the book. *One Hundred Questions Regarding the Three Gorges Project* [M]. Beijing: Science Popularization Press, 2012.

[2] Changjiang Water Resources Commission of the Ministry of Water Resources. *Comprehensive Planning of the Yangze River Basin (2012-2030)* [EB]. 2012.

[3] Changjiang Water Resources Commission. *Encyclopedia of the Three Gorges Project* [M]. Beijing: China Water & Power Press, 1986.

[4] Sun Yat-sen. *Completed Works of Sun Yat-sen* [M]. Beijing: Zhonghua Book Company, 1985.

[5] Editorial Board of *China Three Gorges Construction Yearbook*. *China Three Gorges Construction Yearbook (2000)* [J]. Yichang: China Three Gorges Construction Yearbook Press, 2000.

[6] Second Historical Archives of China. *Photo Collection of Archival Documents Prepared for the Three Gorges Project during the Republic of China Period* [M]. Beijing: China Three Gorges Press, 2016.

[7] Huang Shanzuo. *China's Preparatory Activities in Developing the Three Gorges Hydropower Resources of the Yangtze River in the Republic of China Period* [J]. China Historical Materials of Science and Technology, 1984, 5 (3).

[8] Qian Changzhao. *Memories of Qian Changzhao* [M]. Beijing: Oriental Press, 2011.

[9] Editorial Board of *China Three Gorges Construction Yearbook*. *China Three Gorges Construction Yearbook (1997)* [J]. Yichang: China Three Gorges Construction Yearbook Press, 1997.

［10］国家能源局. 中国水电100年（1910—2010）［M］. 北京：中国电力出版社，2010.

［11］《中国三峡建设年鉴》编纂委员会. 中国三峡建设年鉴（1996）［J］. 宜昌：中国三峡建设年鉴社，1997.

［12］王芳，宋丹. 1947年三峡工程初步规划［M］. 北京：中国水利水电出版社，2003.

[10] National Energy Administration. *A Century of China's Hydropower Development (1910-2010)* [M]. Beijing: China Electric Power Press, 2010.

[11] Editorial Board of *China Three Gorges Construction Yearbook*. *China Three Gorges Construction Yearbook (1996)* [J]. Yichang: China Three Gorges Construction Yearbook Press, 1997.

[12] Wang Fang & Song Dan. *Preliminary Design of the Three Gorges Project in 1947* [M]. Beijing: China Water & Power Press, 2003.

| 阅读提示：|

中华人民共和国成立后，党和国家领导人从人民的长远利益出发、审时度势，积极而又慎重地领导了三峡工程的论证和决策。从国家成立之初开始研究、论证到1992年4月3日全国人大七届五次会议通过《关于兴建长江三峡工程的决议》，其间40多年的论证，记载着无数人的智慧和心血。

三峡工程论证时间之长，论证内容之多、之广泛，参加论证专家之多，为中国大型建设项目中所罕见。在论证的过程中，党和国家领导人遵循"服从多数"和"尊重少数"的民主原则，广泛听取各方意见；同时秉承科学精神、运用科学方法、按照科学程序，对多种行动方案做出合理取舍，从而做出符合客观规律的决策。

After the founding of the People's Republic of China (PRC) in 1949, the leaders of the country and Party led the feasibility work and decision-making about the Three Gorges Project based on the long-term interests of the Chinese nation. An assessment period spanning more than 40 years, starting from the beginning of the PRC to the *Resolution on the Construction of the Three Gorges Project,* approved by the Fifth Session of the Seventh National People's Congress on April 3, 1992, contained the wisdom and hard work of countless participants.

There are very few large-scale construction projects in the history of the PRC, like the Three Gorges Project, which involved an incredible length of time and scale, with the support of countless experts. During the assessment, the leaders of the country and Party solicited advice from all social circles observing the democratic principle of "deferring to the majority" and "respecting the minority". At the same time, they adhered to the spirit of science, applied scientific methods and followed scientific procedures to make reasonable trade-offs among various courses of action. Finally, decisions were made according to objective laws.

Chapter 2 >>>>

三峡工程的研究论证
Research and Feasibility of the Three Gorges Project

第一节 中华人民共和国成立后三峡工程的研究论证

鉴于三峡工程规模空前、技术复杂、施工难度大的特殊性，从中华人民共和国成立之初到三峡工程开工建设，党和国家领导人有计划、有步骤地动员全国各有关方面的力量，对三峡工程采用多种方式、多种途径进行勘察、研究、论证、规划、设计，历时 40 多年。社会各界尤其是自然科学、工程技术等诸多领域的国内外专家学者广泛参与、深入调研，秉承科学精神、爱国热情和历史责任感进行了充分科学论证，为三峡工程顺利建设提供了坚实保障。

一、中华人民共和国成立后对三峡洪水的早期治理

1949 年汛期，长江发生了大洪水。这年大水，长江中下游湘、鄂、赣、皖、苏 5 省受灾农田达 181.4 万公顷，受灾人口 810 万人。洪灾所到之处，摧毁了房屋、道路，人员疾病蔓延、伤亡较大，灾后城乡满目疮痍。同时，洪水造成中下游多处堤防溃决成灾，长江堤防千疮百孔，荆江大堤险象环生。虽然在解放军与沿江老百姓的共同努力下，这次抗洪取得了胜利，但长江中下游特别是荆江河段的防洪形势显得十分严峻而紧迫。

荆江河段上起湖北枝城，下迄湖南城陵矶，全长 347.2 千米，该河段由 9 个滩段、13 个浅滩、33 个水道组成，约占整个长江干线航道的八分之一。该河段九曲回肠、水深流急、崩岸频繁、故道密集，自古就流传着"万里长江，险在荆江"的民谣。荆江两岸人口 1 500 万人，耕地 2 300 万亩，是我国重要的粮棉基地，还有一批重要的大中城市和工矿企业、交通设施、油田等。一旦发生溃坝，广大人民的生命财产均将遭受到巨大损失，严重影响我国经济

Chapter 2 Research and Feasibility of the Three Gorges Project

Section 1 | Research and Feasibility of the Three Gorges Project After the Founding of the PRC

Given the unprecedented scale of the Three Gorges Project and the complicated technologies and enormous difficulties involved in its creation, the leaders of the country and Party mobilized the strength of all relevant departments to survey, research, plan and design the project in various ways from the beginning of the founding of the PRC to the start of its construction. The process lasted for more than 40 years. Domestic and international experts and scholars from the natural sciences to engineering participated in the extensive and in-depth investigation with scientific rigor and a sense of historical responsibility. Their work provided a solid foundation for the smooth progress of the Project's construction.

I. The Early Management of the Flooding in the Three Gorges After the Founding of the PRC

In the flood season of 1949, there was severe flooding in the Yangtze River. Up to 1.814 million hectares of farmland were devastated in the five provinces (Hunan, Hubei, Jiangxi, Anhui and Jiangsu) in the middle and lower reaches of the river. The flooding impacted 8.1 million people. The floods destroyed houses and roads. Disease was rampant. The result was injury and death. Cities and rural areas were met with desolation after the floods receded. Meanwhile, the floods damaged several dykes in the middle and lower reaches of the river. The embankment of the Yangtze River was scarred everywhere; that of the Jingjiang River was dangerous. Although the PLA soldiers and the residents along the river joined hands to combat the floods, finally winning the battle, the need for flood control in the middle and lower reaches of the Yangtze River, especially for the Jingjiang section, was urgent.

The Jingjiang section starts from Zhicheng of Hubei and ends at Chenglingji of Hunan, totaling 347.2 km in length. This section comprises nine beaches, 13 shoals and 33 water courses, accounting for about 1/8 of the whole arterial channel of the Yangtze River. According to a folk rhyme, zigzags, deep water, raging torrents, dense old water courses, and

建设大局。

中华人民共和国成立后，为加快长江治理开发，1950年2月，水利部长江水利委员会（以下简称"长江委"）在武汉成立。1951年12月，长江委主任林一山在《关于治理长江计划基本方案的报告》"治江三阶段"中，首次提出以三峡工程为长江防洪的治本性工程，以荆江分洪工程为当时平原防洪的重要措施。

1952年3月，中南军政委员会①根据中央指示，做出《关于荆江分洪工程的决定》，"一致同意荆江分洪的计划，认为这一计划的方针是照顾全局，兼顾了两省，对两湖人民都是有利的"。同月，中央人民政府政务院发布了《关于荆江分洪工程的决定》，决定指出："为保障两湖千百万人民生命财产的安全起见，在长江治本工程未完成以前，加固荆江大堤并在南岸开辟分洪区乃是当前急迫需要的措施。"至此，荆江分洪工程成为中华人民共和国第一个治理长江的大型工程。

1952年4月5日，荆江分洪工程主体工程北闸、南闸和分洪区围堤的主要组成部分黄天湖大堤全面开工，30万军民和技术人员齐心协力，日日夜夜奋战在工地上，6月20日提前15天完成工程任务。荆江分洪工程的竣工，为防御1954年洪水、保障荆江大堤创造了重要条件。

图2.1 荆江大堤一段（摄影：黄正平）

---

① 中央人民政府在中南地区设立的介于中央和省之间的一级政权机关，机关驻武汉市，下辖河南、湖北、湖南、江西、广东、广西等6个省。

frequent collapse of dykes are characteristic of the section. The population on both sides of Jingjiang is 15 million, and the farmland is 23 million mu (1 hm^2 = 15 mu). In addition to being an important grain and cotton base in China, several important large and medium cities, mines, enterprises, transportation facilities and oil fields exist. Once a dam breach happens, the result is enormous losses in lives and property and a setback to the country's economic development.

After the founding of the PRC, the Changjiang Water Resources Commission of the Ministry of Water Resources (hereinafter referred to as "the Changjiang Commission") was established in Wuhan in February 1950 to accelerate the harnessing and development of the Yangtze River. In December 1951, Lin Yishan, director of the Changjiang Commission, first proposed in the "Three Stages for Harnessing the Yangtze River" in the *Report About the Basic Scheme for the Plan of Harnessing the Yangtze River* that the Three Gorges Project is a primary flood control project for the Yangtze River and the Jingjiang Flood Diversion Project is crucial for flood control on the plains.

In March 1952, the Central South Military and Administrative Committee[①] issued the *Decision About the Jingjiang Flood Diversion Project* by the instructions of the Central Committee of CPC, which states that "We agree to the plan of Jingjiang flood diversion unanimously and believe that the guiding principle of this plan has taken the overall situation of two provinces into consideration and is beneficial to the people both in Hubei and Hunan." In the same month, the Government Administration Council of the Central People's Government released the *Decision About the Jingjiang Flood Diversion Project*, which pointed out that "To ensure the safety of the lives and properties of millions upon millions of people in Hubei and Hunan, we will reinforce Jingjiang dyke and open up a flood diversion zone on the southern bank as an urgent measure before the completion of the essential project for harnessing the Yangtze River." So far, the Jingjiang Flood Diversion Project became the first large-scale project in the PRC's endeavor to harness the Yangtze River.

On April 5, 1952, the construction of the Huangtianhu Dyke, the central part of the main project (the Northern Floodgate, Southern Floodgate and Embankment of the Flood Diversion Zone) of the Jingjiang Flood Diversion Project began. Three hundred thousand PLA soldiers, workers and technicians fought side by side on the construction site day and night. Construction was completed on June 20, 15 days ahead of schedule. Completing the Jingjiang Flood Diversion Project created essential conditions for preventing a flood in 1954 and safeguarding the Jingjiang Dyke.

◐ Figure 2.1　A Section of the Jingjiang Dyke (Photographed by Huang Zhengping)

---

① It is a level of administrative power between the Central Committee of CPC and province set up by the central government in Central South. It is based in Wuhan and has Henan, Hubei, Hunan, Jiangxi, Guangdong and Guangxi under its jurisdiction.

二、中华人民共和国成立初期对三峡工程的研究论证

荆江分洪工程完工后,中央进一步考虑到了长江的治理问题。1953年2月,毛泽东乘坐"长江舰"从武汉到南京,亲自视察长江。在视察期间,毛泽东主要谈的两件事之一就是了解长江流域的情况,并着重研究长江中下游的防洪问题。当他得知费了那么大的力量修这么多的支流水库还是不能解决防洪问题后提出,"毕其功于一役",先修三峡水库的建议。从此,中华人民共和国开始筹划兴建三峡工程。

1954年夏,长江中下游发生了百年少遇的大洪水,虽然经过广大军民的奋力防守、抢救,采取分洪措施,仍淹地4 755万亩,受灾人口1 800多万人,京广铁路100多天不能正常运行。严峻的防洪形势再次敲响了警钟:防洪是治理长江首要而紧迫的任务。于是,中央人民政府要求加快长江流域规划和三峡工程研究。

从1955年开始,长江委在30多个部门和单位的大力协同下,在苏联专家的协助下,全面开展长江流域规划和三峡工程勘测、科研、设计工作。由于长江流域规划涉及中央有关部委和长江流域各省市,1956年3月,国务院决定将长江水利委员会改为长江流域规划办公室(以下简称"长办",1988年又恢复为长江水利委员会,即"长江委")。

1956年6月,当长江流域规划和三峡工程研究有了初步成果时,毛泽东在武汉畅游长江后,写下了《水调歌头·游泳》一词,描绘出"更立西江石壁,截断巫山云雨,高峡出平湖"的三峡工程的壮丽宏图。

1958年1月,中共中央南宁会议期间,毛泽东听取了国内专家对于三峡工程的不同意见,在肯定修建三峡工程必要性的同时,充分吸取不同意见的合理部分,提出"积极准备,充分可靠"的三峡建设方针。同年2月底至3月初,周恩来率100多位党政要员和专家,实地勘察荆江大堤和三峡工程的预选坝址——南津关和三斗坪,并听取了各方意见。

1958年3月,中共中央成都会议通过了《中共中央关于三峡水利枢纽和长江流域规划的意见》,这是中共中央下发的关于三峡工程的第一个重要文件。意见明确提出:"从国家长远的经济发展和技术条件两个方面考虑,三峡水利枢纽是需要修建而且可能修建的,但是最后下决心确定修建及何时开始

II. Research and Feasibility of the Three Gorges Project in the Early Stage After the Founding of the PRC

After completing the Jingjiang Flood Diversion Project, the CPC's Central Committee began to consider further harnessing the Yangtze River. In February 1953, Mao Zedong traveled from Wuhan to Nanjing via the Yangtze River Ship during his inspection of the river. During his tour, one of the two subject matters Mao Zedong discussed was the situation of the Yangtze River basin, with emphasis on flood control of the middle and lower reaches of the river. When learning that several tributary reservoirs were built but could not solve the flooding problem, he proposed building the Three Gorges Reservoir first as a permanent cure. Since then, the PRC planned the construction of the Three Gorges Project.

In the summer of 1954, the middle and lower reaches of the Yangtze River were hit by once-in-a-century heavy floods. Despite all the efforts made by both the PLA soldiers and civilians to contain the floods, 47.55 million *mu* (1 hm^2 = 15 *mu*) of land was flooded. The flood-hit population was more than 18 million, and the regular operation of the Beijing-Guangzhou Railway was disrupted for more than 100 days. The situation was another wake-up call for all in China: Flood control must be primary in harnessing the Yangtze River. The Central People's Government demanded an acceleration in Yangtze River basin planning and research of the Three Gorges Project.

In 1955, the Changjiang Commission started planning the creation of the Yangtze River basin, along with surveying, researching and designing the Three Gorges Project. More than 30 departments and units and experts from the Soviet Union were involved. Because the planning of the Yangtze River basin involves relevant ministries and commissions of the Central Committee of CPC and all the provinces and municipalities along the Yangtze River basin, the State Council decided in March 1956 to transform the Changjiang Commission into the Yangtze River Basin Planning Office (hereinafter referred to as "the Yangtze River Office". In 1988, it was changed back to the Changjiang Water Resources Commission, namely the Changjiang Commission).

In June 1956, preliminary achievements were made in planning the Yangtze River basin and the research of the Three Gorges Project. Mao Zedong wrote the poem *Shui Diao Ge Tou: Swimming*, depicting the grandiose ambition of the project: "The towering rocky cliffs on the river bank can cut off the cloud and rain in Wushan Mountain, and an enormous lake appears amid the gorges."

In January 1958, during the Nanning Conference of the Central Committee of CPC, Mao Zedong entertained the differing opinions of domestic experts on the Three Gorges Project. While affirming the necessity of carrying out this project, he accepted reasonable

图 2.2 葛洲坝工程开工誓师大会

修建，要待各个重要方面的准备工作基本完成之后，才能做出决定。现在应当采取积极准备和充分可靠的方针，进行各项有关工作。"至此，兴建三峡工程正式提上了党和政府的议事日程。

1958 年 11 月，长办基本编制完成《三峡水利枢纽初步设计要点报告》。1959 年 5 月，长办在湖北武昌召开了《三峡水利枢纽初步设计要点报告》讨论会，就坝址选择、正常高水位选择及施工方案进行了讨论。大会一致同意选用三斗坪坝址和正常蓄水位 200 米的方案，这是第一次确定三斗坪坝址。

1960 年，中央考虑到当时国家的经济情况和国际形势，决定放缓三峡工程建设的进程。同年 8 月，周恩来在北戴河主持长江流域规划工作会议时宣布三峡工程延期建设，同时确定"雄心不变，加强科研，加强人防"的方针。1967 年，由于历史的原因，三峡工程建设被搁置下来。

在三峡工程尚未具备上马条件的情况下，1970 年 10 月，武汉军区和湖北省革命委员会向中央和国务院报送《关于兴建宜昌长江葛洲坝水利枢纽工程的请示报告》，建议在三峡工程之前，先兴建葛洲坝水利枢纽。中央在研究

Chapter 2 Research and Feasibility of the Three Gorges Project

◒ Figure 2.2   The Oath-taking Rally for the Start of the Gezhouba Dam

suggestions and put forward the guiding principle of "active preparation and full reliability" for constructing the Three Gorges Project. From the late February to the early March of that year, Zhou Enlai led more than 100 top officials from the Party and government and experts to survey the Jingjiang Dyke as well as Nanjinguan and Sandouping (the chosen site of the Three Gorges Dam) on the spot and listened to the opinions from all parties.

In March 1958, the Chengdu Conference of the Central Committee of CPC approved the *Guideline of the Central Committee of CPC on the Three Gorges Project and Planning of the Yangtze River Basin*. This was the first important document released by the Central Committee of CPC about the Three Gorges Project. The guideline proposed clearly that "In terms of China's long-term economic development and technological conditions, the Three Gorges Project was a necessity. But a final decision to start the project cannot be made until all preparations have been finished. We should proceed with all relevant work based on the principle of 'active preparation and full reliability'". The building of the Three Gorges Project was formally placed on the agenda of the Party and government.

In November 1958, the Yangtze River Office finished the compilation of the *Report on the Key Points of the Preliminary Design of the Three Gorges Project*. In May 1959, the Yangtze River Office held a symposium on the *Report on the Key Points of the Preliminary Design of the Three Gorges Project*, discussing the choice of dam site, choice of high water level and construction program. The participants unanimously agreed on the choice of Sandouping as the dam site and a water storage level of 200 m. It was the first time that Sandouping was selected as the dam site.

In 1960, given China's economic conditions and international situation, the Central Committee of CPC decided to slow down the building of the Three Gorges Project. In August of that year, Zhou Enlai announced the postponement of the Three Gorges Project when presiding over the work meeting about the planning of the Yangtze River basin in Beidaihe; and he affirmed the need to "stay ambitious, strengthen scientific research and strengthen civil air defense" at the same time. In 1967, for the historical reasons, so the Three Gorges Project was shelved.

Because the conditions were not mature to launch the Three Gorges Project, Wuhan Military Area Command and Hubei Revolutionary Committee submitted the *Report for Asking for Instructions From Higher Authorities About the Building of the Gezhouba Dam on the Yangtze River at Yichang* to the Central Committee of CPC and the State Council, which proposed building the Gezhouba Dam before the launch of the Three Gorges Project. The Central Committee of CPC studied the relationship between the Gezhouba Dam and the

了葛洲坝工程与三峡工程的关系，并听取了先兴建葛洲坝工程的不同意见后，于1970年12月26日批准兴建葛洲坝工程，并指出，兴建葛洲坝工程，是有计划、有步骤地为建设三峡工程做实战准备。

1975年葛洲坝工程建设基本走上正轨，1981年年初实现截流，1981年开始发电、通航，1989年全部建成。葛洲坝工程的兴建，一方面满足了华中地区近期用电的迫切需要，另一方面也为三峡工程积累了丰富的经验。

三、改革开放初期对三峡工程的研究论证

1978年12月，党的十一届三中全会及时、果断地把党和国家的工作重心转移到社会主义现代化建设上来。20世纪70年代末，随着大规模经济建设的开展，华中地区甚至全国缺电现象越来越严重，影响了经济建设。针对这个问题，国家提出了"水火并举"的电力工业发展方针，逐步把重点放在水电上。在水电方面，确定重点开发黄河上游、长江中上游干支流和红水河的水力资源，建设一批大型水电站。因此，兴建三峡工程的必要性被再次强调。1979年11月，葛洲坝一期工程基本建成后，水电部向中央、国务院再次提出修建三峡工程的建议。

1980年7月，邓小平视察了三斗坪坝址、葛洲坝工地和荆江大堤，并听取了关于三峡工程的汇报，并指出要抓紧研究。邓小平的这次视察，对三峡工程的兴建是一个极其重要和关键的转折。从此，三峡工程的论证和有关准备工作开始加速进行。

1980年年底，长办编制上报了《三峡水利枢纽论证报告》，后因故论证会没有召开。1981年2月，长办提出了三峡分期开发、初期蓄水位128米、坝顶高程145米的方案。

1982年9月，党的十二大召开，为适应国家经济建设的总要求，三峡工程建设再次成为舆论关注的焦点。国务院提出立即着手建设三峡工程，但考虑到当时的国情，尽量减少水库淹没，建设规模要适当，三峡工程建设采用低坝方案——正常蓄水位为150米。同年11月，邓小平对兴建三峡工程方案果断表态："看准了就下决心，不要动摇。"

Three Gorges Project and listened to different opinions before approving the building of the Gezhouba Dam on December 26, 1970. It pointed out that the construction of the Gezhouba Dam would aid the eventual establishment of the Three Gorges Project.

The construction of the Gezhouba Dam remained on the right track in 1975. The river closure was realized at the beginning of 1981, and it began to generate electricity and was open to navigation that year. The construction of the entire project was completed in 1989. The building of the Gezhouba Dam met the urgent need for electricity in Central China and accumulated plenty of experiences for the Three Gorges' construction.

III. Research and Feasibility of the Three Gorges Project at the Early Stage of Reform and Opening-up

In December 1978, the Third Plenary Session of the 11th CPC Central Committee shifted the country's and Party's focus to building the economy. At the end of the 1970s, with the large-scale economic construction, the lack of electricity in Central China and even the whole country grew more urgent. To solve this problem, the government proposed the policy of developing hydro-power and thermal power concurrently and gradually shifted the focus to hydro-power. In terms of the hydro-power, the authorities were determined to create hydro-power in the upper reaches of the Yellow River, the mainstem and tributaries in the upper and middle reaches of the Yangtze River and the Hongshui River. A group of large-scale hydro-power stations would be built. Therefore, the necessity of making the Three Gorges Project was reiterated. Upon the completion of Phase I of the Gezhouba Dam construction in November 1979, the Ministry of Water Resources and Power Industry once again proposed building the Three Gorges Project to the Central Committee of CPC and the State Council.

In July 1980, Deng Xiaoping inspected the dam site of Sandouping, Gezhouba construction site and Jingjiang Dyke. He listened to the report about the Three Gorges Project and called for speeding up research. Deng Xiaoping's inspection was a turning point for the building of the Three Gorges Project. From then on, preparations were done at a quicker speed.

At the end of 1980, the Yangtze River Office submitted the *Report About the Feasibility of the Three Gorges Project* to the authorities. But the feasibility meeting was aborted. In February 1981, the Yangtze River Office proposed a scheme of phased development of the Three Gorges Project with an initial water storage level of 128 m and dam elevation of 145 m.

In September 1982, the 12th National Congress of CPC was held. To meet national economic requirements, the Three Gorges Project again came under the spotlight. The State

1982年12月，长办开始研究正常蓄水位150米[①]的方案，并于次年3月提交了《三峡水利枢纽150米方案可行性研究报告》。1984年4月，国务院原则批准该方案。这就是通常所说的"150方案"。同年9月，重庆市提出将三峡工程正常蓄水位提高到180米（也称"180方案"），以便万吨级船队可以直达重庆市。

与此同时，社会各界对三峡工程也提出各种意见和建议，还有一些专家和社会知名人士对兴建三峡工程提出异议或者认为应该缓建。中共中央、国务院十分重视这些意见，1986年4月，国务院有关部门赴三峡地区进行实地考察，在听取不同意见后，决定对三峡工程开展进一步论证。

四、1986年对三峡工程的进一步论证

1986年6月2日，中共中央、国务院下发15号文件，即《中共中央、国务院关于长江三峡工程论证工作有关问题的通知》，决定"由水电部广泛组织各方面专家，进一步论证修改原来的三峡工程可行性报告，要注意吸收有不

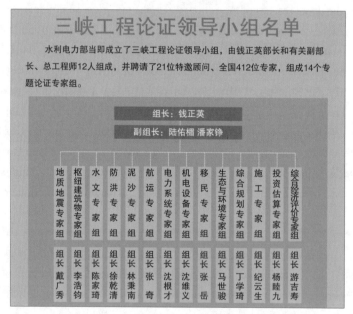

图2.3　三峡工程论证领导小组组成

---

① 指以吴淞口为零点，海拔高程为150米。

Council proposed to set about building the Three Gorges Projects immediately. However, given the national conditions at that time, the inundated area for building reservoirs had to be minimized, and construction was limited. The project was built using a low dam option, with a standard water storage level of 150 m. In November of the same year, Deng Xiaoping took a decisive stance on the building of the Three Gorges Project: "If we are sure of it, we must make up our mind and never waver over it."

In December 1982, the Yangtze River Office began to study the idea of water storage levels at 150 m[①], and it submitted the *Feasibility Study Report About the 150 Meters Scheme for the Three Gorges Project* to the authorities in March of the following year. In April 1984, the State Council approved this scheme in principle. This is what has been called the "150 scheme". In September of the same year, Chongqing City proposed raising the storage level to 180 m (also known as the "180 scheme") so that a fleet of 10,000 tons could reach Chongqing directly.

Meanwhile, people from all sectors of society offered their opinions and suggestions regarding the Three-Gorge Project. Some experts and well-known social figures objected to the project's building or suggested postponing it. The Central Committee of CPC and the State Council attached great importance to these opinions. In April 1986, the relevant departments of the State Council conducted a field investigation at the Three Gorges. After listening to different views, it decided to conduct further feasibility studies.

IV. Further Feasibility Studies of the Three Gorges Project in 1986

On June 2, 1986, the Central Committee of CPC and the State Council released the No. 15 document, namely the *Notice of the Central Committee of CPC and the State Council About Relevant Issues Concerning the Feasibility of the Three Gorges Project*. According to the document, "the Ministry of Water Resources and Power Industry shall extensively organize experts from all fields to further analyze and revise the original feasibility study report about the Three Gorges Project. It should introduce experts holding different viewpoints and then produce a new feasibility study report based on soliciting opinions and conducting in-depth research. The Three Gorges Project Review Committee of the State Council was established to review the feasibility report before submitting it to the Central Committee of CPC and the State Council for approval and finally submitting it to the National People's Congress for deliberation." From then on, the Three Gorges Project entered a stage defined by more

⇐ Figure 2.3 Members of the Three Gorges Project Feasibility Studies Leading Group

---

① It refers to an altitude of 150 m calculated from the zero point at Wusongkou.

同观点的专家参加,在广泛征求意见、深入研究论证的基础上,重新提出三峡工程的可行性报告;成立国务院三峡工程审查委员会负责审查可行性报告,提请中央和国务院批准,最终提交全国人民代表大会审议"。从此,三峡工程的论证工作进入了更广泛、更深入、更全面、更周密、更稳妥的阶段。

1986年6月19日,水电部成立以钱正英为组长、陆佑楣为副组长、潘家铮为副组长兼技术总负责人和杨振怀等12人为组员的三峡工程论证领导小组。这次论证分为10个主题,邀请了全国各行各业的412位专家,组成了14个专家组。专家学者们本着对人民负责的严肃精神和严谨的科学态度,反复分析讨论研究,分别提交了专题论证报告。

在国内进一步论证的同一时期,中国水电部和加拿大国际开发署共同聘请加拿大国际工程扬子江联营公司,与国内平行进行三峡工程可行性研究。1988年8月,加拿大方面按照国际通用标准编制了《中华人民共和国三峡水利枢纽工程可行性研究报告》。他们的总结论是:三峡工程效益巨大,三峡工程技术上是可行的,经济、财务上是合理的,建议早日兴建。这为国内最终方案提供了佐证。

1989年1月17日,长江委重新编制了正常蓄水位为175米的《长江三峡水利枢纽可行性研究报告》,原水电部三峡工程论证领导小组召开第十次会议,审议并原则通过了长江委编制的这份报告(审议稿)。根据近3年的论证成果,重新提出的三峡工程可行性研究报告的总结论是:"三峡工程对四化建设是必要的,技术上是可行的,经济上是合理的,建比不建好,早建比晚建有利。"推荐的建设方案是:"一级开发,一次建成,分期蓄水,连续移民。"至此,三峡工程建设的进一步论证工作全部结束。

extensive, comprehensive, thorough, prudent, and profound work.

On June 19, 1986, the Ministry of Water Resources and Power Industry set up the Three Gorges Project Feasibility Studies Leading Group consisting of Qian Zhengying (group leader), Lu Youmei (deputy group leader), Pan Jiazheng (deputy group leader and concurrently general technology director), Yang Zhenhuai and others (12 group members altogether). The feasibility work was divided into 10 themes. They invited 412 experts from all sectors of society across the country to form 14 expert groups. These experts, with a spirit of responsibility to the people and a rigorous scientific attitude, analyzed, discussed and studied the project before submitting their feasibility reports.

In the same period, the Ministry of Water Resources and Power Industry of PRC and the Canadian International Development Agency jointly engaged CYJV to conduct a project feasibility study. In August 1988, the Canadian side compiled the *Feasibility Study Report About the Three Gorges Project of the People's Republic of China* according to general international standards. Their general conclusion was as follows: The economic benefits of the Three Gorges Project are enormous; technically-speaking, this project is feasible; it is economically and financially reasonable and should be built as soon as possible. The report provided evidence for the project's final domestic plan.

On January 17, 1989, the Changjiang Commission re-compiled the *Feasibility Study Report About the Three Gorges Project* for a standard water storage level of 175 m. The former Three Gorges Project Feasibility Studies Leading Group of the Ministry of Water Resources and Power Industry convened the 10th meeting on which the report (the draft for deliberation) compiled by the Changjiang Commission was reviewed and passed in principle. According to the feasibility study achievements over the past three years, the following general conclusion of the re-proposed feasibility study report about the Three Gorges Project was reached: "The Three Gorges Project is essential to the building of the four modernizations; it is feasible technically and reasonable economically; to build it is better than not to build it; to build it early is more favorable than to build it later." The recommended construction scheme called for "First level development, one-time completion, staged water storage, and continuous resettlement." At this point, a further feasibility study of the construction of the Three Gorges Project was all but completed.

第二节 研究论证的主要内容

三峡工程研究论证的主要内容包括:建设"三峡工程"的必要性和可行性,葛洲坝工程与三峡工程建设的次序,三峡工程坝高、蓄水位的确定,三峡工程上马时间进度表,三峡工程项目管理机制,三峡工程的地质问题,三峡工程的水文问题,三峡工程的泥沙问题,三峡工程的环境问题,三峡工程的安全与人防问题,三峡工程突发事件应对机制的问题,三峡工程的移民问题,三峡工程的经济效益问题,三峡工程的文物保护措施问题,三峡工程的旅游开发措施问题等。下面选取一些典型问题进行阐述。

一、对葛洲坝工程与三峡工程建设次序的论证

在三峡工程论证上马之前,曾经发生过是否要在三峡工程上马之前先修建葛洲坝工程作为三峡工程准备和试点的讨论。1970年10月,修建葛洲坝工程的报告获得批准,将于年底开工。长办主任林一山坚决反对葛洲坝工程先于三峡工程仓促上马,他的理由是:葛洲坝工程在长江总规划中,是作为三峡水利枢纽的一个组成部分、一个反调节电站、一个航运梯级而存在的,它的先期兴建,将对三峡主体工程带来许多困难;而且,其综合效益也远不如三峡工程的好。

针对林一山提出的意见,毛泽东认为一旦发生战争,高坝(三峡)太危险,低坝(葛洲坝)出事则损失小。因此,从全局出发,先上葛洲坝工程优于先上三峡工程。虽然林一山的意见没有被采纳,但是中央下发葛洲坝工程上马的文件时,将林一山的意见一并下发全党,以提醒人们充分重视可能出现的问题。

Chapter 2 Research and Feasibility of the Three Gorges Project

Section 2 Main Content of Research and Feasibility

The main content of the research and feasibility of the Three Gorges Project includes the necessity of building the project, the construction sequence of the Gezhouba Dam and the Three Gorges Project, the determination of the dam height and storage level of the Three Gorges Project, the schedule of the Three Gorges Project, the project management of the Three Gorges Project, and problems concerning the geology, hydrology, sediment, environment, safety and civil air defense, emergency response mechanism, relocation, economic benefits, cultural relics preservation measures and tourism development. The following problems emerged:

I. Feasibility of the Construction Sequence of the Gezhouba Dam and the Three Gorges Project

Before the feasibility study of the Three Gorges Project was launched, authorities discussed whether the Gezhouba Dam should be built as the preparation and a pilot project of the Three Gorges Project in advance. In October 1970, the report on the construction of the Gezhouba Dam was approved, and the building was set to begin at the end of the year. Lin Yishan, the then director of the Yangtze River Basin Planning Office, resolutely opposed the idea of starting the Gezhouba Dam in such a hurry before the Three Gorges Project. He held that, in the general planning of the Yangtze River, the Gezhouba Dam was an integral part of the Three Gorges Project and served as a reverse regulating reservoir and shipping cascade of the former; to build the Gezhouba Dam earlier would bring many difficulties to the main work of the Three Gorges Project; moreover, it was much inferior to the Three Gorges Project in terms of comprehensive benefits.

In response to Lin Yishan's opinions, Mao Zedong believed that, in case of war, the high dam (Three Gorges) would be too dangerous, and the low dam (Gezhouba) would suffer less loss. Therefore, based on the overall situation, it was better to launch the Gezhouba Dam before the Three Gorges Project. Although Lin Yishan's opinions were not adopted, the CPC

事实证明,葛洲坝工程的建成不仅本身具有巨大的防洪、发电、航运效益,更对三峡工程上马具有重要而积极的意义:第一,葛洲坝工程为三峡工程积累了丰富的经验,为三峡工程做了重要的"实战准备"。第二,葛洲坝工程的实践证明了三峡工程论证的科学性、合理性(如泥沙论证、生态环境论证等),使各种分歧意见更趋统一。第三,葛洲坝工程的建成,也使国内大多数水电专家相信,中国水电科技队伍完全能够设计包括三峡工程在内的重大水利工程。第四,实践证明了葛洲坝工程的巨大效益,而三峡工程的论证表明,三峡工程的综合效益将远远超过葛洲坝工程。葛洲坝工程的成功激发了人们对三峡工程的巨大热情,使三峡工程的论证和建设进行得更快更好。

图 2.4　葛洲坝工程全景

二、三峡工程大坝选址的论证

万丈高楼平地起,必须有牢固的地基。三峡大坝的坝址选在三斗坪,是经过大量的地质勘探,在两个坝区、15 个坝段、数十个坝轴线中,历时 24 年,由专家充分论证后确定的。

1954 年春天,长江委对长江上游进行了地质勘察,目的是为三峡大坝选个理想的坝址。1958 年,周恩来总理召集了 100 多位专家考察三峡坝址。经

Central Committee issued them along with the documents on the launch of the Gezhouba Dam to the whole Party to remind people to pay full attention to problems that might occur.

Completing the Gezhouba Dam not only created great benefits in flood control, power generation and shipping but also provided plenty of momentum for the launch of the Three Gorges Project. First, the Gezhouba Project allowed those who worked on it to accumulate rich experience, making critical practical preparations for the Three Gorges Project. Second, the outcome of the Gezhouba Project proved that the selected sediment and ecological environment for the Three Gorges Project were scientific and reasonable and divergent opinions were more unified as a result. Third, the completion of the Gezhouba Dam also convinced most domestic hydropower experts that China was fully capable of designing major water conservancy projects, including the Three Gorges Project. Fourth, practice proved the great benefits of the Gezhouba Project, and the feasibility work of the Three Gorges Project showed that the comprehensive benefits of the Three Gorges Project would far exceed the Gezhouba. The success of the Gezhouba Project aroused people's great enthusiasm for building the Three Gorges Project and laid a more solid foundation for the feasibility and construction of the Three Gorges Project.

⊂ Figure 2.4   Panorama of the Gezhouba Dam

II. Feasibility of dam site selection for the Three Gorges Project

A high-rise building must have a solid foundation. The site of the Three Gorges Dam is located at Sandouping and was chosen after 24 years of geological exploration in two dam areas, 15 dam sections and dozens of dam axes.

In the spring of 1954, the Changjiang Water Resources Commission (CWRC) conducted a geological survey on the Yangtze River's upper reaches to select an ideal dam site for the Three Gorges Dam. In 1958, the then Premier Zhou Enlai convened more than 100 experts to investigate the site of the Three Gorges Dam. After repeated investigation, study, feasibility study and comparison, the dam site was finally located on the small island under the jurisdiction of Sandouping in the middle section of the Xiling Gorge of the Yangtze River in 1979, which is a little more than 40 km away from Yichang City, Hubei Province, the boundary point between the middle and lower reaches of the Yangtze River.

Sandouping almost rolls into one all the advantages of high dam sites at home and abroad.

First, granite is the ideal geological rock mass for building high concrete gravity dams. The granite in the Sandouping dam site is solid and dense, with complete rock mass. It can withstand ton-level pressure per square centimeter and is almost impervious, which is rarely

图 2.5　三斗坪江段中堡岛岩芯

过多次勘察研究和论证比选，最终于 1979 年把三峡大坝的坝址选定在长江西陵峡中段小岛三斗坪，这里距长江中、下游分界点——湖北省宜昌市 40 余千米。

三斗坪几乎集中了国内外高坝坝址的所有优点。

第一，三斗坪拥有建筑混凝土高坝最理想的地质岩体——花岗岩。三斗坪坝址的花岗岩，质地坚密，岩体完整，抗压强度达到了每平方厘米能经得起吨重的压力，且几乎不透水，在国内外的高坝坝址中非常少见。

整个长江干流上，类似三斗坪这样的好基岩还有不少，但并非所有拥有好基岩的江段都适合建坝。在宜宾以上建坝不能有效地解决长江中下游的防洪问题，自宜昌以下进入江汉平原，修坝又会淹没大片地区，所以只有从宜宾到宜昌这一段适宜建坝。而在这 1 000 多千米的江段上，仅有三斗坪拥有良好的花岗岩基岩，条件可谓得天独厚。

第二，三斗坪位于一个稳定性较高的地块上。以坝址为中心，半径为 320 千米范围内近 2 000 年的历史记载证明，三斗坪属典型弱震环境。经国家地震权威部门鉴定、核准，坝址区地震基本烈度为Ⅵ度，相当于对房屋造成轻微破坏的强度。

第三，三斗坪坝址河谷开阔，河床右侧有个中堡岛，便于大坝水电站船闸等枢纽建筑物布置，有利于施工中的导流和截流，还能够保证施工期长江干流航运不断航。

三、坝高、蓄水位的反复论证

1958 年以前，长江委的中苏专家曾拟定过 260 米、235 米、220 米、210 米、200 米、190 米等 6 个正常蓄水位方案，进行比较研究。结果表明：三峡正常蓄水位越高，则技术经济指标越优越，防洪、发电、航运的效益也越大。

◐ Figure 2.5   Drill Cores on the Zhongbaodao Island of the Sandouping Section

seen among high dam sites at home and abroad.

There are many good bedrock sites like Sandouping in the main channels of the Yangtze River, but not all river sections with good bedrock are suitable for building dams. Building a dam above Yibin will not effectively control floods in the middle and lower reaches of the Yangtze River, while large areas would be flooded if dams were built on the Jianghan Plain below Yichang. Therefore, the section from Yibin to Yichang was the only site suitable for dam construction. Along this river section of more than 1,000 km, only Sandouping enjoys good granite bedrock.

Second, Sandouping is located on a highly stable plot. According to historical records of the past 2,000 years, Sandouping is a typical weak earthquake environment within a radius of 320 km with the dam site at the center. National seismological authorities have verified that the basic seismic intensity in the dam site area is VI, equivalent to the intensity of slight damage to homes.

Third, the river valley of the Sandouping dam site is wide, and there is a small island named Zhongbaodao on the right side of the riverbed, making it convenient for the layout of the dam, hydropower station, ship lock and other hub structures. It was also conducive to water diversion and damming during construction and could ensure uninterrupted navigation of the main channels of the Yangtze River during the construction period.

III. Repeated Feasibility Studies of the Dam Height and Water Storage Level

Before 1958, the experts from China and the Soviet Union on the Changjiang Commission once prepared six standard water storage level schemes, including 260 m, 235 m, 220 m, 210 m, 200 m and 190 m, for comparison. The results indicated that: The higher the standard water storage level in the Three Gorges Project, the more favorable the technical and economic indices are, and the bigger the benefits from flood control, power generation and shipping. But if the water storage level were higher than 200 m, it would cause heavy losses from inundation in the urban areas of Chongqing and neighboring countryside; as a result, the number of people to relocate is enormous. After weighing the pros and cons repeatedly, the Chengdu Conference of the Central Committee of CPC in 1958 decided the following: "The height of the normal water storage level for the Three Gorges Dam should not be higher than

但正常蓄水位高于 200 米以后，重庆市区及邻近的农村都将造成较大的淹没损失，移民人数巨大。经过反复审慎的利弊权衡，1958 年的中央成都会议决定："三峡大坝正常水位的高程应当控制在 200 米，不能高于这个高程；同时，在规划设计中还应当研究 190 米和 195 米两个高程，提出有关的资料和论证。"

1980 年至 1982 年，长办先后提出了初期蓄水位 128 米、坝顶高程 145 米的低坝方案和最高蓄水位为 200 米的高坝方案等多个方案。1984 年 4 月，国务院原则批准长江委的《三峡水利枢纽 150 米方案可行性研究报告》，确定 150 米水位的低坝方案。后来因为重庆市向中央提出，将蓄水位提高到 180 米，以使万吨级船队直达重庆。邓小平当时表态认为中坝方案相对而言属于最佳方案，它既可以保障三峡水电站多发电，又能够使得万吨级船队通过大坝直达重庆。从蓄水防洪、发电、通航、库区移民等多个因素综合考虑，175 米是能够使各方面利益相对极化的边际决策，体现了三峡工程在坝高决策上的科学性。

在 1986 年进一步论证阶段，14 个专家组对兴建三峡工程的任务和技术经济条件，又做了多方面的论证，最后初选"坝顶高程 185 米，最终正常蓄水位 175 米，初期蓄水位 156 米"作为可行性研究阶段的基本方案。

四、三峡移民安置规划的长期探索

三峡移民工程是我国乃至世界水利建设史上的奇迹，也是三峡工程最大的难点。在工程总投资中，用于移民安置的经费就占 45%，尚不包括后期规划中所追加的投资在内。三峡工程蓄水后，由于淹没的面积之大、涉及移民搬迁的地域范围之广、产生的移民数量之多、持续的时间之长、涉及事项之多，以及与之相关的决策环节之复杂程度，在世界水利水电建设史上都是绝无仅有的。

在 1986 年进一步论证时成立的 14 个专家组中，专门设置了移民专家组。移民专家组总的结论是：三峡工程移民安置任务艰巨，但有解决的途径和办法，只要切实加强领导，紧紧依靠群众，精心做好规划，提前进行开发，并实行科学管理，经过努力可以把移民安置好。如果三峡工程推迟兴建，则不仅严重影响库区有关县市的经济发展，而且由于人口增加，淹没实物日益增

200 m; at the same time, the height of 190 m and 195 m should be studied in the planning and design, and relevant materials and feasibility work should be presented."

From 1980 to 1982, the Yangtze River Office proposed several schemes in succession, including the low dam scheme (the early-stage water storage level was 128 m, and the height of the dam was 145 m) and the high dam scheme (the highest water storage level was 200 m). In April 1984, the State Council approved the *Feasibility Study Report for the 150 Meters Scheme of the Three Gorges Project* submitted by the Changjiang Commission in principle, thus confirming the low dam scheme with the water storage level being 150 m Later, Chongqing City proposed to the Central Committee of CPC that the water storage level be raised to 180 m to allow fleets holding 10,000 tons to reach Chongqing directly. Deng Xiaoping took a stand at that time by opting for a middle dam scheme. This scheme could ensure the Three Gorges Project hydropower station generates more electricity and make fleets of 10,000 tons reach Chongqing directly. Taking into consideration many factors like water conservancy and flood prevention, power generation, shipping and relocation of residents in the reservoir area, people thought 175 m was a marginal decision that comparatively maximizes the interests of all parties, reflecting the scientific basis for determining the dam height for the Three Gorges Project.

In 1986, 14 expert groups conducted further feasibility work and examined the technological and economic conditions involved. At last, a preliminary "dam height of 185 m, the final standard water storage level of 175 m and the early-stage water storage level of 156 m" was decided upon.

IV. Long-term Exploration of How to Relocate Local Residents at the Three Gorges

The project of relocating residents from the Three Gorges area is a miracle in the history of water conservancy construction in our country and even in the world. It is also the project's greatest challenge. The funds used for relocation accounted for 45% of the project's total investment, in addition to the added investment at later stages. The water storage of the Three Gorges Project will inundate a large area, resulting in the need to relocate a staggering number of residents over a long period. The entire process involves extremely complicated decision-making. All these challenges are unprecedented in the world's history of water conservancy and hydropower stations.

There was an expert relocation group among the 14 expert groups set up in 1986. The group concluded that relocating residents for the Three Gorges Project is arduous but possible through effective leadership, close reliance on the masses, careful and advanced planning,

多，标准逐渐提高，移民安置费用将会急剧增长，安置难度也将显著加大。根据库区各县市的迫切要求和经济发展的需要，建议对三峡工程的兴建尽快做出决策。

为确保决策的科学性，三峡移民工作首先从试点工作开始，通过开发试点、广泛听取各方的意见和建议，积极探索有效途径，并在此基础上总结经验，制定详细的移民规划。同时在总结过去40年水利建设移民搬迁安置工作经验的基础上，国家还提出了开发性移民的新思路。

第三节 科学论证形成决策

三峡工程论证史同时也是三峡工程论辩史，通过正反两方面意见的相互碰撞和驳难，使三峡工程的规划、设计一步步走向成熟。三峡工程的上马和建设是按照"服从多数"和"尊重少数"的民主原则进行决策，经过几十年的广泛深入论证，凝聚了最广泛的智慧和力量，充分尊重和吸纳不同意见，形成最终决策。

一、广泛深入的论证

1986年6月19日，成立的三峡工程论证领导小组，为了接受各方面的监督、指导，论证领导小组聘请21人为特邀顾问。论证工作分为10个专题：地质地震、水文与防洪、泥沙与航运、电力系统规划、水库淹没与移民、生态与环境、综合水位方案、施工、工程投资估算、经济评价。

论证领导小组聘请了参加过由国家计委、国家科委组织的论证三峡工程正常蓄水位的106位专家，还聘请了国务院所属17个部委、中国科学院所属12个院所、29所高等院校和8个省市专业部门的各有专长的专家、学者306位，共计412位专家，分成14个专家小组参加论证工作。其中，中国科学院

and scientific management. If the building of the Three Gorges Project is postponed, the economic development of relevant counties and cities would suffer. What's more, the funds for relocation would rise sharply due to the increase in population and the increasing number of submerged objects. Given the urgent demand of all counties and cities in the reservoir area and the need for economic development, they called for an early decision on constructing the Three Gorges Project.

To guarantee the scientificalness in the decision-making, the relocation program in the Three Gorges area started as a pilot. The group solicited the opinions and suggestions of all parties and explored the most effective means to relocate residents. All these efforts were directed toward a detailed relocation plan. At the same time, our state proposed something new based on the experiences of water conservation resettlement over the past 40 years.

Section 3 | Decision-making Based on Scientific Feasibility Work

The history of the Three Gorges Project is a history of debate. Through collision and refutation, the Three Gorges Project matured step by step. The launch and construction of the Three Gorges Project were decided according to the democratic principle of "deferring to the majority" and "respecting the minority". Based on the extensive and in-depth discussions over several decades, a final decision was made by pooling together the most extensive wisdom and respecting different opinions.

I. Extensive and In-depth Feasibility Work

On June 19, 1986, the Three Gorges Project feasibility studies leading group was set up. To receive supervision and guidance from various parties, the group hired 21 people as special consultants. The feasibility work was divided into 10 special topics: earthquakes, hydrology and flood control, silt and navigation, planning of the electric power system, reservoir flooding and relocation of residents, ecology and environment, comprehensive water level scheme, construction, project investment estimation and economic evaluation.

The leading feasibility studies group engaged 106 experts who had participated in

学部委员15人，全国各学科学会理事长、副理事长和各专业委员会主任、副主任60余人，有高级职称的359人。专家中既有赞成兴建三峡工程的专家，也有不赞成或在某一方面有不同意见的专家；水利电力系统以外的专家占51.7%。上述数据充分说明参加论证的专家、学者，既具有很高的权威性，又具有广泛的代表性。

  三峡工程论证是一个复杂而庞大的系统工程，论证程序采用先专题、后综合，综合与专题互相交叉的论证方法。从流域、地区和全国经济发展三个不同层次分别考虑，首先审查各专题的基本资料，各专家组制定专题论证纲要，进行初步论证工作，在初步论证的基础上，综合择优选出一个各方面都可以接受的有代表性的正常蓄水位方案，再由各专家组深入论证。在有严格科学依据的基础上，提出专题论证报告。

  为了使专题论证符合科学化、民主化的原则，论证领导小组还对论证工作提出三点要求：第一，要充分利用过去的工作成果，包括长江委和其他部门、单位的大量工作成果和国家计委、国家科委组织进行的论证成果；同时又不局限于过去的结论。第二，要注意听取和重视各种不同意见，包括少数乃至个别专家的意见，对这些问题和意见要深入论证，实事求是地做出回答，对原来的可行性报告该修改的修改，该补充的补充。第三，论证的结论一定要有严格的科学依据，经得起检验。

  为了贯彻以上三点要求，各专家组根据论证工作的需要，在全国范围内委托高等院校和科研、勘测、设计、机电设备制造等单位，补充进行了大量调查、试验、计算等研究工作；同时，国家科委配合论证工作，组织了科技攻关。

Chapter 2 Research and Feasibility of the Three Gorges Project

studying the standard water storage level of the Three Gorges Project organized by the State Planning Commission and the State Scientific and Technological Commission, as well as 306 experts and scholars of different specialties from 17 ministries and commissions under the State Council, 12 institutes under the Chinese Academy of Sciences, 29 institutions of higher learning eight specialized departments at the provincial or municipal level. There were 412 experts altogether, and they were divided into 14 expert groups. Among them were 15 members of the Chinese Academy of Sciences, more than 60 people who were presidents or vice presidents of the societies of their respective disciplines or directors or deputy directors of different special committees, and 359 people having senior professional titles. Some favored building the Three Gorges Project while others did not; experts without a background in water conservancy or electric power accounted for 51.7%. The above data illustrates that the experts and scholars who participated in the feasibility studies were of high authority and broad representativeness.

The Three Gorges Project is gigantic and complicated. The expert groups needed to consider the situation from three dimensions: the river basin, the region and national economic development. First, they reviewed the basic materials of every topic, created thematic feasibility studies and carried out preliminary feasibility work. Based on the initial feasibility work, a representative standard water storage level plan acceptable to all parties would be selected based on comprehensive merit and further demonstrated by the expert groups. Feasibility reports on topics were presented and supported on a strictly scientific basis.

For feasibility work to fall in line with the scientific and democratic principles, the leading group also put forward three requirements for the feasibility work: First, to make full use of the results of past work, including by the Changjiang Commission and other departments and units and the State Planning Commission and the State Scientific and Technological Commission, without being confined to previous conclusions. Second, we should listen to and emphasize all kinds of different viewpoints; factual answers to the original feasibility report should be modified where necessary. Third, the conclusion must be scientifically sound and stand examination.

To meet the three requirements mentioned above, all expert groups entrusted institutions of higher learning and units in scientific research, surveying, design and manufacturing of electromechanical equipment nationwide with the task of conducting additional investigations, experiments and calculations. Meanwhile, the State Scientific and Technological Commission organized the workforce to tackle critical scientific and technical problems.

According to the above-mentioned principles and methods, the expert groups determined the construction scheme of "developed in one scale, completed at one go, water-storage levels rising in stages and relocation of local residents in succession," with the standard

根据上述论证原则和论证方法，专家组确定了"一级开发，一次建成，分期蓄水，连续移民"和正常蓄水位175米、坝顶高程185米的建设方案，各专家组又进行了深入细致科学的论证，最终分别提出了论证报告。

二、充分吸纳不同意见

尊重并吸纳少数人的反对意见既是民主决策的体现，更是科学决策的要求。对于三峡工程这样一个涉及地质、水利、电力、交通、环境、人防、移民等广泛领域，工艺极端复杂、史无前例的工程，更应该慎之又慎，要求决策必须做到充分、可靠、万无一失，对少数人的反对意见必须引起足够的重视并认真加以研究。三峡工程之所以经过几十年旷日持久的反复论证后才上马，正是因为充分重视了少数反对意见。

从中华人民共和国成立初期三峡工程提出伊始，毛泽东就指出要重点听取反对三峡工程的少数人的意见。1966年4月10日，毛泽东在时任长办主任林一山关于修建三峡的报告上批示："需要一个反面报告。"

1958年1月，中共中央在南宁召开政治局扩大会议，毛泽东决定把三峡工程问题正式提交政治局讨论，还决定让长办主任林一山、电力部水力发电建设总局局长李锐参加会议，他们是当时国内对三峡工程的两种不同意见的代表。林一山等人认为，"长江流域规划中必须首先解决防洪问题"，三峡是建设大坝的最好地址，在技术上也是可行的。李锐等人则认为，用三峡防洪是不必要的，三峡发的电用不了，会造成投资浪费，三峡工程所涉及的工程技术问题也无法解决。

毛泽东在耐心听取和认真研读双方观点后，并没有倾向于大多数人的意见，而是表示"三峡问题并没有最后决定要修建"。会议即将结束时，毛泽东在肯定修建三峡工程必要性的同时，提出"积极准备，充分可靠"的方针，并委托周恩来亲自抓长江流域规划和三峡工程研究。1958年2月底至3月初，周恩来率领中央和有关地方负责同志、中苏专家100多人，实地考察荆江河势和两岸大堤、三峡工程坝址和库区，并听取了各方面的意见。

1984年4月，国务院原则批准了三峡工程正常蓄水位为150米的可行性研究报告。然而，重庆市提出了180米蓄水位的建议，一些全国人大代表、

water storage level being 175 m and the dam height 185 m. All the expert groups worked thoroughly, carefully and scientifically before finally presenting their feasibility reports.

II. Embracing Different Opinions

Respecting and accepting the objections of minorities is not only the embodiment of a democratic decision-making process but also the requirement of scientific decision-making. The Three Gorges Project is an unprecedented project requiring extremely complicated technology and involving a wide range of fields like geology, water conservancy, electricity, transport, environment, civil air defense and relocation of residents. Extreme caution was exercised in handling such a project. Our decisions had to be reliable and perfectly safe. Our emphasis on the objections of minority opinions made the feasibility study of the Three Gorges Project a long-drawn-out process spanning several decades.

When the Three Gorges Project was first proposed at the beginning of the founding of the PRC, Mao Zedong pointed out that we should focus on the opinions of minorities against the move. On April 10, 1966, Mao Zedong gave instructions to the report about the building of the Three Gorges Project submitted by Lin Yishan, the then director of the Yangtze River Office: "A report about objections is demanded."

In January 1958, the Central Committee of CPC held the enlarged meeting of the Political Bureau in Nanning. Mao Zedong decided to submit the proposal of the Three Gorges Project to the Political Bureau for formal discussions, and he also decided to allow Lin Yishan, director of the Yangtze River Office, and Li Rui, director-general of the Hydropower Generation Construction Bureau of the Ministry of Electricity, to attend the meeting. Both were representatives of the two opposing sides of the Three Gorges Project at that time in China. Lin Yishan and his supporters believed that the top priority in the planning of the Yangtze River basin is the issue of flood control, the Three Gorges are the best site for building the dam, and it is technically feasible. Li Rui and his supporters believed that it was unnecessary to use the Three Gorges to prevent floods; the electricity the Three Gorges generates is more than needed; thus, the project would lead to a waste of investment; it would not be possible to solve the engineering and technological problems of the project.

After patiently listening to and carefully reading the viewpoints of both sides, Mao Zedong didn't favor the majority's views. Still, he claimed that "we have not finally made up our mind to build the Three Gorges Project yet." At the end of the meeting, Mao Zedong put forward the guiding principle of "well-prepared and fully reliable" while affirming the necessity for building the Three Gorges Project. He entrusted Zhou Enlai with planning the Yangtze River basin and researching the Three Gorges Project. From the end of February to

全国政协委员及专家学者提出了一些不同意见，引起了党中央、国务院的高度重视，于1986年6月2日下发《中共中央、国务院关于长江三峡工程论证工作有关问题的通知》，要求"注意吸收有不同观点的专家参加，发扬技术民主"。

为了广泛听取不同意见，14个专家组中都聘请了持有不同意见的专家。论证领导小组办公室收集国内外报刊上发表的对三峡工程的各种不同意见，汇编成《对建设三峡工程不同意见文章选编》1～8辑；同时还收集了第六届全国人大、全国政协第六届第二、第三、第四、第五次会议上代表议案、委员提案中对三峡工程的不同意见，分别汇编成册，印发给参加论证工作的每位专家。

在论证过程中，各专家组在本专业范围内独立负责地进行论证，不受行政干预，经反复调研、充分讨论且认真对待不同意见后，提出本专题论证报告，每位专家要签名负责。仍然有不同意见的专家，可以不签名，提出的书面意见作为专题论证报告的附件一并上报。14个专家组中，有9个专家组的论证报告一致通过；有5个专家组的10位专家11人次没有在论证报告上签名，有9位专家附上了他们个人的不同意见。1990年7月，时任国务院总理李鹏等在听取三峡工程论证成果汇报时，把这9位专家全部请到了中南海，并安排他们发言，认真听取他们的意见。

正是因为有了不同意见，才使得根据论证成果重新提出的长江三峡工程可行性研究报告更加细致、精确和稳妥。在三峡工程建设过程中，针对不同意见，设计、科研部门都进行了论证、计算和复核。正是因为各界人士的不同意见和认真对待这些不同意见，才有了今天的三峡工程。

特别要强调的是，正是因为听取了各方不同的意见，现在实施的方案较原方案相比，在防洪、发电、航运等方面的效益大大提高。三峡工程蓄水运行至今，主要工程问题均在初步设计预计范围内，工程的实际表现已在一定程度上回答了规划论证阶段的各种疑虑。

Chapter 2 Research and Feasibility of the Three Gorges Project

the beginning of March 1958, Zhou Enlai led more than 100 people, including the comrades on the Central Committee of CPC and from local authorities in charge of this project and experts from China and the Soviet Union to carry out on-the-spot survey at Jingjiang River and the dykes on both sides, the dam site of the Three Gorges Project and the reservoir area. He listened to the opinions of all parties.

In April 1984, the State Council approved the feasibility study report supporting a standard water storage level of 150 m for the Three Gorges Project in principle. However, Chongqing proposed to raise the water storage level to 180 m. Some deputies to the National People's Congress, members of the national committee of CPPCC, experts and scholars put forward different viewpoints, to which the Central Committee of CPC and the State Council paid great attention. They released the *Notice of the Central Committee of CPC and the State Council About Relevant Issues Concerning the Feasibility of the Three Gorges Project* on June 2, 1986, requiring people to engage experts with different viewpoints and carry forward technological democracy.

To listen to different opinions extensively, all 14 expert groups engaged those with differing viewpoints. The office of the leading feasibility studies group collected different views about the Three Gorges Project published in domestic and international newspapers and periodicals and compiled into the 1-8 Parts of the *Selection of Articles Having Different Opinions About the Three Gorges Project*. At the same time, different viewpoints from the proposals of the deputies or members of the sixth People's Congress and the second, third, fourth and fifth sessions of the sixth National Committee of CPPCC were collected and compiled into books before being released to each expert participating in the feasibility work.

The expert groups carried out their work without administrative interference. The feasibility report concerning special topics was presented after repeated surveys, sufficient discussion and pondering over different opinions in earnest. Each expert needed to sign his or her name on the report. Those experts still holding different opinions could refuse to sign. The proposed written suggestions would be submitted as the enclosure of the feasibility report about the particular topic. Among the 14 expert groups, the feasibility reports of nine expert groups were passed unanimously; 10 experts in five groups didn't sign their names on the feasibility reports on 11 occasions, and nine experts enclosed their different personal opinions. In July 1990, Li Peng, the then Premier of the State Council, invited all nine experts to Zhongnanhai while listening to the feasibility results report about the Three Gorges Project. Li Peng asked them to make speeches and listened to their opinions carefully.

These different opinions made the rewritten feasibility study report more meticulous, accurate and reliable. The design and scientific research departments engaged with different ideas during the project's construction. Because the diverse views from all walks of life were

三、形成最终决策

1990年7月,国务院召开了三峡工程论证汇报会,决定将重新编制的《长江三峡水利枢纽可行性研究报告》提交国务院三峡工程审查委员会审查。审查工作自1990年12月开始,1991年8月结束,共召开了三次会议,并组织专家赴现场进行了实地考察。在审查期间,审查委员会还组织163名专家进行预审,提出预审意见。在此基础上,形成审查意见。

审查意见的主要结论是:兴建三峡工程的效益是巨大的,特别是对防御长江荆江河段的洪水灾害是十分必要和迫切的;技术上是可行的,经济上是合理的;我国国力是能够承担的,资金是可以筹措的;无论是从发挥三峡工程巨大的综合效益,还是从投资费用和移民工程的需要来看,早建都比晚建有利。

在第七届全国人大第五次会议期间,李鹏总理提交了《国务院关于提请审议兴建长江三峡工程的议案》,邹家华副总理作了《关于提请审议兴建长江

图2.6 第七届全国人大第五次会议投票现场

treated seriously, the Three Gorges Project is where it is today.

It is imperative to emphasize that listening to different opinions from all sides made the program currently in place much more effective in terms of flood control, power generation and shipping compared to the original program. To a certain extent, the project's actual performance has answered all doubts in the planning and feasibility work stage.

III. The Final Decision

In July 1990, the State Council held a meeting on the feasibility work and decided to submit the recompiled *Feasibility Study Report About the Three Gorges Project* to the Three Gorges Project Review Committee of the State Council for examination. The review started in December 1990 and ended in August 1991. During that period, three meetings were held, and experts carried out on-the-spot investigations. In addition, the Review Committee organized 163 experts to conduct a preliminary review and propose suggestions. On this basis, the review comments were formed.

The main conclusion was as follows: The economic benefits brought by the Three Gorges Project are enormous, and the project is vital and urgent in the prevention of the flooding at the Jingjiang section of the Yangtze River; it is feasible technically and reasonable economically; the national power of our country can afford it, and the funds can be raised; building it at an earlier date is more favorable than at a later date no matter in terms of giving play to the enormous comprehensive benefits of the Three Gorges Project or in terms of investment and resident relocation program.

During the fifth session of the seventh National People's Congress, Premier Li Peng submitted the *State Council's Proposal for Submitting the Building of the Three Gorges Project for Review*. Vice-premier Zou Jiahua presented his *Explanation About the Proposal for Submitting the Building of the Three Gorges Project for Review*. On April 3, 1992, the meeting put the resolution on building the Three Gorges Project to the vote. The voting results were: 1,767 votes in favor, 177 votes against and 664 abstentions; 25 representatives declined to vote. The affirmative votes accounted for 67.1% of all votes, exceeding 50%. On the same day, Wan Li, the NPC Standing Committee chairman, announced solemnly at 15:21 pm: The *Resolution on Building the Three Gorges Project* was passed. The passing of this resolution

○ Figure 2.6　The Ballot Casting Scene of the Fifth Session of the Seventh National People's Congress

三峡工程的议案的说明》。1992年4月3日，会议对关于兴建三峡工程的决议进行表决。表决结果为：赞成1767票，反对177票，弃权664票，有25名代表未按表决器。赞成票占全部票数的67.1%，超过半数。同日，15时21分，全国人大常委会委员长万里庄严宣布：《关于兴建长江三峡工程的决议》通过。兴建三峡工程决议的通过，充分体现了决策的科学化、民主化。

三峡工程在论证和表决过程中，将几十年来不同意见的争论都严格定性在学术和业务的范畴。三峡工程议案审议过程中的少数人意见，以及表决中的反对票、弃权票都得到宪法和法律的保护，不受任何法律追究。潘家铮院士曾说过，对三峡工程贡献最大的人是那些反对者。"正是他们的追问、疑问甚至是质问，逼着你把每个问题都弄得更清楚，方案做得更理想、更完整，质量一期比一期好。"

在新华社关于决议表决通过的新闻报道中这样写道：

掌声里，许多人流下了激动的泪水。

掌声里，人们知道，在共和国年轻的历程里，她的最高权力机关专门对某项建设工程做出决策，这是第一次。它意味着一个中国人做了70年的梦会在不远的将来成为现实。

投赞成票的，投反对票的，投弃权票的，包括未按表决器的，都在这神圣的殿堂里表示了自己的意见，他们都有一颗中国心，都是对国家、对民族、对历史负责的。

本章小结：

三峡工程是科学决策和民主决策的高度统一。科学决策贯穿三峡工程论证和建设管理全过程，民主决策凝聚了最广泛的智慧共识，为党和国家准确认识事物内在规律、及时把握历史发展机遇、清醒判断社会主要矛盾、果断进行集中统一决策提供了坚实保障。从20世纪50年代毛泽东同志批示"积极准备、充分可靠"，到1982年邓小平同志提出"看准了就下决心，不要动摇"，到1986年国家组织412位专家对三峡工程进行重新论证得出"建比不建好，早建比晚建有利，建议早作决策"的论证结论，再到1992年七届全国人大五次会议审议通过《关

fully reflected that the decision-making process was scientific and democratic.

In the feasibility work and voting of the Three Gorges Project, the debates over different opinions lasting for several decades were strictly confined to the scope of academia and business. In reviewing the proposal about building the Three Gorges Project, the views of the minority and the votes against it or abstentions were protected by the constitution and law and were not subject to any legal action. Academician Pan Jiazheng once said that those who made the most significant contributions to the Three Gorges Project were the people objecting to it. "It was exactly their questioning, doubt or even interrogation that forced us to clarify each matter and to produce a more ideal and complete scheme. The next scheme was always better than the previous one in quality."

Xinhua News Agency's report on the voting about the resolution read:

> Many people burst into tears with excitement in applause.
>
> In applause, people understood for the first time that our young republic's highest organ of state power had decided on a specific and vital project. It meant the dream the Chinese people had pursued for 70 years would soon become a reality.
>
> Those who voted for it or against it or abstained or didn't press the voting machine all expressed their viewpoints in this holy hall. They love China and are responsible for our country, nation, and history.

Chapter Summary:

The Three Gorges Project reflects the deep integration of science and democracy in the process of its decision-making. Science was applied in the whole process of the feasibility work, construction and management of the Three Gorges Project. Democratic decision-making pooled the wisdom from the broadest spectrum of society, laying a solid foundation for the Party and government to understand the inherent law of the things accurately, seize historical opportunities for development in a timely manner, judge the the principal contradiction in Chinese society soberly, and carry out unified and centralized decision-making resolutely. From comrade Mao Zedong's instruction of "well-prepared and fully reliable" in 1950s to the idea of "If we are sure of it, we must

于兴建长江三峡工程的决议》，都是在不同历史阶段，三峡工程科学决策和民主决策辩证统一的生动体现。

参考文献：

［1］不尽的思念［M］. 北京：中央文献出版社，1987.

［2］陈夕. 中国共产党与长江三峡工程［M］. 北京：中共党史出版社，2014.

［3］长江流域规划办公室. 三峡水利枢纽初步设计要点报告［EB］，1958.

［4］李兵. 邓小平长江三峡行［J］. 党史博览，2004（1）.

［5］《中国三峡建设年鉴》编纂委员会. 中国三峡建设年鉴（2000）［J］. 宜昌：中国三峡建设年鉴社，2000.

［6］《中国三峡建设年鉴》编纂委员会. 中国三峡建设年鉴（2001）［J］. 宜昌：中国三峡建设年鉴社，2001.

［7］《中国三峡建设年鉴》编纂委员会. 中国三峡建设年鉴（1995）［J］. 北京：中国三峡出版社，1996.

make up our mind and never waver over it" proposed by comrade Deng Xiaoping in 1982, to the feasibility conclusion of "To build it is better than not to build it; to build it early is more favorable than to build it later. We advise an early decision" drawn from a new round of feasibility study of the Three Gorges Project by 412 experts organized by the government in 1986, and to the *Resolution on Building the Three Gorges Project* was passed by the fifth session of the seventh National People's Congress in 1992, the dialectical unity of science and democracy in the decision-making process of the Three Gorges Project at different historical stages was vividly illustrated.

References:

[1] *Endless Longing* [M]. Beijing: Central Party Literature Press, 1987.

[2] Chen Xi. *The Chinese Communist Party and the Three Gorges Project on the Yangtze River* [M]. Beijing: Chinese Communist Party History Publishing House, 2014.

[3] The Yangtze River Basin Planning Office. *Report About the Key Factors in the Preliminary Design of the Three Gorges Project* [EB], 1958.

[4] Li Bing. *Deng Xiaoping's Trip to the Three Gorges on the Yangtze River* [J]. General Review of the History of CPC, 2004 (1).

[5] Editorial Board of China Three Gorges Construction Yearbook. *China Three Gorges Construction Yearbook (2000)* [J] Yichang: China Three Gorges Construction Yearbook Press, 2000.

[6] Editorial Board of China Three Gorges Construction Yearbook. *China Three Gorges Construction Yearbook (2001)* [J] Yichang: China Three Gorges Construction Yearbook Press, 2001.

[7] Editorial Board of China Three Gorges Construction Yearbook. *China Three Gorges Construction Yearbook (1995)* [J] Yichang: China Three Gorges, 1996.

> 阅读提示：

　　三峡枢纽工程是一座具有防洪、发电、航运、水资源利用等巨大综合效益的大型水利水电工程，由拦江大坝、水电站和通航建筑物等部分组成。它不仅有世界上最大规模的混凝土建筑物，而且有世界上装机容量最大的水电站、规模最大的升船机、泄洪能力最大的泄洪闸、级数最多且总水头最高的内河船闸，导流截流之规模与难度堪称世界之最。

　　这项工程采用"一级开发，一次建成，分期蓄水"的方针，"三期导流、明渠通航"的施工方案，在整个建设过程中"三围长江，两改江流"，历经十七载风雨，三峡枢纽工程终于破浪而出，壁立西江。这其中，攻克了多少技术难关，又凝聚了多少人的智慧与汗水。

　　The Three Gorges Project is a large-scale multipurpose hydro-development project designed to yield comprehensive benefits in flood control, power generation, navigation and water resource utilization. The project includes a river dam, power generators, navigation structures and more. It is not only the world's largest concrete structure, but also the world's largest hydropower station in terms of installed capacity. In addition, it boasts the world's largest ship lift, the world's largest flood discharge sluice gate, and the world's largest inland ship lock in terms of the number of steps and gross water head. Furthermore, it is home to the world's largest scale and most challenging water diversion and closure.

　　The project was undertaken under the official briefs of "developed on one scale, completed in one go, with water-storage levels rising by stages" and "diversion in three phases and navigation through open channels". To ensure the quality of the concrete placement, temporary cofferdams were built three times to enclose the dam site as the foundation pit, and the water was diverted twice to the pre-built drainage channels. After 17 years, and thanks to the arduous efforts of numerous experts and workers and in spite of complex technical difficulties, the dam was finally completed.

Chapter 3 >>>>

第三章
三峡枢纽工程的建设
Construction of the Three Gorges Project

第一节 三峡枢纽工程的构成

人们常常把"三峡枢纽工程"简称"三峡工程",实际上三峡工程包括枢纽工程、输变电工程和移民工程。

三峡枢纽工程是长江中上游段建设的大型水利工程项目,它位于长江三峡西陵峡中段湖北省宜昌市境内的三斗坪,距下游的葛洲坝水利枢纽38千米。

三峡枢纽工程的主要建筑物包括三大部分:

- 挡水泄洪建筑物:三峡拦河大坝(通常简称"三峡大坝")。
- 发电建筑物:三峡水电站。
- 通航建筑物:包括双线五级连续梯级船闸(通常简称"三峡船闸")和垂直升船机。

从长江左岸至右岸三峡枢纽建筑物的布置是:三峡船闸,垂直升船机,左岸非溢流坝段和电源电站、左岸电站,泄洪坝段,右岸非溢流坝段和右岸电站、右岸地下电站。

三峡枢纽工程规模巨大,是当今世界上最大的水利水电工程。从工程规模来看,三峡枢纽工程的混凝土浇筑总量达2 807万立方米,是世界上最大规模的混凝土建筑物。枢纽工程最大泄洪流量超过

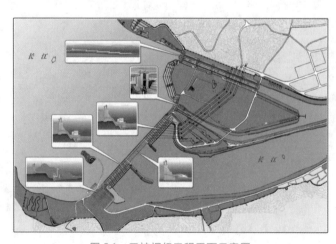

图 3.1 三峡枢纽工程平面示意图

Chapter 3 Construction of the Three Gorges Project

Section 1 | Composition of the Three Gorges Project

The Three Gorges Water Conservancy Pivotal Project, more commonly known as the Three Gorges Project, consists of the pivotal dam project, a power transmission and transformation project and a resettlement project.

The Three Gorges pivotal dam project is a large-scale water conservancy project in the middle and upper reaches of the Yangtze River. It is located in Sandouping, Yichang City, Hubei Province, in the middle section of Xiling Gorge, which is part of the Three Gorges, 38 km away upstream from Gezhouba Dam.

The main structures of the Three Gorges pivotal dam project include:
• Water retention and flood release structure: the Three Gorges Dam.
• Power generation structures: the Three Gorges Hydropower Station.
• Navigation structures: the double-line continuous five-step ship lock (commonly known as the Three Gorges ship lock) and the vertical ship lift.

Figure 3.1 shows the layout of the dam. From top to bottom (left to right banks of the Yangtze River) it shows: The ship lock, vertical ship lift, left-bank non-overflow section and power source station, left-bank power station, spillway section, right-bank non-overflow section and right-bank power house, right-bank underground power station.

The Three Gorges Dam is the world's largest hydropower project. In terms of project scale, the total volume of concrete placement for the Three Gorges Project was about 28.07 million m^3, making it the world's largest concrete structure. The maximum discharge capacity of the dam exceeds 100,000 m^3/s, and the total flood discharge flow exceeds 100 million kW, making it the largest in the world. With 34 sets of water-turbine generators, the hydropower station has a generating capacity of 22.5 million kW, the world's biggest hydropower plant in terms of installed capacity. The Three Gorges ship lock is the world's largest inland ship lock in terms of the number of steps, gross water head (113 m), and the water head borne by the water conveyance system (45.2 m). With 24 large miter gates (12 pairs), the dam is colloqui-

◐ Figure 3.1　Plan of the Three Gorges Pivotal Dam Project

10万立方米每秒，泄洪水流总功率逾1亿千瓦，是世界上规模最大的。三峡水电站共安装水轮发电机组34台，总容量达2 250万千瓦，是当今世界上装机容量最大的水电站。三峡船闸是目前世界上级数最多、上下游水位落差最大（113米），输水系统承担的水头最大（45.2米）的内河船闸。24扇大型人字闸门，号称"天下第一门"。垂直升船机提升高度113米、提升重量15 500吨，可通过3 000吨级船舶，也是世界上最大的。

一、三峡大坝

1. 三峡大坝的概况

三峡大坝是一座混凝土重力坝，坝顶轴线总长2 309.47米，坝顶高程185米，泄洪坝段建基面最低高程4米，最大坝高181米，相当于60层楼房的高度；坝顶最大宽度22.60米，坝底最大宽度126.73米。大坝横剖面的形状近似直角梯形。

混凝土重力坝是依靠自身重量及其与地面的摩擦力抵抗水库上游水的压力荷载，以维持自身稳定的大坝。经过计算，三峡大坝正常蓄水175米时，整个大坝将承受约2 000万吨的水压力。

三峡大坝的重量，主要包括混凝土、发电机组设备、各种金属埋件和闸门等金属结构的重量，总计约为4 000万吨，相当于5个半埃及胡夫金字塔的重量。

浩荡东去的长江水，是通过布设在三峡大坝上的许多孔洞，流向大坝下游的。

三峡大坝共布设89个孔洞，不同位置的孔洞，其功能各不相同。其中，泄洪坝段分两层布设45个孔洞，包括22个表孔和23个深孔，主要用于泄洪。大坝左、右侧的厂房坝段布设38个孔洞，包括26个进水孔与发电机组对应，用于水轮发电机组发电；3个排漂孔、2个冲沙孔和7个排沙孔，主要用于各种漂浮物的下泄和冲沙。右岸的地下电站也布设6个孔洞，为6台发电机组的进水孔。

为了利用长江水流的能量，通过三峡大坝的水流一般都要经过水轮发电机组发电后，再流向大坝下游。根据三峡水电站的设计方案，每台机组正常

ally known as the "World's No. 1 Sluice Gate". The vertical ship lift is also the biggest in the world with maximum lifting height of 113 m and a maximum lifting weight of 15,500 tons, allowing the transportation of ships weighing as much as 3,000 tons.

I. The Three Gorges Dam

1. Overview of the Three Gorges Dam

The Three Gorges Dam is a concrete gravity dam. The axial length of the dam is 2,309.47 m, the crest height is 185 m, the minimum height of the foundation surface of the spillway section is 4 m, and the maximum height of the dam is 181 m, equal to the height of a 60-story building. The maximum crest width is 22.60 m and the maximum base width is 126.73 m. The cross section of the dam is approximately a right trapezoid.

Concrete gravity dams rely on their own weight and friction with the ground to resist the pressure load of the reservoir's upstream water and ensure stability. It has been calculated that at a normal water storage level of 175 m, the Three Gorges Dam bears the water pressure of approximately 20 million tons.

The weight of the Three Gorges Dam is mainly due to the concrete, generator sets, and the various embedded metal parts, gates and other metal structures. This amount to 40 million tons, 5.5 times that of the Great Pyramid of Giza (a.k.a. the Pyramid of Khufu).

The Yangtze water flows downstream through the numerous holes distributed throughout the dam.

There are totally 89 holes in the Three Gorges Dam, covering various different functions. Among them, the spillway section has 45 holes in two layers, including 22 surface holes and 23 deep holes, which are mainly for flood discharge. The power plant section has 38 holes on the left and right sides of the dam, including 26 inlet holes corresponding to the generator sets, which are used for generating power through the water-turbine generator sets. In addition, it has three drift holes, two sand flushing holes and seven sand discharge holes, which are used for discharging various floating objects and flushing the sand. The underground power station on the right bank is also equipped with six holes, which function as inlet holes for six of the generator sets.

In order to make use of the energy of the Yangtze, the water flowing through the Three Gorges Dam usually passes through the water-turbine generator sets for power generation before moving downstream. According to the design scheme of the Three Gorges Hydropower Station, each set needs approximately 950 m^3/s of water flow for normal power generation, and the water flow required for all the generator sets to generate power at the same time is

发电时约需过水流量 950 立方米每秒，所有机组同时发电所需的水流流量约为 3.1 万立方米每秒。

每年 10 月至来年 5 月是枯水期，长江上游的来水流量一般均小于 3.1 万立方米每秒。此时，除通过三峡船闸的极少部分水量外，上游来水全部通过水轮发电机组发电后流向下游。

每年 6 月至 9 月的汛期，三峡水库下泄流量要服从防洪调度需要。当上游来水流量大于 3.1 万立方米每秒时，部分流量通过水轮发电机组发电后流向下游，超过机组发电所需的流量，要在保证下游防洪安全的前提下，根据防洪调度指令，或蓄在水库内，或通过泄洪坝段泄向下游。

2. 三峡大坝的建筑材料

三峡大坝是混凝土重力坝，其原材料都是经过严格优选出来的，这也是三峡大坝坚固可靠的秘诀。这些优选的原材料主要有：骨料、水泥、粉煤灰、拌和水、外加剂。

骨料：是混凝土中的主要材料，约占混凝土体积的 80%，是大坝的主体。骨料分为粗骨料——石头、细骨料——沙。三峡大坝的骨料，来源于大坝基坑、船闸和料场开挖出来的新鲜花岗岩，具有较高的抗压强度。由人工砂石料开采加工系统用机械破碎成不同粒径的人工骨料，提供三峡工程使用。

水泥：是大坝骨料的黏合剂，只有它的加入才使骨料由一盘散沙，成为坚如磐石的混凝土坝体。三峡大坝使用的水泥，很有讲究。它是低热或中热

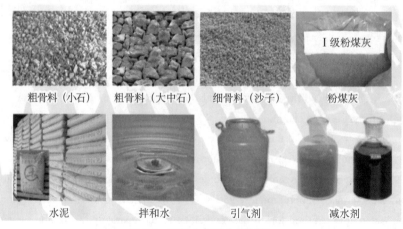

图 3.2　三峡大坝的主要材料

approximately 31,000 m³/s.

During the dry season from October to May, the inflow of water from the upper Yangtze is generally less than 31,000 m³/s. At this time, except for a small proportion of the water that passes through the Three Gorges ship lock, all the upstream water from the upstream goes through the water-turbine generator sets for power generation before flowing downstream.

During the flood season from June to September, the discharge of the Three Gorges Reservoir is subject to the flood control requirements. When the upstream water flow is more than 31,000 m³/s, part of the flow will go through the water-turbine generator sets for power generation before heading downstream, while the excessive flow shall, in order to ensure downstream flood control safety, be either stored in the reservoir or discharged downstream through the spillway section according to flood control operational procedures.

2. Building materials of the Three Gorges Dam

The Three Gorges Dam is a concrete gravity dam built with carefully selected materials that ensure the strength and stability of the dam, largely comprised of aggregates, cement, fly ash, mixing water and admixtures.

Aggregates: Aggregates are the main materials in concrete, accounting for approximately 80% of its total volume, and are used in the main body of the dam. The aggregates can be divided into coarse aggregate, namely crushed stone, and fine aggregate, namely sand. These materials come from fresh granite, which has a high compressive strength, excavated from the dam foundation pit, ship lock and stockyard. The granite was crushed into artificial aggregates of different sizes by an artificial aggregate mining and processing system, and then used to build the dam.

Cement: Cement is the adhesive for the dam's aggregate that binds the loose aggregates into the solid concrete dam body. The cement used in the Three Gorges Dam was strictly selected. It is comprised of both low-heat and moderate-heat cement, which has the following advantages: First, the low-heat nature can reduce the hydration heat temperature of the concrete inside the dam; second, the micro-expansive nature can reduce shrinkage cracks that can be a problem for concrete structures.

Fly ash: Fly ash is an admixture in dam concrete and comes from the waste generated after fuel combustion in thermal power plant power generation. Fly ash can take the place of part of cement and thus reduce the calorific value in the hardening process of cement. Fly ash microspheres can also greatly improve the workability of the concrete mixture, reducing the

⊂ Figure 3.2　Main Building Materials of the Three Gorges Dam

水泥，这类水泥具有以下优点：一是低发热性，可降低大坝内部混凝土的水化热温度；二是具有微膨胀性，可减少混凝土干缩裂缝。

粉煤灰：是大坝混凝土中的掺和物。粉煤灰本是火电厂燃烧发电后生成的废料，掺加粉煤灰可取代部分水泥并减少水泥硬化过程中的发热量。利用粉煤灰的微珠效应，可大大改善混凝土拌和物的和易性，而且还可降低混凝土的用水量，方便混凝土的振捣施工。

拌和水：也是混凝土成分之一。没有水的参与就不会使水泥发生化学变化，水泥就不会成为黏合剂。水在混凝土中的作用，很像一句成语"成也萧何败也萧何"，水少了拌和不成混凝土，水多了混凝土强度降低。如何少用拌和水而且不会影响混凝土的强度与和易性呢？这就需要掺加外加剂。

外加剂：三峡大坝主要采用了减水剂和引气剂两种外加剂。别看外加剂掺加的分量仅占混凝土水泥重量的百分之几，甚至千分之几，但是作用非凡。三峡大坝选用了与其他原材料有良好适应性，且减水率在18%以上，其他指标满足国标一等品的高效减水剂。这是降低混凝土用水量的一个非常重要的措施，为配制高强度性能的大坝混凝土创造了条件。引气剂的作用，是利用在混凝土中生成互不贯通的微小气泡，增强抗冻融系数来提高混凝土的耐久性，可以达到延长大坝使用寿命的效果。

三峡大坝的不同部位，还会根据功能进行强化处理，比如大坝泄洪坝面的混凝土中掺入了钢纤维，增加抗冲耐磨性能；通过大坝的发电水流采用压力钢管进入水轮发电机组发电，避免了有压力的江水直接冲刷大坝混凝土本体，保证大坝的安全。

二、三峡水电站

1. 三峡水电站的概况

三峡水电站包括左、右岸2个坝后式水电站，1个地下水电站和1个电源电站。2个坝后式水电站分别位于左、右两侧厂房坝段紧靠大坝的下游，地下水电站位于河道右岸的山体中，电源电站位于河道左岸的山体中。三峡水电站总装机容量2 250万千瓦，是目前世界上装机容量最大的水电站，其中，单

concrete's water consumption, and facilitating its vibration operation.

Mixing water: Mixing water is also a component of concrete. Without water, the required chemical change in the cement will not occur, and it will not become adhesive. The role of water in concrete is very similar to the Chinese idiom that "the key to one's success is also one's undoing". Insufficient water will result in an inability to mix the cement into concrete, while too much water will reduce the strength of concrete. In order to reduce the amount of mixing water without affecting the strength and workability of concrete admixtures are added.

Admixtures: In addition to fly ash, the Three Gorges Dam adopts two other admixtures, namely a water reducing agent and an air entraining agent. Although the quantity of these admixtures accounts for only a few hundredths or even thousandths of the cement's weight, their effect is considerable. The Three Gorges Dam employs a high-performance water reducing agent that is very adaptable to other raw materials; it has a water reduction rate of over 18%, and its other indicators meet the national first-class standard. This is a very important measure to reduce the concrete's water consumption, creating the conditions needed to prepare high-strength dam concrete. The role of the air entraining agent is to improve the concrete's durability by generating tiny, discrete bubbles in the concrete and enhancing the frost thaw resistance coefficient, thus extending the dam's service life.

Different parts of the Three Gorges Dam have also been consolidated according to their functions. For example, steel fibers were mixed into the concrete of the structure's flood discharge dam surface to increase its impact and wear resistance. The water for power generation flowing through the dam enters the water-turbine generator sets through penstocks to generate electricity, which avoids the pressure water from directly washing against the dam concrete body and ensures the dam's safety.

II. Three Gorges Hydropower Station

1. Overview of the Three Gorges Hydropower Station

The power generation structures consist of two powerhouses at the dam toes on the left and right banks, an underground powerhouse, and a power source station. The two powerhouses at the dam toes are located on the left and right sides of the power plant section, close to the downstream section of the dam. The underground powerhouse is housed in the mountain on the right bank of the river and the power source station is located in the mountain on the river's left bank.

The Three Gorges Hydropower Station has a total installed capacity of 22.5 million kW,

图 3.3　三峡左岸电站厂房内景（摄影：张祖新）

机容量 70 万千瓦的水轮发电机组 32 台（左岸电站安装 14 台，右岸电站安装 12 台，地下电站安装 6 台）；电源电站安装 5 万千瓦水轮发电机组 2 台。

三峡水电站建成前，世界上装机容量最大的水电站是伊泰普水电站，该电站位于南美洲巴拉那河上，由巴西、巴拉圭两国合建。1991 年该水电站建成时，安装了 18 台单机容量 70 万千瓦的水轮发电机组，设计总容量 1 260 万千瓦，年均发电量 750 亿千瓦时。2001 年，该水电站扩建了 2 台单机容量 70 万千瓦的水轮发电机组，总装机容量达 1 400 万千瓦。

2. 三峡水电站的水轮发电机组

三峡水电站共安装了 32 台额定容量 70 万千瓦的水轮发电机组，该机组是当今世界最大的水轮发电机组。那么，70 万千瓦水轮发电机组的规模到底有多大？

从发电效益看，一台水轮发电机组的额定出力为每小时发电 70 万千瓦时（每秒钟发电 486 千瓦时），如果一个家庭一年使用 1 500 千瓦时电量，那么一台机组一个小时的发电量相当于 467 个家庭一年的用电量。

从机组重量来看，机组由进水口压力钢管、蜗壳、导水机构、基础环、

○ Figure 3.3　Interior of the Left-bank Powerhouse of the Three Gorges Hydropower Station
(Photographed by Zhang Zuxin)

making it the world's biggest hydropower plant in terms of installed capacity. Within it, there are 32 water-turbine generator sets generating 700 MW each (14 sets on the left bank, 12 sets on the right, and 6 sets underground). The power source station has a further two 50 MW water-turbine generator sets.

Before the completion of the Three Gorges Hydropower Station, Itaipu Hydroelectric Dam was the world's largest operational hydroelectric power plant in terms of installed power. Constructed as a cooperation between Brazil and Paraguay, it is located on the Paraná River. Upon completion in 1991, the dam had 18 water-turbine generator sets of 700 MW capacity each, with a total designed capacity of 12.6 million kW and an average annual power generation of 75 billion kWh. In 2001, two new units, each with a single unit capacity of 700 MW were added to the plant, increasing its total installed capacity to 14 million kW.

2. The Three Gorges Hydropower Station's water-turbine generator sets

32 water-turbine generator sets, with a per-unit generation capacity of 700 MW, have been installed in the Three Gorges Hydropower Station, the world's largest in terms of rated capacity. Such a vast capacity is difficult to conceptualize.

In terms of power generation efficiency, the rated output of one water-turbine generator set is 700 MW/h (486 kWh/s). If the average family consumes 1,500 kWh of electricity in a year, the electricity generated by one generator set in one hour can supply the annual power consumption of 467 families.

Each set consists of a water turbine that is composed of an inlet penstock, volute chamber, water distributor, foundation ring, seating ring, runner, draft tube among other things. It is also composed of a generator that is composed of upper and lower frames, a generator stator and rotor and the principal axis of the set that connects it to the water-turbine generator. These components are extremely large: the maximum weight of a single piece is up to 460 tons, and the total weight of a set is 7,000 tons, equivalent to the total weight of all the steel in the Eiffel Tower. To hoist the generator rotor, for example, required the synchronous operation of two 1,000-ton bridge cranes.

In terms of the size of the sets, the diameter of the inlet penstock is 12.4 m, which is equivalent to the height of a three-story building. The diameter of the principal axis of the connector between the turbine and the generator is 3.8 m, and the inner diameter of the gen-

座环、转轮、尾水管等构成的水轮机，和由上下机架、发电机定子与转子等组成的发电机，以及将水轮机、发电机连接起来的机组主轴构成，这些部件个个都是"巨无霸"，最大单件的总重量达460吨，一台机组的总重量达7 000吨，相当于法国巴黎埃菲尔铁塔所用钢材重量的总和。仅发电机转子的吊装，就需要2部1 000吨以上的桥式起重机同步起吊才能胜任。

从机组的尺寸来看，机组的进水钢管直径达12.4米，相当于三层楼房的高度。水轮机和发电机的连接主轴的直径达到了3.8米，发电机定子内直径超过18米。2011年4月15日，三峡建设者在右岸地下电站正在施工中的28号机组的基坑内，举办了一场由160人参加，历时1个小时的庆祝"五一国际劳动节"的合唱音乐会。

外形和重量如此巨大的庞然大物，其安装精度要求却是以零点几毫米甚至微米计算。如此高的安装精度对我国大型水电机电安装建设者的技术水平和能力是一个重大考验。以水轮机和发电机的连件主轴的安装为例，主轴安装垂直度的设计要求为0.2毫米，然而三峡水电站左、右岸电站全部机组主轴的垂直度精度达到0.01毫米，小于一根头发丝的直径(0.06～0.1毫米)。

高精度的机组安装得到的回报是机组较高标准的运行稳定性和机组运行寿命的延长。以机组顶盖振动为例，在机组稳定性试验中，机组在稳定运行区运行时的水轮机顶盖垂直振动大约50微米，在国际允许值40～140微米范围内，水平振动值约25微米，在国际标准20～40微米范围内。数字是枯燥的，举一个形象例子，在厂房中，如此巨大的设备按照常理，机组运转起来，应该机声隆隆，厂房颤动；然而，三峡水电站水轮发电机组运行时，声音却不大，几乎感觉不到振动，曾经有一个试验，将一枚1元的硬币竖立在运行中的水轮机顶盖上，硬币纹丝不动。

三、三峡船闸和垂直升船机

1. 三峡船闸和垂直升船机的概况

三峡水库蓄水至正常蓄水位175米时，三峡大坝上游和下游的水位最大落差达113米，客货轮船必须经过三峡船闸或垂直升船机，才能往返于上游或下游。

erator stator exceeds 18 m. To give a sense of the scale, on April 15, 2011, builders of the Three Gorges Project held a one-hour chorus concert to celebrate the upcoming International Labor Day in the foundation pit of the No. 28 generator set that was under construction in the underground power station on the right bank at the time; it was attended by 160 people.

For such an enormous structure with such a huge shape and weight, the calculation unit of installation accuracy needs to be to the millimeter or even micrometer. This was a serious test of the technical level and ability of large hydropower electromechanical installation builders in China. Taking the installation of the principal axis of the connector between the turbine and the generator as an example, the designed perpendicularity of the principal axis installation was 0.2 mm; in the end, the actual perpendicularity accuracy of the principal axes of all sets in the left and right banks of the Three Gorges Hydropower Station is 0.01 mm, much smaller than the diameter of a hair (0.06-0.1 mm).

High installation accuracy yields higher-standard operation stability and a longer service life of the sets. In a stability test of the vertical vibration of the water turbine's head cover, when the set was operating in the stable operation area, was approximately 50 μm, within the international allowable value of 40-140 μm; the horizontal vibration value was about 25 μm, within the international standard of 20-40 μm. The operation of such huge equipment in the powerhouse could be expected to be noisy enough to vibrate the building, but this is not the case with the water-turbine generator sets in the Three Gorges Hydropower Station, where they generate much weaker sound and almost no vibration. In one experiment, a coin lay absolutely still on the head cover of a running water turbine.

III. Ship Lock and Vertical Ship Lift

1. Overview of the ship lock and vertical ship lift

When the Three Gorges Reservoir is at its normal water storage level of 175 m, the maximum water head between the upstream and downstream of the Three Gorges Dam will reach 113 m. Passenger and freight ships must pass through the ship lock or the vertical ship lift to travel between the upstream and downstream sections.

The Three Gorges ship lock is the world's largest inland ship lock in terms of the number of steps, the gross water head, the size of lock chamber (280 m × 34 m × 5 m), and the water head borne by a single-step lock. The main section of the ship lock is 1,621 m long, with five lock chambers on a single line, totaling ten lock chambers on the double lines, all of which were excavated from the granite mountain, one of the key technologies in the construction process of the Three Gorges Project. The highest slope, 175 m, is located on the south slope of the third lock head. There are five lock chambers in the single-line five-step

⊃ 图 3.4　三峡船闸全景（摄影：黄正平）

　　三峡船闸是目前世界上级数最多、上下游水位落差最大、闸室尺寸最大（长 280 米 × 宽 34 米，槛上水深 5 米），单级船闸承担的水头最大的内河船闸。三峡船闸主体结构段总长 1 621 米，单线 5 个闸室，双线共 10 个闸室，完全是在花岗岩山体中开挖出来的，是三峡枢纽工程施工中的关键技术之一。最大高边坡位于三闸首南坡，坡高达 175 米。单线五级船闸有 5 个闸室，闸室之间有输水廊道相通，像连通器一样可以使相邻两个闸室中的水位相齐平。

　　垂直升船机主要作用是为 3 000 吨级客货轮和特种船舶提供快速过坝通道，并与三峡船闸联合运行，加大枢纽的航运通过能力和保障枢纽通航质量。其提升高度为 113 米、提升重量为 15 500 吨，是目前世界上提升高度最高和提升重量最大的垂直升船机；主体承船厢长 132 米、宽 23.4 米、高 10 米，有效水域长 120 米、宽 18 米、最小水深 3.5 米，一次可装运一艘 3 000 吨级客货轮。

　　三峡垂直升船机对安全性要求极高，设计、制造和安装的难度都非常大。为了给科研、设计多一点时间，国务院三峡工程建设委员会（以下简称"国务院三峡建委"）于 1995 年 5 月做出升船机缓建的决定。2007 年 7 月，原"钢丝绳卷扬提升"方案改为"齿轮齿条爬升"方案，螺母柱安全机构可以随时锁定，可保证升船机运行过程中"万无一失"。垂直升船机实际上就是一个特大型"电梯"，设置了与承船厢满载 15 500 吨相等的平衡铁块；提升或者下降承船厢时，只需给予 400 吨的驱动力就可以了，不会消耗大量电能。垂直升船机是船舶来往于大坝上下游的快速通道。三峡船闸设计通过能力

图 3.5　三峡船闸超大规模人字闸门
（单扇闸门高 38.5 米、宽 20.2 米、厚 3.0 米，重 850 吨）

Chapter 3 Construction of the Three Gorges Project

Figure 3.4 Panorama of the Three Gorges Ship Lock (Photographed by Huang Zhengping)

ship lock, which are linked by the water delivery gallery like a connector to ensure the water in the two adjacent lock chambers is at the same level.

The main function of the vertical ship lift is to allow 3,000-ton passenger or cargo vessels and special boats to quickly pass through the dam. It operates jointly with the Three Gorges ship lock, so as to increase the navigation capacity and ensure the navigation quality of the dam. The vertical ship lift is the biggest in the world, with maximum lifting height of 113 m and maximum lifting weight of 15,500 tons. The ship chamber has external dimensions of 132 m×23.4 m×10 m and effective water area in it is 120 m×18 m×3.5 m (minimum water depth), which can allow the passage of a 3,000-ton ship.

The vertical ship lift of the Three Gorges Project requires extremely high safety, and presented complex difficulties in its design, manufacturing and installation. To give more time for scientific research and design, the Three Gorges Project Construction Committee of the State Council (hereinafter referred to as "the TGPCC") decided, in May 1995, to postpone the construction of the ship lift. In July 2007, the original plan of having a steel wire rope-winching evaluation was changed to a gear and racking climbing plan. The nut column

⊂ Figure 3.5 The Towering Miter Gates of the Three Gorges Ship Lock
(A single gate is 38.5 m high, 20.2 m wide, 3.0 m thick and weighs 850 tons.)

图 3.6 运行中的三峡垂直升船机（摄影：陈臣）

为年单向 5 000 万吨，垂直升船机每次可通过一艘 3 000 吨级船舶。船舶从下游驶往上游（或从上游驶往下游），必须通过三峡船闸或垂直升船机。人们形象地比喻船舶通过三峡船闸过坝是"爬楼梯"，船舶乘垂直升船机过坝是"坐电梯"。

2. 船舶如何通过三峡船闸和垂直升船机

船舶是如何通过三峡船闸的呢？假如船舶要从下游驶向上游——船舶驶入第五闸室后，关闭下游人字门，通过地下输水系统，第四闸室往第五闸室内充水，当两个闸室内水面齐平时，打开上游人字门，船舶驶入第四闸室。依此类推，船舶就好像爬过一级又一级楼梯一样，通过第一闸室，驶向大坝上游航道。船舶下行的过程相反。三峡船闸单个闸室净宽 34 米、长 280 米，是世界上最大的闸室，可同时容纳 6 艘 3 000 吨级船舶。一艘船通过三峡船闸大约要 3 个小时。

三峡垂直升船机是客轮和特种船舶的快速过坝通道，垂直升船机就像我们日常乘坐的电梯，容纳船舶的承船厢就像电梯的轿厢。承船厢内部净长 120 米、宽 18 米、水深 3.5 米，可容纳一艘 3 000 吨级船舶，承船厢自重加上船舶和厢内水体的总重达 15 500 吨，是当今世界上提升高度和总重量最大的垂直升船机。

船舶是如何通过垂直升船机过坝的呢？假如船舶要从大坝下游驶向上游——当船舶驶入承船厢，关闭承船厢下游闸门，承船厢通过爬升装置上升，直到承船厢内水位与上游水库水位齐平时，打开承船厢上游闸门，船舶驶出承船厢进入上游航道。船舶从上游向下游行驶，与其过程相反。一艘船"乘坐"垂直升船机过坝，约需 30 分钟。

Figure 3.6 Vertical Ship Lift of the Three Gorges Project in Operation (Photographed by Chen Chen)

safety mechanism can be locked at any time to ensure absolute safety during ship lifting. The vertical ship lift is actually a very large elevator, which is equipped with a balancing iron block equal to the ship chamber's full load of 15,500 tons. The lifting or lowering of the ship chamber only needs 400 tons of driving force, therefore its electricity consumption is low. The vertical ship lift is a fast channel for ships traveling between the dam's upstream and downstream. The design navigation capacity of the Three Gorges ship lock is 50 million tons/single line/year, and the vertical ship lift can elevate one 3,000-ton ship at a time. Ships traveling on this stretch of the river must pass through the Three Gorges ship lock or the vertical ship lift. The operation of passing through the former has been likened climbing stairs, and that of the latter as "taking the elevator".

2. How ships pass through the Three Gorges ship lock and the vertical ship lift

If a ship needs to travel from the downstream to the upstream of the dam it enters the fifth lock chamber, at which point the lower miter gate is closed and the fifth lock chamber fills with water from the fourth lock chamber through the underground water conveyance system. The upstream miter gate is opened when the water in the two lock chambers is at the same level, and then the ship then enters the fourth lock chamber. In this way, the ship is moves as if it is climbing one stair after another until it eventually passes through the first lock chamber and travels to the dam's upstream. The process of traveling downstream is the same but in reverse. The available area of each lock chamber is 280 m × 34 m, and is the largest lock chamber in the world, with the capacity for six 3,000-ton ships. The average time it takes for ships to pass through the ship lock is three hours.

The vertical ship lift is a fast channel for passenger ships and other vessels requiring faster transit times. The vertical ship lift operates like an elevator with the ship chamber being like the elevator car. The effective water area within it is 120 m ×18 m × 3.5 m (minimum water depth), which can allow the passage of one 3,000-ton ship. The total maximum weight of the ship chamber, the ship and the water is 15,500 tons and it is the world's largest vertical ship lift in terms of both lifting height and total weight.

After the ship enters the ship chamber from downstream, the lower gate of the ship chamber is closed, and the ship chamber rises through the climbing device until the water level in the ship chamber is equal to that of the upstream reservoir. Then, the upstream gate of the ship chamber is opened and the ship can move out of the ship chamber and enter the upstream channel. The process can be done in reverse for vessels traveling downstream. It takes the average ship only 30 minutes to cross the dam by the vertical ship lift.

第二节　三峡枢纽工程的建设过程

在水利水电工程建设中，为了修建大坝，一般需先用围堰把将要建大坝的位置全部或部分围起来（在水利水电行业中，围堰是指临时挡水的坝，围堰围护成的封闭范围称为基坑），同时让江水"改道"，流向预先建好的泄水通道，这一过程在水利水电行业中被称为"施工导流"。在围堰基坑内将存留的江水抽干，开挖到新鲜坚硬的岩石，作为大坝的地基，再在地基上浇筑混凝土大坝；如果大坝需要分期施工时，必须经过多次导流，才能建成完整的大坝。为了统筹解决好大坝施工导流过程中的江水宣泄、通航等问题，一般需进行施工导流方案选择和规划。

三峡大坝坝址河谷开阔，宽达 1 000 多米，中堡岛将长江分为左侧的大江和右侧的后河，方便江水改道，有利于进行分期导流；由于长江为我国的水运交通动脉，施工期间通航问题至关重要，不能断航；另外三峡大坝坝址位于葛洲坝水库中，水深达 60 多米，江底有厚达 10 余米的粉细砂。这些都是三峡大坝施工导流规划中必须考虑的重点和难点。经过多种技术方案比较，三峡工程施工导流采用"三期导流、明渠通航"的施工方案，在整个建设过程中"三围长江，两改江流"。总工期 17 年，即施工准备和一期工程 5 年，二期工程和三期工程各 6 年。

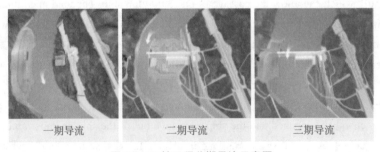

图 3.7　三峡工程分期导流示意图

Section 2　Construction Process of the Three Gorges Project

In the construction of water conservancy and hydropower projects, cofferdams are usually built to enclose part of or the entire location of the dam before it is built; in the water conservancy and hydropower industry, a cofferdam is defined as a dam that temporarily retains water, and the area enclosed by cofferdam is called the foundation pit. This allows the water to be diverted to the pre-built drainage channels. a process that is called "construction diversion". Dam builders first drain the water in the foundation pit, excavate it to expose fresh, hard rock for the dam's foundation, and then pour the concrete dam on to this foundation. If the dam needs to be constructed in stages, the river is diverted multiple times before the dam is completed. In order to solve the problems of water discharge and navigation during construction diversion, a construction diversion scheme must be made.

At the Sandouping dam site, the river valley is more than 1,000 m wide, with a small islet known as Zhongbaodao in the river, which separates the Yangtze into the main river course on the left and the Houhe River on the right; this location is favorable for a river diversion scheme with phased construction. As the Yangtze River is China's water transport artery, the navigation issue during the construction period was of great importance as it could not be interrupted. In addition, the dam site of the Three Gorges Dam is located in the Gezhouba Reservoir, with a water depth of more than 60 m and silty fine sand that is more than 10 m deep on the river's bed. These were the key points and difficulties that had to be considered in the construction diversion scheme for the Three Gorges Dam. After the comparison of several technical schemes, the Three Gorges Project was organized by a scheme of "diversion in three phases and navigation through open channels". To ensure the concrete placement quality, temporary cofferdams were built three times to enclose the dam site as the foundation pit, and the water was diverted twice to the pre-built drainage channels. The dam took a total of 17 years to be finished: five years of preparatory and phase I projects, six years for the phase II project, and six years for the phase III project.

◐ Figure 3.7　Diagram of the Phased Diversion of the Three Gorges Dam

一、施工准备和一期工程

1993年至1997年,进行施工准备和修建一期工程。准备工程主要项目有:对外专用公路、场内道路、人工砂石料场、混凝土拌和系统、水电气供应系统和西陵长江大桥等同时展开。经过一年多的施工准备,三峡枢纽工程于1994年12月14日正式开工。

三斗坪坝址处的长江,天然情况下,自右岸至左岸有后河、中堡岛、大江。一期工程在后河上、下游及沿中堡岛左侧修筑一期土石围堰,形成一期基坑,把基坑内的江水抽干后,在围堰保护下挖除中堡岛、修建混凝土导流明渠、浇筑混凝土纵向围堰,并浇筑三期碾压混凝土围堰基础和部分堰体混凝土。一期土石围堰束窄河道约30%,大江仍然承担着泄流及通航任务。以如期实现规模和难度位于世界之最的大江截流为完成标志。

所谓大江截流,就是以纵向围堰和左岸滩地为物料基地,从左、右两岸向大江抛投石块和砂砾料,不间断地向江心逼拢,最后"斩断"大江,迫使江水从导流明渠流向下游。1997年11月8日,经过六个半小时的鏖战,15

图3.8 大江截流上游戗堤龙口即将合龙现场

I. Preparatory and Phase I Projects

Preparations and phase I construction were carried out from 1993 to 1997. The preparations included the simultaneous construction of external dedicated roads, on-site roads, artificial aggregate stock ground, a concrete mixing system, water, electricity and gas supply systems, and the Xiling Yangtze River Bridge. After more than one year of preparations, official construction commenced on December 14, 1994.

The Yangtze at the Sandouping dam site is naturally divided right to left into the Houhe River, Zhongbaodao islet and the main river course. During the phase I project, phase I earth-rock cofferdams were built at on the upper and lower Houhe River and along the left side of Zhongbaodao to form the phase I foundation pit. After the water in the foundation pit was drained, under the protection of the cofferdams, builders excavated Zhongbaodao, built concrete open channels for diversion, poured the lengthways concrete cofferdam, and poured the phase III roller compacted concrete (RCC) cofferdam foundation as well as the concrete for part of the cofferdam body. Phase I earth-rock cofferdams confined the riverway by about 30%, and the main river course still undertook the tasks of discharge and navigation. The completion of this stage was marked by the completion of closing the Yangtze River as scheduled, the world's largest project in terms of scale and difficulty.

The so-called river closure in fact used the lengthways cofferdams and left-ban bottomland as the material base, on to which rocks and sand-gravel were dumped into the river from the left and right banks to close the middle of the river, thus forcing the water to flow downstream via the open diversion channel. At 15:30 on November 8, 1997, after six and a half hours of hard work, the river closure of the main channel for the Three Gorges Project was carried out, constituting the largest scale closure in terms of flow (8,480 m^3/s) and water depth (60 m) in the world.

II. Phase II Project

The successful river closure forced the Yangtze water to flow from the open diversion channel to the downstream reaches, and ships travelled from the open diversion channel and the temporary ship lock on the left bank both upstream and downstream. The phase II project lasted approximately six years, from November 1997 to June 2003. Builders built upstream and downstream earth-rock horizontal cofferdams in the river, which formed the phase II foundation pit together with the phase I concrete lengthways cofferdam. After the water in

◐ Figure 3.8　Site of the Closure Gap Nearing Completion in the Upper River

时30分世界上截流水深最大（60米）、截流流量最大（8 480立方米每秒）的三峡工程大江截流成功。

二、二期工程

大江截流成功后，迫使长江水从导流明渠流向下游，船舶从导流明渠和左岸临时船闸驶向上下游。1997年11月至2003年6月近6年的时间里，修建二期工程。在大江上修筑二期上、下游土石横向围堰，与一期修建的混凝土纵向围堰形成二期基坑；把基坑中的江水抽干后，在围堰保护下，修建大坝的泄洪坝段、左岸厂房坝段及电站厂房、临时船闸等。与此同时，修建左岸五级永久船闸和右岸茅坪溪防护坝。上述建筑物具备挡水条件后，拆除二期上、下游土石围堰，进行导流明渠截流，长江水改由左岸大坝泄洪坝段的导流底孔和泄洪深孔流向下游，船舶先从临时船闸、后从五级船闸通行。修筑三期上、下游土石围堰，在上游土石围堰保护下浇筑三期碾压混凝土围堰，利用该围堰及左岸大坝挡水，以如期实现三峡水库135米蓄水、五级船闸试通航、首批机组发电为完成标志。

导流明渠是一条宽350米的人工河道，大江截流后，江水从此流向下游，船舶从此驶向上下游。导流明渠截流后，完全截断了长江，迫使江水通过左岸大坝底部的22个导流底孔流向下游。船舶从临时船闸驶向上游或下游。在导流明渠内填筑土石围堰，将围堰中的江水抽干，修建碾压混凝土围堰及右岸电站。导流明渠截流的成功是三峡二期工程将要转向三期工程建设的重要标志。

◌ 图3.9 导流明渠截流图

the foundation pit was drained, and under the protection of the cofferdams, builders built the spillway section, the left-bank power plant section as well as the powerhouse, temporary ship lock, and other structures. At the same time, they built the left-bank five-step permanent ship lock and right-bank Maopingxi protective dam. After these structures met the required water retention standards, the phase II upstream and downstream earth-rock cofferdams were removed, and river closure was carried out through the open diversion channel. The Yangtze water then flowed downstream from the diversion bottom outlets and the deep outlets of the spillway section of the left bank. Ships passed through the section using first the temporary ship lock and then the five-step ship lock. After this, the builders constructed the phase III upstream and downstream earth-rock cofferdams, and poured the phase III RCC cofferdam under the protection of the upstream earth-rock cofferdam. The cofferdam and the left-bank dam were used to retain water. The completion of this stage was marked by the completion of the 135-meter water storage of the Three Gorges Reservoir, the trial navigation of the five-step ship lock and the power generation of the first batch of generator sets as scheduled.

The open diversion channel was a 350-meter artificial watercourse, through which water flowed downstream and ships travelled upstream or downstream after the river was closed. After the closure of the open diversion channel, the Yangtze River was completely dammed, forcing the water to flow downstream through the 22 diversion outlets at the bottom of the dam on the left bank, and requiring ships to travel upstream or downstream using the temporary ship lock. Builders filled the open diversion channel with earth and rocks to build the cofferdam, drained the water in the cofferdam, and built the RCC cofferdam and the power station on the right bank. The successful closure of the open diversion channel was an important landmark in the transition from phase II to phase III.

Figure 3.9 River Closure of the Open Diversion Channel

2003年6月1日，三峡工程如期下闸蓄水，至6月10日22时坝前水位由106米上升至135米。6月16日，世界上最大的内河船闸——三峡船闸开始试通航。7月18日，左岸电站2号和5号两台机组并网发电，投入商业运行。至此，三峡二期工程如期高质量完成。

三、三期工程

2003年7月至2009年7月的6年，修建三期工程。当二期基坑内的左侧大坝建好后，拆除二期上、下游土石围堰，开始由左侧这部分大坝挡水。再在右岸的导流明渠内进行三期明渠截流，修筑三期上、下游土石围堰，在上游土石围堰保护下迅速浇筑较高的三期碾压混凝土围堰。由三期碾压混凝土围堰、混凝土纵向围堰与三期下游土石围堰形成三期基坑。把基坑中的江水抽干后，在三期基坑内修建大坝右岸厂房坝段及右岸电站厂房和右岸非溢流坝段。此期长江水改由左岸电站机组和泄洪坝段预留的导流底孔、泄洪深孔宣泄，船舶先从临时船闸、后从五级船闸通行。以实现三峡水库156米初期蓄水、正常蓄水位175米试验性蓄水、双线五级永久船闸通航、26台70万千瓦水轮发电机组全部投入商业运行为完成标志。

2003年7月，三峡枢纽工程进入三期工程施工。2005年9月16日，左岸电站14台机组提前1年全部投入商业运行。2006年5月20日，右岸大坝全线浇筑到设计高程185米；汲取左岸大坝出现极少数温度裂缝的教训，进一步完善了混凝土温控防裂成套技术，使右岸大坝没有出现裂缝，打破了"无坝不裂"的"魔咒"，创造了世界混凝土大坝浇筑史上的奇迹。

经国务院三峡三期工程验收委员会批准，2006年10月27日，三峡水库蓄水至设计初期蓄水位156米，提前1年进入初期运行，开始初步发挥防洪、发电、航运等综合效益。2008年10月30日，右岸电站12台机组全部投入商业运行。至此，三峡工程初步设计规定的建设任务（国家批准缓建的垂直升船机除外），提前1年全部完成。

2010年10月26日，三峡枢纽工程试验性蓄水至正常蓄水位175米，这是该工程第一次蓄水至正常蓄水位，各建筑物的各项监测数据正常，开始全

The dam's sluice gates were closed on June 1, 2003 as scheduled, allowing the reservoir behind it to begin filling. By 22:00 on June 10, the water level had risen from 106 m to 135 m. On June 16, the Three Gorges ship lock, the world's largest inland ship lock, began its trial operation. On July 18, the No.2 and No.5 generator sets of the powerhouse on the left bank were connected to the grid and put into commercial operation. This marked the completion of phase II construction as scheduled.

III. Phase III project

Phase III of the project lasted six years from July 2003 to July 2009. After the left dam section in the phase II foundation pit was completed, the upstream and downstream earth-rock cofferdams of phase II were removed and the left dam section was used for retaining water. Builders then completed the phase III open channel closure in the diversion channel on the right bank, built phase III upstream and downstream earth-rock cofferdams, and, under the protection of upstream earth-rock cofferdams, quickly poured the phase III RCC cofferdam that was relatively high. The phase III foundation pit was formed by the phase III RCC cofferdam, concrete lengthways cofferdam and phase III downstream earth-rock cofferdam. After draining the water from the foundation pit, builders constructed the right-bank power plant section, right-bank powerhouse and right-bank non-overflow dam section in the phase III foundation pit. In this phase, the Yangtze's water was discharged by the left-bank generator sets and the diversion bottom outlets and the deep outlets reserved in the flood discharge dam section, and ships passed through the section using first the temporary ship lock and then the five-step ship lock. The completion of this stage was marked by the attainment of the initial water storage level of 156 m, the trial normal water storage level of 175 m, and the commercial operation of 26 water-turbine generator sets of 700 MW capacity each.

Phase III construction had begun in July 2003. On September 16, 2005, all the 14 generator sets of the left-bank power station were put into commercial operation one year ahead of schedule. On May 20, 2006, the dam section on the right bank was poured to the design elevation of 185 m. Drawing lessons from some temperature cracks that had appeared in the left-bank dam section, builders further improved the concrete temperature control and crack prevention technology, which ensured there were no cracks in the right-bank dam section, disproving the rule that "there is no dam has no cracks", and creating a world-first in the history of concrete dam pouring.

With the approval of the Three Gorges Project Phase III Acceptance Committee under the State Council, the Three Gorges Reservoir was filled to the designed initial water level of 156 m on October 27, 2006, and started operation one year ahead of schedule, creating

面发挥其巨大的综合效益。截至2021年,连续12年每年汛后都蓄水至正常蓄水位175米。

需要特别一提的是两次截流和二期围堰防渗墙工程。三峡大江截流最大水深60米,截流流量8 480～11 600立方米每秒,截流施工最大日抛投强度19.40万立方米;同时,截流期间长江仍需通航,这在世界截流工程中是少有的。明渠截流设计流槛为9 010～10 300立方米每秒,总落差达4.11米,最大流速逾8米每秒;同时明渠底部为人工浇筑的混凝土河床,特别光滑,不利截流材料的稳定。这些因素导致两次截流施工难度非常大。

二期围堰防渗墙最大深度达73.5米,截水面积8.345万平方米,施工时间短,围堰基础地质条件复杂,存在许多难以攻克的地质问题。围堰在上游水压力作用下会产生变形,设计采用塑性混凝土以适应变形。施工采用综合方法,取得了圆满的效果。

三峡枢纽工程三期导流和两次截流,其规模、难度和复杂程度堪称世界之最。三期导流的顺利实施,保证了三峡枢纽工程所有建筑物都能在干地上施工,顺利建成三峡大坝,确保了施工期长江航运通畅。以上的工作使三峡大坝横贯长江,让"截断巫山云雨,高峡出平湖"的愿望变成现实,这是中华民族治水史上的一次壮举,多少代中国人开发和保护长江的梦想成为现实。

comprehensive benefits for flood control, power generation navigation and more. On October 30, 2008, all 12 generator sets of the right-bank power station were put into commercial operation. By then, all the construction tasks specified in the preliminary design of the Three Gorges Project (with the exception of the construction of the vertical ship lift, a postponement approved by the state) had been completed one year ahead of schedule.

On October 26, 2010, the dam was experimentally filled to its normal water level of 175 m for the first. The monitoring data of all the structures was normal, and the project began creating wide-ranging comprehensive benefits. By 2021, the flood-filling water level in the flood season had reached this normal level for 12 consecutive years.

The two river closures and the cut-off wall construction of the phase II cofferdams also deserve mention. The greatest water depth of the river closure was 60 m, the closure discharge volume was 8,480-11,600 m^3/s, and the maximum dumping intensity reached 194,000 m^3 in 24 hours. Moreover, uninterrupted navigation during the closure process has rarely been seen in river closure projects from around the world. The design discharge of the open diversion channel closure was 9,010-10,300 m^3/s, with a gross head of 4.11 m and maximum flow velocity exceeding 8 m/s. In addition, during the closure of the diversion channel, it was not easy to maintain the steadiness of the fill material on the slippery riverbed formed by a manually-placed concrete base. These factors presented great challenges to the two river closures.

The maximum depth of the cut-off walls of the phase II cofferdam was 73.5 m, with a water cut-off area of 83,450 m^2. With a short construction period and the complex geology of the cofferdam foundation, there were many geological problems that were difficult to overcome. The cofferdam would deform under the action of upstream water pressure, so plastic concrete was used in the design to compensate. With this comprehensive method, satisfactory results were achieved.

The three phases of river diversion and two river closures of the Three Gorges Project were the world's greatest in terms of scale, difficulty and complexity. The smooth implementation of the three phases of river diversion ensured that all the structures of the Three Gorges Project could be constructed on dry land, that the Three Gorges Dam was successfully completed, and that the Yangtze River navigation was uninterrupted during the construction period. The above work has enabled the Three Gorges Dam to span the Yangtze River, making the dream "to hold back Wushan's clouds and rain, till a smooth lake rises in the narrow gorges" (quoted from Chairman Mao's poem) a reality. This is a feat in the history of flood control of the Chinese nation: the dream of generations of Chinese people to develop and protect the Yangtze River has finally come true.

本章小结：

三峡枢纽工程位于西陵峡的三斗坪，是长江流域治理开发的关键工程，也是目前世界上规模最大的水利水电工程，具有防洪、发电、航运、水资源利用等巨大综合效益。三峡枢纽工程主要由拦河坝、水电站和通航建筑物组成，在建设过程中采用分期导流方式，分三期施工完成：一期工程以1997年11月8日实现大江截流为标志；二期工程以2003年实现水库初期蓄水、第一批机组发电和永久船闸通航为标志；三期工程实现全部机组发电和枢纽工程完建，2009年完工。三峡枢纽工程以其规模宏大、效益显著、技术要求高而举世瞩目，成为中华民族治水史上的一次壮举。

参考文献：

[1] 本书编委会. 百问三峡［M］. 北京：科学普及出版社，2012.

[2]《中国三峡建设年鉴》编纂委员会. 中国三峡建设年鉴（1998）［J］. 宜昌：中国三峡建设年鉴社，1998.

[3]《中国三峡建设年鉴》编纂委员会. 中国三峡建设年鉴（1999）［J］. 宜昌：中国三峡建设年鉴社，1999.

[4]《中国三峡建设年鉴》编纂委员会. 中国三峡建设年鉴（2007）［J］. 宜昌：中国三峡建设年鉴社，2007.

[5]《中国三峡建设年鉴》编纂委员会. 中国三峡建设年鉴（2009）［J］. 宜昌：中国三峡建设年鉴社，2009.

[6]《中国三峡建设年鉴》编纂委员会. 中国三峡建设年鉴（2011）［J］. 宜昌：中国三峡建设年鉴社，2011.

第三章 | 三峡枢纽工程的建设
Chapter 3 Construction of the Three Gorges Project

Chapter Summary:

The Three Gorges Project is located in Sandouping, the middle section of Xiling Gorge. It is a key project for China's development and control of the Yangtze River, and is currently the world's largest hydropower project. It produces vast and comprehensive benefits such as flood control, power generation, waterway transport, and water-resource utilization. The project encompasses the dam, hydropower station and navigation structures. River diversion was carried out in three phases: the completion of the phase I project was marked by the river closure on November 8, 1997; the completion of phase II was marked by the completion of the initial water storage, the power generation of the first batch of generator sets and successful navigation using the permanent ship lock; the completion of the phase III project was marked by the power generation of all the generator sets and the completion of the key project. All the three phases were completed in 2009. As a feat in the history of flood control of the Chinese nation, the Three Gorges Project has attracted worldwide attention for its large scale, remarkable benefits and high technical requirements.

References:

[1] Editorial Board of the book. *One Hundred Questions Regarding the Three Gorges Project* [M]. Beijing: Science Popularization Press, 2012.

[2] Editorial Board of *China Three Gorges Construction Yearbook*. *China Three Gorges Construction Yearbook (1998)* [J]. Yichang: China Three Gorges Construction Yearbook Press, 1998.

[3] Editorial Board of *China Three Gorges Construction Yearbook*. *China Three Gorges Construction Yearbook (1999)* [J]. Yichang: China Three Gorges Construction Yearbook Press, 1999.

[4] Editorial Board of *China Three Gorges Construction Yearbook*. *China Three Gorges Construction Yearbook (2007)* [J]. Yichang: China Three Gorges Construction Yearbook Press, 2007.

[5] Editorial Board of *China Three Gorges Construction Yearbook*. *China Three Gorges Construction Yearbook (2009)* [J]. Yichang: China Three Gorges Construction Yearbook Press, 2009.

[6] Editorial Board of *China Three Gorges Construction Yearbook*. *China Three Gorges Construction Yearbook (2011)* [J]. Yichang: China Three Gorges Construction Yearbook Press, 2011.

> 阅读提示：

三峡输变电工程是三峡工程的重要组成部分，承担着三峡水电送出的重要任务。该工程于1997年开工建设，2011年竣工投产，以三峡水电站为中心，向华东、华中、南方电网送电，供电范围包括九省两市，惠及半数国民，是世界上覆盖面积最广、受惠人口最多、建设规模最大的输变电工程。

三峡输变电工程的建设不仅确保了三峡电力"送得出、落得下、用得上"，而且促进了以三峡电网为中心的全国电网互联格局的形成，对加速实现西电东送通道建设目标、全面提高我国输变电工程建设水平都起到重要作用。

As an important part of the Three Gorges Project (TGP), the Three Gorges Power Transmission and Transformation Works (TGPTTW) undertakes the important task of transmitting the hydropower generated by the Three Gorges Hydropower Station (hereinafter referred to as "the TGHS"). Commenced in 1997 and completed and commissioned in 2011, the TGPTTW centers on the TGHS and transmits power to the East China, Central China, and China Southern Power Grids. It provides power to nine provinces and two municipalities, benefiting half of the population in China. It is the largest power transmission and transformation system in the world, having the widest coverage, the largest beneficiary population, and is the largest in terms of construction scale.

The TGPTTW not only ensures the transmission, acceptance, and utilization of the power generated by the TGHS, but also promotes the formation of a national grid interconnection centered on the Three Gorges Power Grid. It plays an important role in speeding up the construction of the west-east power transmission works project and upgrades China's power transmission and transformation works in general.

Chapter 4 >>>>

第四章
三峡输变电工程的建设
Construction of Three Gorges Power Transmission and Transformation Works

第一节 三峡输变电工程的综述

三峡输变电工程是三峡工程的重要组成部分，是三峡水电站电力送出及其效益实现的根本保证，是世界上覆盖面积最广、受惠人口最多、建设规模最大的输变电工程。

一、三峡输变电工程的概况

三峡输变电工程系统设计方案于 1995 年获得批复，1997 年开工建设，2011 年全部建成投产。

三峡输变电工程建设规模巨大、建设周期长、施工难度大、影响范围广，在我国输变电工程建设史上史无前例。

（1）建设规模巨大

截至 2009 年，三峡输变电系统共有单项工程 91 项。其中，交流输变电工程 88 项，包括线路工程 55 项（线路总长度 6 519 千米）、变电工程 33 项（变电总容量 2 275 万千伏）；直流工程 3 项，线路总长度 2 965 千米，换流站总容量 1 800 万千伏。为了保证三峡电力安全、稳定、可靠地送出，相应建设了涉及全国涵盖 9 省 2 市电力系统的通信、调度等二次系统，共五大类 26 个单项工程，以及与输变电工程配套建设的二次系统工程。

（2）建设周期长

从 1995 年三峡输变电系统设计获得批复算起，至 2011 年全部建成，工程建设历时 16 年；从 1997 年第一个单项输变电工程开工算起，工程建设也有 14 年之久。

Chapter 4 Construction of Three Gorges Power Transmission and Transformation Works

Section 1 Overview of the TGPTTW

As an important part of the Three Gorges Project (TGP), the Three Gorges Power Transmission and Transformation Works (TGPTTW) is a fundamental guarantee of the transmission of power and the benefits delivery of the TGHS. The station is the largest power transmission and transformation works in the world, with the widest coverage, the largest beneficiary population, and it has the largest construction scale.

I. Introduction to the TGPTTW

The system design scheme of the TGPTTW was approved in 1995. Its construction commenced in 1997 and was completed and commissioned in 2011.

The TGPTTW had an enormous construction scale, long construction period, and was a challenging project with an extensive range of influence, which was unprecedented in China's history of power transmission and transformation works construction.

(1) An enormous construction scale

By 2009, there were 91 single works in the Three Gorges Power Transmission and Transformation System. Among them, there were 88 AC power transmission and transformation works, including 55 line projects (with a total length of 6,519 km), 33 power transformation works (with a total capacity of 22.75 million kV), and 3 AC power transmission and transformation works, with a total length of 2,965 km and a total capacity of converter stations of 18 million kV. In order to ensure the safe, stable, and reliable transmission of the power generated by the TGHS, 26 single works of five categories were constructed, covering communication, dispatching, and other secondary systems within the power systems of nine provinces and two municipalities in China, as well as secondary system works constructed for supporting power transmission and transformation works.

(2) A long-running project

The construction of the TGPTTW lasted for 16 years from 1995, when the design of the

(3)施工难度大

三峡输变电工程沿线山地约占60%,地形、地貌、气象、环境条件复杂。跨越县级行政区划163个,跨越长江、汉江等大型通航河流164次,跨越高速公路、铁路等骨干交通线路和高等级电力、通信线路500余次。

(4)影响范围广

三峡输变电工程的供电范围涉及华中、华东、川渝、南方电网,覆盖区域包括湖北、湖南、河南、江西、安徽、江苏、浙江、四川、广东、上海和重庆等9省2市,覆盖面积超过182万平方千米,人口超过6.7亿人。

二、三峡输变电工程的作用

三峡输变电工程的建成投产,对于三峡电力外送、促进全国电网联网、能源资源优化配置、推进我国电网建设水平提高等方面,发挥了重要作用。

1. 确保三峡电力送出

在整个三峡工程投资中,三峡输变电工程投资所占比重不到四分之一,但工程的如期建设并保证电力的顺利送出,对实现三峡枢纽工程的发电、防洪、航运、水资源利用等综合效益,都具有重要的意义。根据三峡国际专家组最终报告的评估结论,三峡工程的发电效益在综合效益中所占比例最大,达到70%。三峡输变电工程保证三峡水电站全部电力安全、稳定、可靠地送出,并在有关地区合理消

图4.1 三峡输变电工程施工

Three Gorges Power Transmission and Transformation System was approved, to 2011, when the works were completed. The construction still lasted for 14 years from the first power transmission and transformation single project's commencement in 1997.

(3) Large construction difficulties

Mountainous areas along the TGPTTW account for about 60% and the topography, landforms, meteorological and environmental conditions are complex. The TGPTTW crosses 163 county-level administrative regions, 164 large navigable rivers such as the Yangtze River and the Hanjiang River, also crossing backbone transportation routes such as expressways and railways, as well as high-grade power and communication lines over 500 times.

(4) A wide range of impacts

The power supply areas of the TGPTTW involve the Central China, East China, Sichuan and Chongqing, and China Southern Power Grids, and cover nine provinces and two municipalities, namely Hubei, Hunan, Henan, Jiangxi, Anhui, Jiangsu, Zhejiang, Sichuan, Guangdong, Shanghai, and Chongqing, covering an area of more than 1.82 million km^2 and serving a population of more than 670 million.

II. Role of the TGPTTW

Since being commissioned, the TGPTTW has played a major role in the power transmission of the Three Gorges, promoting the national power grid network, optimizing the allocation of energy resources, and improving the level of China's power grid construction.

1. Ensuring the power transmission from the TGHS

The investment in the TGPTTW accounts for less than one-fourth of the total investment of the TGP. However, the completion of the TGPTTW on schedule, along with its smooth power transmission is of great significance to realizing the comprehensive benefits of the TGP, including power generation, flood control, shipping, and water resource utilization. According to the Three Gorges International Expert Panel's assessment, found in its final report, the power benefit of the TGP accounts for the largest proportion, up to 70%. The TGPTTW ensures safe, stable, and reliable transmission of all power generated by the TGHS and reasonable power consumption in relevant regions. Realizing the transmission, acceptance, and utilization of such power is necessary to realize the benefit of power generation of

⊆ Figure 4.1 Construction of the TGPTTW

纳，实现三峡电力"送得出、落得下、用得上"，是实现三峡工程发电效益、保障枢纽工程投资回收的必要条件。

表 4.1 三峡水电站 2014 年以来发电量统计表　　（单位：亿千瓦时）

年份	2014	2015	2016	2017	2018	2019	2020	2021
三峡水电站全年累计发电量	988	870	935	976	1 016	968	1 118	1 036

说明：表中统计数据来自中国长江三峡集团有限公司官网、历年三峡水电站运行报告、央广网、中国新闻网等官方媒体报道。

2. 促进全国联网的实现

从地理位置上看，三峡水电站恰好位于西电东送、南北互供的中心，可以东联上海、西接川渝、北达京津、南至广州，是全国联网的中枢和最为理想的能源基地。三峡输电系统则横贯东西，沟通南北，在全国电网互联格局中处于中心位置。通过三峡水电站 5 个大规模交直流送出工程，直接形成了华中—川渝、华中—华东、华中—南方的联合电网，并促进了华中—西北与华中—华北之间的电网互联，为全国联网打下了重要的基础。

三峡输变电工程建成投产后，一个仅次于北美联合电网（约 8 亿千瓦）、与欧洲联合电网（约 5 亿千瓦）规模相当的世界级特大型电力系统已基本形成。电网横跨黄河、长江、珠江三大水系，覆盖多个大型煤电基地与水电基地，东西相距超过 2 000 千米。

由于电源结构、负荷特性、季节气候等方面差异较大，电网的互联、大电网的形成对于获取地区电网之间潜在的巨大错峰效益、水电站群补偿调节效益，以及水火互补容量效益等成为可能。从电力特性上看，三峡水电站发出的电力清洁环保、价格低廉、保证率高。因此，三峡输变电工程对于提高电网运行的经济性、安全性、灵活性和可靠性都具有重要意义。同时，由此建立起的更大范围内的能源资源优化配置交易平台，也为推动全国范围的电力市场建设和电力交易，构建更加开放、共赢的电力市场体系提供了良好的平台，为取得更大的能源与市场资源优化配置的经济效益、社会效益和环境效益创造了条件。

the TGP and guarantee the investment recovery of the project.

Table 4.1 Power Generation Volume of the TGHS Since 2014 (Unit: 100 million kWh)

Year	2014	2015	2016	2017	2018	2019	2020	2021
Total Annual Power Generation of Three Gorges Hydropower Station	988	870	935	976	1,016	968	1,118	1,036

Note: The data in the table are from official media reports such as the official website of the China Three Gorges Corporation, the operation reports of the TGHS over the years, www.cnr.cn, and www.chinanews.com.cn.

2. Promoting the national power grid networking

Geographically, the TGHS is located just at the center of the west-to-east power transmission works. It mutually transmits between the south and the north and connects Shanghai in the East, Sichuan and Chongqing in the West, Beijing and Tianjin in the North, and Guangzhou in the South. So it is at the center of China's national power grid network, and the most ideal location for an energy base. The Three Gorges power transmission system is in the center of the national power grid's interconnection network because it runs from east to west and connects north and south. Through the five large-scale AC and DC transmission works of the TGHS, an interconnected power grid of Central China – Sichuan and Chongqing, Central China – East China, and Central China – China Southern Power Grids has been formed, and the power grid interconnection between Central China – Northwest China and Central China – North China has been enhanced, laying an important foundation for national power grid networking.

After the TGPTTW was completed and commissioned, a world-class, super-large power system, second only to the North American Supergrid (with a capacity of about 800 million kW), and comparable in scale to the European Network of Transmission System Operators for Electricity (ENTSO-E, with a capacity of about 500 million kW), has been formed. The power grid spans the three major water systems – the Yellow River, the Yangtze River, and the Pearl River- covering several large coal-fired power bases and hydropower bases, with a distance of more than 2,000 km from east to west.

Due to the large differences in power supply structures, load characteristics, and seasonal climates, the interconnection of power grids and the formation of large power grids make it possible to obtain the potential huge peak-shifting benefits between regional power grids, the complementary, regulative benefits of hydropower stations, and the complementary capacity benefits of hydropower and coal-fired power. In terms of electric power characteristics, the electric power generated by the TGHS is clean and environment-friendly, with low cost and

自2003年以来,中国电力负荷持续快速增长,煤、电、油、运全面紧张。三峡水电站发电以后,国家电网公司充分发挥三峡水电站的作用,利用三峡输变电系统西电东送的能力,合理调度三峡及四川水电,与华东、南方电网火电充分进行了互补交换,有效缓解了电力供应的紧张局面,三峡水电站作为中国电网骨干电源的重要地位也得到了验证。

3. 提升了输变电工程建设水平

三峡输变电工程示范了电网规划审批、基金筹资、输变电工程招投标与工程监理及公司法人负责建设运营等管理制度,实现了与国际先进管理经验的接轨;开展了工程稽察与国家验收工作,形成了完备的国家监管和企业实施管理体系。通过三峡输变电工程建设探索并实现的体制创新和一系列管理创新、技术创新,不仅保证了三峡电力系统建设任务的全面完成,而且极大地提高了中国输电系统的规划设计水平,推动了中国超高压输变电工程建设的总体水平进入到世界先进行列,尤其是在直流技术引进及国产化方面取得的成效,使中国跻身于世界直流输电设备生产和建设运行的大国,并进入世界先进行列。

根据国际大电网会议(CIGRE)对世界22项直流输电系统2002年运行指标的统计分析表明,三峡输变电工程的三常直流工程(三峡—常州)、三广直流工程(三峡—广东)的能量利用率可分别排名到世界第一、第五,能量可用率指标也处于先进行列。2007年9月,三峡输变电工程三沪工程(三峡—上海)从多个国家推荐的40多个项目中脱颖而出,获得2007年度"亚洲输变电工程年度奖"。

a high reliability. Therefore, the TGPTTW is of great significance in terms of improving the economic efficiency, security, flexibility, and reliability of the power grid's operation. Meanwhile, the trading platform for the optimization and allocation of energy resources on a wider scale derived from it also provides a means for the development of the national electricity market and electricity trade. It also establishes a more open and mutually beneficial electricity market system, helping obtain the economic, social, and environmental benefits of better energy and market resources allocation.

Since 2003, China's power load has continued to grow rapidly, with tensions on coal, electricity, oil, and transmission. After the TGHS started generating power, the State Grid Corporation of China (State Grid) has given full play to the role of the TGHS, utilized the west-to-east power transmission capacity of the Three Gorges Power Transmission and Transformation System, and reasonably dispatched the hydropower of the Three Gorges and Sichuan to complementarily exchange with the thermal power of the East China Power and China Southern Power Grids. This effectively eased tensions in the power supply. The important position of the TGHS as the backbone power source of China's power grids has been confirmed.

3. Improving the construction of power transmission and transformation works

The TGPTTW has set a good example of the planning and approval of the power grid, fund financing, tendering and bidding of power transmission and transformation works, works supervision, and construction and operation by corporate legal persons, realizing its integration with advanced international management experience. The TGPTTW has been inspected and accepted by the State and a complete management system of supervision by the State and implementation by enterprises has been formed. The system, management, and technological innovations explored and realized through the TGPTTW not only ensure the full completion of the Three Gorges power system, but also greatly improve China's planning and design of power transmission systems. It also enhances China's overall construction skills in EHV transmission and transformation works, and it is now among the world's most advanced. In particular, the achievements made in the import and localization of DC technology have made China one of the world's major players, and ranks at the top among other countries in terms of the production, construction, and operation of DC power transmission equipment.

According to statistical analysis of the operation indexes of 22 DC transmission systems in the world in 2002, carried out by the International Conference on Large High Voltage Electric System (CIGRE), the energy utilization rates of the Three Gorges-Changzhou

三、三峡输变电工程的特点

要想将三峡水电站发出的强大电力通过三峡输变电网络送往半个中国,超高压输电技术(包括直流和交流)是最好的选择。根据电工理论,为了输送一定的功率,输电线路的输电损耗与输电线路电流的二次方成正比,与输电电压的二次方成反比;输电线路电压降与输电线路电流成正比,与输电电压成反比。因此,为了提高电力传输功率、减少输送线路的功率损耗和线路电压降,大功率超高压远距离输电技术已成为电力输送的首选。

在实际情况中,三峡输变电工程最远输电距离达1 300千米,而三峡发电机组发出的电力是20千伏的交流电,电压等级较低。要想远距离输送,必须通过变电站把电压升高,并采用超高压输电技术。交流电升压后直接输送的技术是超高压交流输电技术。为进一步减少远距离输电的线路损耗,提高输电系统稳定性,三峡输变电工程采用的是超高压直流输电,即通过换流站把超高压交流电转换成超高压直流电,在用电地区再通过换流站将直流电重新转换成交流电。

根据国际大电网会议的数据,340千伏和500千伏线路推荐输送容量为600兆瓦和1 200兆瓦。经过我国电力系统权威专家的论证,三峡输变电工程采用了当时具有国际先进水平、我国已完全掌握的500千伏超高压交流输电技术和±500千伏超高压直流输电技术。三峡输电工程共通过18回500千伏超高压输电线路将三峡水电站的电力送往华中电网的龙泉、江陵、葛洲坝、宜都、荆门5个交直流变电站,除部分电能在华中电网、川渝电网消纳以外,大部分电能通过换流技术将交流转变成直流后,经过±500千伏超高压直流输电线路送往华东电网和南方电网。

送往华中电网和川渝电网的电力,通过500千伏超高压交流输电线路,

图4.2 三峡输变电工程电力输送示意图

Chapter 4 Construction of Three Gorges Power Transmission and Transformation Works

DC Transmission Works and the Three Gorges-Guangdong Transmission Works under the TGPTTW can rank first and fifth in the world respectively, and the ratio of energy availability is also among the first. In September 2007, the Three Gorges-Shanghai DC Transmission Works under the TGPTTW stood out among more than 40 works recommended by many countries and won the Asian Power Transmission Project Award in 2007.

III. Characteristics of the TGPTTW

The EHV transmission technology (including DC and AC) is the best option for transmitting the TGHS's power to half of China through the Three Gorges transmission and transformation network. According to electrical engineering theory, in order to transmit certain power, the transmission loss of the transmission line is proportional to the quadratic of the transmission line current and inversely proportional to the quadratic of the transmission's voltage. A voltage drop of the transmission line is proportional to the line's current and inversely proportional to its voltage. Therefore, high-power EHV long-distance transmission technology has become the foremost option for improving transmission power, and reducing power loss and voltage drops in power transmission lines.

In practical terms, the TGPTTW transmits as far as 1,300 km, but the power generated by the generating units in the TGHS is 20 kV AC, which is considered low. For long-distance transmission, voltage must be boosted through substations and EHV transmission technology shall be employed. The direct transmission technology after AC voltage boosting is called EHV AC transmission technology. In order to further reduce transmission loss over long distances, and to improve the stability of a transmission system, EHV DC transmission is employed in the TGPTTW. That is, a converter station converts EHV AC into EHV DC and DC is converted back into AC through a converter station at the power consumption locations.

According to the data of CIGRE, the recommended transmission capacities for 340 kV and 500 kV lines are 600 MW and 1,200 MW, respectively. Through the demonstration of authoritative power system experts in China, the TGPTTW adopted the 500 kV EHV AC transmission technology and the ±500 kV EHV DC transmission technology, which were internationally advanced at that time and were fully mastered by China. The TGPTTW transmits the power generated by the TGHS to five AC and DC substations of the Central China Power Grid, namely Longquan, Jiangling, Gezhouba, Yidu, and Jingmen, through 18 circuits of 500 kV EHV transmission lines. Except for that consumed through the Central China

◐ Figure 4.2　Schematic Diagram of Power Transmission of the TGPTTW

直接送往用电地区。送往华东和南方电网的电力，要先送往换流站，把超高压交流电转换成±500千伏超高压直流电，再通过直流输电线路，远距离送往用电地区。

三峡水电站输送电力的地区，多为经济发达地区，长期承受缺电之苦。三峡输变电工程的建设，将源源不断的三峡电能输往这些地区，极大地缓解了电力供应的紧张局面，为国民经济发展提供了强有力的能源支撑，对实现我国能源合理配置具有显著的经济效益和社会效益。截至2021年年底，三峡水电站累计发电量达15 028亿千瓦时，源源不断的三峡电力有力地支持了华东、华中、广东等地区电力供应，成为我国重要的大型清洁能源生产基地。

第二节 三峡输变电工程的建设过程

一、三峡输变电工程的建设内容

三峡输变电工程的建设内容包括直流输电工程、交流输电工程和二次系统。

（1）直流输电工程

三峡输变电工程建设的三项直流输电工程均为：额定直流电压等级±500千伏，交流母线额定电压500千伏，每项直流输电工程的双极额定输送功率为300万千瓦；三项直流输电线路长度合计2 854千米、换流站6座，输送容量900万千瓦。其中：

Power and Sichuan and Chongqing Power Grids, most of the power is transmitted to the East China and China Southern Power Grids via ±500 kV EHV DC transmission lines after being converted into DC by converter technology.

The power transmitted to the Central China Power and the Sichuan and Chongqing Power Grids is directly transmitted to destinations for consumption through 500 kV EHV AC transmission lines. The power transmitted to the East China and the China Southern Power Grids is transmitted to converter stations to convert EHV AC into ±500 kV EHV DC, and is then transmitted to destinations for consumption over a long distance through DC transmission lines.

Most of the destination regions where TGHS power is transmitted are well-developed and suffer from long-term power shortages. The TGPTTW transmits power continuously generated by the TGHS to these regions, greatly easing power supply shortages and providing strong energy support for national economic development. This brings remarkable economic and social benefits in terms of rational allocation of energy resources in China. By the end of 2021, the cumulative power output of the TGHS reached 1,502.8 billion kWh. The continuous supply of such power has strongly supported the regions such as East China, Central China, and Guangdong, and this station has become an important large-scale clean energy production base in China.

Section 2 | Construction Process of the TGPTTW

I. Construction Scope of the TGPTTW

The construction scope of the TGPTTW involves DC transmission works, AC transmission works, and secondary systems.

(1) DC transmission works

The three DC transmission works of the TGPTTW are all rated at ±500 kV DC voltage and 500 kV AC busbar voltage, and the rated transmission power of each DC transmission works is 3 million kW. The three DC transmission works feature a total line length of 2,854 km and six converter stations with a transmission capacity of 9 million kW, including:

- 三峡—常州 ±500千伏直流输电工程（简称"三常直流工程"）是三峡电力送往华东地区的主干通道之一，直流输电线路860千米、换流站2座，三峡侧为龙泉换流站，常州侧为政平换流站，因此也称为"龙政直流工程"。
- 三峡—广东 ±500千伏直流输电工程（简称"三广直流工程"）是三峡电力送往广东的主干道，直流输电线路941千米、换流站2座，三峡侧为江陵换流站，广东侧为鹅城换流站，因此也称为"江城直流工程"。
- 三峡—上海 ±500千伏直流输电工程（简称"三沪直流工程"）是三峡电力送往华东地区的又一主干通道，直流输电线路1 053千米、换流站2座，三峡侧为宜都换流站，上海侧为华新换流站，因此也称为"宜华直流工程"。
- 荆门—沪西 ±500千伏直流输电工程（团林—枫泾）和葛沪直流综合改造工程是三峡电力送往华东地区的又一主干通道，新建换流站2座，每座换流站容量300万千瓦。直流输电线路全长1 024.9千米，其中922.9千米与葛南直流同塔双回架设，是地下电站送出工程，2009年开工，2011年建成投产。

灵宝直流背靠背换流站工程是西北电网与华中电网互联工程，作为三沪直流工程国产化中间试验项目，纳入三峡输变电工程管理，工程规模不计入三峡输变电工程建设规模。

（2）交流工程

截至2009年，三峡输变电工程建成500千伏交流输变电工程88项，其中：线路工程55项，共61条，线路总长6 519千米；变电工程33项，共建成500千伏变电站22座，开关场2座，变电总容量2 275万千伏。截至2011年全部（含地下电站送出等工程）建成，三峡输变电工程最终建成规模为：500千伏交流变电总容量2 275万千伏，交流输电线路7 280千米；±500千伏直流换流容量2 400千瓦，直流输电线路4 913千米。

• The Three Gorges-Changzhou ±500 kV DC Transmission Works (Three Gorges-Changzhou DC Works) is one of the main channels of the TGHS to East China, with 860 km of DC transmission lines and two converter stations (the Longquan Converter Station on the Three Gorges side and the Zhengping Converter Station on the Changzhou side), therefore also known as "Longquan-Zhengping DC Works".

• The Three Gorges-Guangdong ±500 kV DC Transmission Works (Three Gorges-Guangzhou DC Works) is the main channel of the TGHS to Guangdong, with 941 km of DC transmission lines and two converter stations (the Jiangling Converter Station on the Three Gorges side and the Echeng Converter Station on the Guangdong side), therefore also known as "Jiangling-Echeng DC Works".

• The Three Gorges-Shanghai ±500 kV DC Transmission Works (Three Gorges-Shanghai DC Works) is another main channel of the TGHS to East China, with 1,053 km of DC transmission lines and two converter stations (the Yidu Converter Station on the Three Gorges side and the Huaxin Converter Station on the Shanghai side), therefore also known as "Yidu-Huaxin DC Works".

• The Jingmen-Western Shanghai ±500 kV DC Transmission Works (Tuanlin-Fengjing) and the Gezhouba-Shanghai DC Comprehensive Renovation Project form another main channel of the TGHS to East China, with two new converter stations and a capacity of 3 million kW each. The total length of the DC transmission lines is 1,024.9 km, of which 922.9 km is erected on the same tower with double circuits with the Gezhouba-Nanqiao DC Transmission System. It is the transmission works of an underground hydropower station, commenced in 2009 and completed and commissioned in 2011. The Lingbao DC Back-to-back Converter Station is the interconnection works between the Northwest China Power Grid and the Central China Power Grid. As the intermediate pilot works for localization of the Three Gorges-Shanghai DC Works, it is included in the TGPTTW management, while its project scale is not included in the construction scale of the TGPTTW.

(2) AC transmission works

By 2009, 88 500 kV AC power transmission and transformation works had been completed for the TGPTTW, including 61 lines in 55 line works totaling 6,519 km, and 33 transformation works, with 22 500 kV substations and two switchyards, with a total capacity of 22.75 million kW. By 2011, the whole works (including such works as power transmission of the underground hydropower station) were completed and the final completed scale of the TGPTTW included 22.75 million kV of 500 kV AC transformation capacity, 7,280 km of AC transmission lines, 2,400 kW of ±500 kV DC converter capacity, and 4,913 km of DC transmission lines.

（3）二次系统

三峡输变电工程二次系统包含国家电力调度控制中心与华中、华东、四川、重庆等相关网省（市）电力调度通信中心的电能量管理系统、电能量计费系统、交易管理系统、继电保护及故障信息管理系统、安全稳定控制装置、系统通信六个大类的37个子项工程。其中二次系统调度自动化项目19个子项，系统通信类项目18个子项。建成了"三纵一横"的网状主干电力通信网络，由跨省主干光纤通信电路、微波通信电路、综合网管系统、同步网、国家电力数据网等组成。

二、三峡输变电工程的建设阶段

由于三峡输变电工程不仅直接关系到三峡电力电量能否"送得出、落得下、用得上"，也是获得三峡工程投资效益、实现长江中下游水能资源滚动开发的重要渠道。因此，三峡输变电工程项目的建设安排与三峡水电站装机进度同步：2003年三峡水电站首批机组并网发电，2009年26台机组全部建成发电。

根据三峡水电站的装机计划与工程实际进度，结合整个三峡输变电工程总体系统规划设计，三峡输变电工程的建设安排分成三个阶段：第一阶段为1997—2003年，配合三峡水电站首批机组投入运行后的电力外送；第二阶段为2004—2006年，配合三峡左岸电站14台机组全部投入运行及其电力外送；第三阶段为2006—2007年，配合三峡右岸12台机组投入

图4.3　三峡大坝旁边的输电线路

(3) Secondary systems

The secondary systems of the TGPTTW include 37 sub-item works of six major categories of the National Electric Power Dispatching and Control Center and the electric power dispatching communication centers of relevant network provinces (municipalities), such as Central China, East China, Sichuan, and Chongqing. These works include the electric energy management system, energy metering system, transaction management system, relay protection and fault information management system, safety and stability control devices, and system communication. Among them, there are 19 sub-item works within the secondary system dispatching automation works and 18 sub-item works within the system communication works. They create a "three vertical and one horizontal" main power communication network, which is composed of cross-province main optical fiber communication circuits, microwave communication circuits, an integrated network management system, synchronous networks, and a national power data network.

II. Construction Stages of the TGPTTW

The TGPTTW relates to the transmission, acceptance, and utilization of the power generated by the TGHS and it is also an important channel to obtain the investment benefits of the TGP and realize the progressive development of hydropower resources in the middle and lower reaches of the Yangtze River. Therefore, synchronized construction was arranged for the TGPTTW with the TGHS: In 2003, the first-generation units of the TGHS were put into operation and connected to the grid, and in 2009, all 26 units were put into operation.

According to the installation plan of the TGHS and the actual project progress, the construction of the TGPTTW was divided into three stages in combination with its overall system planning and design: The first stage, from 1997 to 2003, supported the power transmission after the first generating units of the TGHS were put into operation. The second stage, from 2004 to 2006, supported the operation of 14 generating units of the Three Gorges Left Bank Hydropower Station and its power transmission. The third stage, from 2006 to 2007, supported the operation and power transmission of all 12 generating units of the Three Gorges Right Bank Hydropower Station.

According to the original plan, the TGPTTW would be completed by 2008. Due to reasonable and effective management, the TGPTTW was completed one year ahead of schedule. By the end of 2007, the TGPTTW was completed in general.

⊂ Figure 4.3   Transmission Lines Next to the Three Gorges Dam

运行及电力外送。

按照原计划，三峡输变电工程建设到2008年全部完成。由于管理合理、措施有效，已经批复的三峡输变电工程建设项目提前一年完成。2007年年底，三峡输变电工程基本建成。

2008年至2011年，三峡输变电工程开工建设后续工程。

1. 第一阶段（1997—2003年）的建设实施情况

第一阶段的建设目标是确保三峡工程2003年首批投产机组发电量的送出。其中关键是建成三常工程，以及加强华中电网的主网架和湖北与河南、湖北与湖南、湖北与江西的联网工程，加强直流工程落点到常州、政平之后的交流500千伏输变电工程的配套送出。为配合川电东送，安排三峡—万县输电线路，建设500千伏万县变电站。河南、湖南、江西3省三峡输变电工程的重点是建设负荷中心的项目。

为解决三峡库区建设和移民用电需要，支持万县地区的经济发展，以及验证三峡输变电建设管理体系运行，1997年，国家电力公司向国家计委提出《关于三峡输变电工程1997年开工的请示》，将长寿—万县500千伏输电线路工程作为三峡输变电工程的第一个开工项目，万县变电站初期暂降压220千伏运行。该请示得到国家计委的批准后，工程于1997年3月开工，标志着三峡输变电工程从此开始正式进入建设实施阶段。

第一阶段从三峡输变电工程开工至2003年6月底三峡水电站首批机组投入运行，共建成交流线路2 599千米，交流变电容量775万千伏；三峡至华东第一条直流线路——三常工程全线架通，两端换流站实现双极投产。经国家验收，所有项目全部合格，确保了三峡水电站首批机组发电送出。

2. 第二阶段（2004—2006年）的建设实施情况

第二阶段的建设目标是确保三峡左岸电站14台机组发电全部送出。其中重中之重是建成三广工程，以及相应的交流输变电工程。同时，为全面检验和提高我国直流换流站国产化能力，并作为三峡右岸直流工程的中间试验项目，建成灵宝直流背靠背换流站工程。

第二阶段共计投产交流线路39个单项工程，线路长4 964千米；交流变

From 2008 to 2011, the follow-up works of the TGPTTW commenced.

1. Construction of the first stage (1997-2003)

The construction goal of the first stage was to ensure the power transmission of the first batch of generating units, which were commissioned in the TGP in 2003. The key was to complete the Three Gorges-Changzhou Works, strengthen the main grid structure of the Central China Power Grid and the networking works between Hubei and Henan, Hubei and Hunan, and Hubei and Jiangxi. It was also a goal to enhance the supporting transmission of 500 kV AC transmission and transformation works after DC works were placed in Changzhou and Zhengping. In order to cooperate with the Sichuan-East China Electricity Transmission Works, the Three Gorges-Wanxian Transmission Line was arranged and a 500 kV Wanxian Substation was constructed. The focus of the TGPTTW in Henan, Hunan, and Jiangxi was to build load centers.

In order to meet the needs of the construction of the Three Gorges Reservoir Area, the power needs of the resettlers, and to support the economic development of Wanxian County, and verify operation of the management system of the Three Gorges power transmission and transformation construction, the State Power Corporation submitted to the State Development Planning Commission the *Request for Commencement of the Three Gorges Power Transmission and Transformation Works in 1997* to take the Changshou-Wanxian 500 kV Transmission Line Works as the first works for commencement of the Three Gorges Power Transmission and Transformation Project and reduce 220 kV at the Wanxian Substation initially. After the Request was approved by the State Development Planning Commission, the TGPTTW commenced in March 1997, which marked the TGPTTW officially entered the construction stage.

In the first stage, from the commencement of the TGPTTW to the operation of the first batch of units of the TGHS at the end of June 2003, a total of 2,599 km of AC lines were completed, with an AC converter capacity of 7.75 million kV. The first DC line from Three Gorges to East China – the Three Gorges-Changzhou Works – was fully connected and the respective converter stations were commissioned. After the acceptance check by the State, all the works were confirmed and power transmission of the first generating units of the TGHS was ensured.

2. Construction of the second stage (2004-2006)

The construction goal of the second stage was to ensure the transmission of all power generated by the 14 units of the Three Gorges Left Bank Hydropower Station. In this stage, the most important were the Three Gorges-Guangdong Works and the corresponding AC

电 25 个单项工程，变电总容量 1 675 万千伏。直流输电工程三广工程建成，直流线路长 975 千米，换流站 3 座（包括灵宝直流背靠背换流站工程）、容量 672 万千瓦。

3. 第三阶段（2006—2007 年）的建设实施情况

第三阶段的建设目标是确保三峡右岸电站 12 台机组发电全部送出。期间将完成葛洲坝换流站改接到三峡水电站右一母线工程。此阶段重点工程为三沪工程及与此相应的交流输变电工程。

2006 年 12 月 9 日，三沪工程建成投产，工程全长约 1 100 千米，直流额定电压 ±500 千伏，额定电流 3 000 安，额定功率 300 万千瓦。

同时，建设完成系统二次及通信工程相应的安全控制及调度自动化工程，湖北荆州到湖南益阳、湖北荆门到孝感、湖北咸宁到凤凰山、湖北潜江到咸宁等交流线路工程，万县、长寿、双林、宜兴、吴江、荆州的变电站扩建工程。

2007 年 9 月，三峡到荆州的双回输电线路建成投运，标志着我国三峡电力外送输变电网络基本建成。

4. 第四阶段（2008—2011 年）的建设实施情况

2008 年，三峡输变电工程开始建设三峡输电线路优化完善工程、融冰工程、地下电站送出工程、荆门至沪西直流输电工程、葛沪直流线路改造工程、宜都至江陵改接潜江兴隆工程等。截至 2011 年年底，上述工程全部按预定进度建成投产。

三峡输变电工程的按期、高质量、高水平建成，成功地把三峡水电站源源不断发出的巨大电能安全可靠地输送到华中、华东、华南和川渝地区，惠及半数国民，为国民经济建设和社会发展提供了强有力的能源支持，奠定了全国电网互联的基本格局，促进了更大范围的能源资源的优化配置。

transmission and transformation works. In addition, the Lingbao DC Back-to-back Converter Station was completed in order to comprehensively verify and improve the localization capacity of DC converter stations in China and serve as the intermediate pilot works of the DC transmission works of the Three Gorges Right Bank Hydropower Station.

In the second stage, 39 AC lines were commissioned, with a total length of 4,964 km. 25 AC converters with a total capacity of 16.75 million kV were completed. In addition, the Three Gorges-Guangzhou DC Transmission Works was completed, with 975 km of DC lines, three converter stations (including Lingbao DC Back-to-back Converter Station), and a capacity of 6.72 million kW.

3. Construction of the third stage (2006-2007)

The construction goal of the third stage was to ensure successful transmission of all power generated by the 12 units of the Three Gorges Right Bank Hydropower Station. In this stage, the Gezhouba Converter Station would be connected to the right first busbar of the TGHS. The key works of this stage were the Three Gorges-Shanghai Works and the corresponding AC power transmission and transformation works.

On December 9, 2006, the 1,100 km Three Gorges-Shanghai Works was completed and commissioned, with a rated voltage of DC ±500 kV, rated current of 3,000 A, and rated power of 3 million kW.

Meanwhile, the completed works included the corresponding safety control and dispatching automation works for the secondary systems and communication works. Also included was AC line works from Jingzhou in Hubei to Yiyang in Hunan, Jingmen to Xiaogan in Hubei, Xianning to Fenghuangshan in Hubei, and Qianjiang to Xianning in Hubei. And substation expansion works were completed in Wanxian, Changshou, Shuanglin, Yixing, Wujiang, and Jingzhou.

In September 2007, the double-circuit transmission line from Three Gorges to Jingzhou was completed and put into operation, completing China's Three Gorges power transmission and transformation network.

4. Construction of the fourth stage (2008-2011)

The TGPTTW works commenced in 2008 included the Three Gorges transmission line optimization works, ice melting works, transmission works of the underground hydropower station, Jingmen-Western Shanghai DC Transmission Works, Gezhouba-Shanghai DC Line Renovation Works, and works on Changing Yidu-Jiangling to Yidu-Xinglong in Qianjiang. By the end of 2011, all the above works were completed and commissioned as scheduled.

The TGPTTW, completed on schedule, was a work of high quality, which safely and

本章小结：

三峡输变电工程是三峡工程的重要组成部分，是三峡水电站电力送出及其效益实现的根本保证。该工程从1997年开工建设至2011年全部建成投产，分为三个阶段建设完成，最终共建成：三峡水电站送出50万伏交流线路18回，线路总长7 280千米，变电总容量2 275万千伏，直流线路总长4 913千米，换流站总容量2 400万千瓦，并相应建设了涉及全国、涵盖9省2市的通信、调度自动化及与输变电工程配套的二次系统工程。三峡输变电工程的建成投产，对于三峡电力外送、促进全国联网、能源资源优化配置、推进我国电网建设水平提高等方面，发挥了重要作用。

参考文献：

［1］本书编委会. 百问三峡［M］. 北京：科学普及出版社，2012.

［2］国家电网公司. 中国三峡输变电工程·综合卷［M］. 北京：中国电力出版社，2008.

［3］《中国三峡建设年鉴》编纂委员会. 中国三峡建设年鉴（2009）［J］. 宜昌：中国三峡建设年鉴社，2009.

［4］《中国三峡建设年鉴》编纂委员会. 中国三峡建设年鉴（2011）［J］. 宜昌：中国三峡建设年鉴社，2011.

reliably transmits power generated by the TGHS to Central China, East China, South China, Sichuan, and Chongqing. It benefits half of the people in China, provides strong energy support for national economic and social development, sets a basic pattern for the interconnection of national power grids, and promotes the optimal allocation of energy resources over a wider range of the country.

Chapter Summary:

As an important part of the TGP, the TGPTTW is the fundamental guarantee for the power transmission of the Three Gorges Hydropower Station and the realization of its benefits. The Project commenced in 1997, was implemented in three stages, and commisioned in 2011, and the completed works include 18 circuits of 500kV AC lines of the Three Gorges Hydropower Station, a total line length of 7,280km, a total converter capacity of 22.75 million kV, a total length of DC lines of 4,913km, and a total capacity of converter stations of 24 million kW, as well as the secondary system works supporting the communication, dispatching automation, and transmission and transformation works involving entire China and covering nine provinces and two municipalities. After being commissioned, the TGPTTW has played an important role in the power transmission of Three Gorges, the promotion of national networking, the optimization of energy resources allocation, and the promotion of China's grid construction.

References:

[1] Editorial Board of This Book. *One Hundred Questions about Three Gorges* [M]. Beijing: Popular Science Press, 2012.

[2] State Grid Corporation of China. *China's Three Gorges Power Transmission and Transformation Works*. Comprehensive Volume [M]. Beijing. China Electric Power Press, 2008.

[3] Editorial Board of Construction Yearbook of China's Three Gorges. *Construction Yearbook of China's Three Gorges* (2009) [J]. Yichang: Publishing House of Construction Yearbook of China's Three Gorges, 2009.

[4] Editorial Board of Construction Yearbook of China's Three Gorges. *Construction Yearbook of China's Three Gorges* (2011) [J]. Yichang: Publishing House of Construction Yearbook of China's Three Gorges, 2011.

阅读提示：

移民工程是三峡工程建设的重要组成部分，关系三峡工程的成败。三峡移民工程涉及的地域之广、数量之多、持续时间之长、需协调的利益关系之复杂等，在水利水电工程建设史上是独一无二的，被国内外舆论称为"世界级难题"。

三峡移民工程实行的开发性移民方针是我国水库移民工作的一次重大变革，它从根本上改变了中华人民共和国成立以后实施了30多年的补偿性、被动性移民的基本思路，使移民安置与移民后续发展、经济结构调整、经济社会发展相结合。三峡工程百万移民的成功安置是一项古今中外罕见、民生所系的伟大壮举。

Resettlement is an important part of the construction of the Three Gorges Project (TGP), which is very vital to the project's success. The Three Gorges Resettlement is unique in the history of water conservancy and hydropower development because of the wide and many areas it involved, the long duration, and complicated interest relations that need to be coordinated. Thus domestic and international public opinions deem it a "world-class problem".

The development-driven resettlement policy carried out for the Three Gorges Resettlement is a great change in terms of reservoir resettlement work in China. It fundamentally changes the basic thinking of compensatory and passive resettlement that was implemented for over 30 years since the People's Republic of China's founding. It combines resettlement with the subsequent development, economic restructuring, and economic and social development. The successful resettlement of millions of resettlers in the TGP is a great feat that is rare in both China and abroad, in both ancient and modern times.

Chapter 5 >>>>

第五章
三峡移民工程的实施
Implementation of the Three Gorges Resettlement

第一节　三峡移民工程的难点

三峡工程涉及的百万移民搬迁是"世界级难题"。"难"在库区社会经济发展滞后、自然禀赋较差、安置容量有限;"难"在影响范围大,政策平衡、协调困难;"难"在移民传统观念较强、技能水平偏低,自我调整、重建家园的能力较弱;"难"在地处当时的国家级集中连片贫困区,移民后续发展乏力等。因此,移民动迁在三峡工程决策阶段就被摆在了重要位置,推行开发性移民政策,举全国之力实施移民搬迁安置,工程浩大,任务艰巨,亘古未见。

一、搬迁安置难

三峡工程涉及搬迁安置移民约131万人,其中农村移民55万余人,城集镇移民75万余人,根据移民意愿,规划报告中主要采取就地后靠、县内安置等方式。但是,在搬迁安置实施过程中,有限的安置容量给三峡库区移民安置造成了巨大困难。主要表现在:一是库区各县自然环境恶劣,山高坡陡,质量较好的耕地主要集中在淹没线下,而淹没线上土地贫瘠、分散、坡度大,不宜开垦;二是库区人口比较密集,人地矛盾突出,当地人均耕园地面积不足1亩,通过调整土地安置移民困难较大;三是库区生态环境脆弱,地质破碎,水土流失严重,自然灾害频发,开发改造土地不仅容易引发滑坡、泥石流等灾害,部分新建居民点和开垦的耕园地也存在安全隐患;四是受封库及自然条件的影响,当地基础设施建设落后,产业发展滞后,经济基础薄弱,无法吸纳大量移民就业。

Chapter 5　Implementation of the Three Gorges Resettlement

Section 1　Difficulties of Three Gorges Resettlement

The relocation of millions of people involved in the TGP was a "world-class problem". The difficulties lay in the lagging social and economic development, poor natural conditions, and a limited nearby resettlement capacity of the surroundings of the project, and lay in the large impact range, difficult policy balancing and coordination, as well as the strong traditional thinking, low level of skills, the weak ability to adjust and build new homes among settlers. In addition, there would be weak follow-up development of resettlers as they were in poor areas at that time, etc. Therefore, resettlement was placed in an important position in the decision-making stage of the TGP. It was unprecedented, and was a huge project and arduous task, which marshals the full resources of the country to carry out.

I. Difficult Relocation and Resettlement

The TGP involved the relocation and resettlement of about 1.31 million people, including more than 550,000 from rural areas and 750,000 from cities and towns. According to the wishes of resettlers, the planning report mainly adopted the methods of moving backwards, and resettlement within local county. However, during the Three Gorges Reservoir Area resettlement, the limited resettlement capacity had caused great trouble. The main problems were: First, the natural environment of every county in the reservoir area was abominable, with high mountains and steep slopes, and cultivated land with better quality was mostly under the flood line, while the land above the flood line was barren and scattered, with a big slope, and was not suitable for cultivation. Second, the reservoir area was densely populated, and the local per capita area of arable land was less than 1 *mu* (about 666.7 m^2). It was difficult to resettle people through land rearrangement. Third, the reservoir area featured a fragile ecological environment, broken zones, serious soil erosion, and frequent natural disasters. Land development and transformation could easily cause landslides, debris flows, and other disasters. Some new residential areas and reclaimed farmland also had potential safety hazards. Fourth, it was impossible to absorb a large number of resettlers for employment due

二、组织协调难

三峡工程移民工作组织协调难主要体现在：一是影响范围大，三峡水库20年一遇回水水面面积为1 084平方千米，其中淹没陆域面积632平方千米，涉及湖北省和重庆市的20个县（区）（期间，1997年省级机构改革，重庆为直辖市，机构重组）。省（直辖市）级之间、县级之间，甚至是乡镇之间需要开展大量的组织协调工作，地区之间的政策措施平衡难度较大；二是影响类别多，不仅涉及移民搬迁安置，还涉及城市（县城）迁建、集镇迁建、工矿企业搬迁和专业项目（如公路）建设，以及文物古迹保护等，涉及的利益主体多、参建单位多，需要做大量的组织协调工作；三是持续时间长、政策跨度大，按照三峡工程"一级开发，一次建成，分期蓄水，连续移民"的建设方案和"先淹先搬、移民进度与枢纽工程相衔接"的原则，三峡工程移民搬迁安置至2009年全部完成，时间长达17年，分四期进行。同时，这段时期又是我国市场经济发展的重要阶段，社会经济环境、物价水平、相关政策等变化较快，造成移民政策和安置方式需要根据实际情况不断进行调整，给移民工作造成了政策平衡、衔接、协调难等诸多困难。

三、社会重建难

大型水利水电工程形成的大水库不同于线性工程，其淹没范围较广，甚至整个县城、集镇、村组需要搬迁和重建，移民的生产生活设施全部受到影响。由于搬迁安置过程也是社会经济活动解构与重构的过程，所以水库移民相比其他工程移民难度更大，措施更复杂。移民个体往往有很强的乡土观念、安土重迁的情结，担心原有的亲情网络、社会关系网络的丧失，不愿意改变原有的生产和生活方式，受限于自身素质、风险承受能力、对今后生存的担忧等，仅凭一己之力难以重建家园、开启新生活，也难以做出动迁决定。无论是动员移民搬迁，制定有力措施帮助移民恢复并发展生产，还是保证移民生产生活资料配置到位，都需要做大量的工作。而各级地方政府还要统筹考虑移民搬迁安置、工矿企业处理、基础设施建设和区域经济发展，以及移民后续安稳致富、利益相关方的平衡等问题，以利于重构、恢复并提升库区的

to the impact of the reservoir closure and natural conditions, backward local infrastructure, lagging industrial development, and a weak economic base.

II. Difficult Organization and Coordination

Main difficulties in resettlement organization and coordination of the TGP: First, the impact range was large. The backwater area of the Three Gorges Reservoir once-in-20-years is 1,084 km^2, of which the flooded land area is 632 km^2, involving 20 counties (districts) in Hubei Province and Chongqing Municipality (Chongqing was changed into a Municipality with institutional reorganization during Provincial Institutional Reform in 1997). A lot of organization and coordination needed to be carried out between provinces (municipalities), counties, and even townships, and it was difficult to balance policies and measures between regions. Second, there was much following work to be done, which not only involved the relocation and resettlement of the people, but also involved the relocation of cities (county towns), market towns, industrial and mining enterprises, the construction of specialized projects (such as highways), and the protection of cultural relics and historical sites. There were many stakeholders and construction participants, and a lot of organization and coordination needed to be done. Third, it was a long period, and involved many policies. According to the construction plan of the TGP of "Grade-I development, completion at one go, staged water storage, and successive resettlement" and the principle of "the affected in the flooded areas moving first and carrying out the resettlement along with the project", the TGP relocation and resettlement was completed in 2009, which lasted 17 years and was carried out in four phases. In addition, this period was also an important stage for the development of China's market economy, in which social and economic environments, price levels, and related policies changed rapidly, resulting in the need for constant adjustment of resettlement policies and methods according to the actual conditions, and causing such difficulties in policy balance and coherence, and coordination to resettlement.

III. Difficult Social Reconstruction

A large reservoir constructed by a large-scale water conservancy and hydropower project was different from the linear projects. Its flooding range was wider, even the entire county, towns, and villages needed to be relocated and reconstructed, and the production and living facilities of the resettlers were all affected. Since the process of relocation and resettlement was also a process of deconstruction and reconstruction of social and economic activities, the reservoir resettlement was more difficult and the measures were more complicated than other

社会经济。

四、后续发展难

三峡库区大多数淹没县属于当时的国家级或省级贫困县，社会经济发展相对滞后，特别是能够大量吸纳移民就业的二、三产业不够发达，大部分县在移民安置前期的产业结构仍以农业为主；库区基础设施建设滞后，制约当地招商引资、经济发展，部分具有地方特色的产业未能得到很好的发展，特别是部分特色农产品未能形成规模、创立自己的品牌；当地土地资源匮乏，绝大多数后靠安置移民人均耕地面积不足 0.8 亩，且土地质量和农田水利设施较以前的河滩地更差，改造难度较大。由于历史原因，移民搬迁安置前，库区教育水平发展滞后，移民受教育程度普遍较低，导致移民搬迁安置后适应新环境、新生产方式的能力不足，从事非农行业的竞争力不强。以上所有因素，造成部分移民搬迁安置后，发展后劲不足，要实现"稳得住、能致富"的目标面临较大挑战。

project resettlement. Resettlers often feel deeply attached to their native land and are worried about leaving their original community and social relationship network. They were unwilling to change their work and lifestyle, and were concerned about future survival due to limited personal capabilities, risk tolerance, etc. It was difficult for them to build a new homeland and start a new life, thus making the decision to relocate. A lot of work needed to be done, whether it be to mobilize the resettlers, take effective measures to help them restore and develop production, or ensure that new production and living materials were in place. Except for the resettlement of people, local governments at all levels also had to consider issues such as relocation of industrial and mining enterprises, infrastructure construction, and regional economic development, as well as the follow-up support to resettlers, and deal with the balance of all stakeholders in order to facilitate reconstruction, restoration, and improvement of the local social economy.

IV. Difficult Follow-up Development

Most of the flooded counties in the area were poverty-stricken at the national or provincial level at that time, and their social and economic development was relatively lagging. In particular, the secondary and tertiary industries that could absorb a large number of resettlers were not developed enough, and the industrial structure of most counties in the early stage of resettlement was still dominated by agriculture. The infrastructure construction in the reservoir area was lagging behind, which restrained local investment and economic development. Some industries with local characteristics were not well-developed, especially some characteristic agricultural products were not developed on a large scale, and had no brands of their own. Local land resources were scarce, and most of the resettlers who moved backwards had less than 0.8 *mu* (about 533.3 m^2) of cultivated land per capita. The land quality and farmland water conservancy facilities were worse than the previous flood land, making it more difficult to transform the land. Due to historical reasons, education in the area had lagged behind before the relocation and resettlement, which made them ill-prepared to adapt to the new environment and new ways of work, especially lacking competitiveness in non-agricultural industries. All the factors above resulted in poor ability to advance in life, and there was a large challenge to achieve the goal of "becoming stable and wealthy".

第二节 三峡移民工程的政策体系

国家和三峡库区各地政府根据我国宪法和法律规定，结合库区实际，先后制定并颁发、实施了一系列政策法规。其中，1993年8月，国务院公布施行《长江三峡工程建设移民条例》，这是我国首部为特定工程项目移民而颁布的行政法规。《长江三峡工程建设移民条例》借鉴了以往水库移民经验和教训，并在长期实践探索中逐步完善。2001年2月，国务院修订颁布了《长江三峡工程建设移民条例》。2006年7月，国务院修订颁布了《大型水利水电工程建设征地补偿和移民安置条例》。以这些条例为基础，配合专项法规和地方法规，形成了移民政策体系和法律体系，使三峡移民走上了法制化的道路。这些法律法规对于规范开展和圆满完成三峡库区移民工作任务有巨大的保障作用，可以保障移民合法权益，促进库区移民、经济、社会、资源和环境协调可持续发展，确保百万移民"搬得出、稳得住、逐步能致富"搬迁安置目标的顺利实现。

一、三峡移民工程政策的基本方针

所谓基本方针，就是对移民工作具有长期稳定的指导意义，能够从根本上起到规范和推动作用的政策。三峡移民工程政策的基本方针主要包括以下几个方面。

1. 开发性移民的方针

1984年，党中央、国务院在审议三峡工程150米建设方案的可行性报告时，提出要探索开发性移民的新路子。为此，我国政府提出了开发性移民方

Section 2 | Policy System of the Three Gorges Resettlement

In accordance with the Constitution and other laws of China, the State and local governments of the Three Gorges Reservoir Area have formulated, promulgated, and implemented a series of policies and regulations in consideration of the conditions in the reservoir area. In August 1993, the State Council promulgated and implemented the *Regulations on Residents-Resettlement for the Yangtze River Three Gorges Project Construction*, which was the first administrative regulation promulgated by the State for the resettlement of a specific project. The experience and lessons of previous reservoir resettlements were drawn upon, and the Regulations were gradually improved over the long-term. In February 2001, the State Council revised and promulgated these regulations. In July 2006, the State Council revised and promulgated a set of regulations called the *Regulations on Land Requisition Compensation and Resettlement for Construction of Large and Medium-sized Water Conservancy and Hydropower Projects*. Based on these regulations and in conjunction with special and local regulations, a resettlement policy system and a legal system were formed, bringing the Three Gorges resettlement into a legal framework. These laws and regulations played a huge role in standardizing and successfully completing the resettlement, safeguarding the legitimate rights and interests of resettlers, promoting the coordinated and sustainable development of resettlement, economy, society, resources, and the area's natural environment, and ensuring that millions of resettlers could attain the goal of smooth relocation and resettlement and providing them with stability and the means to better themselves.

I. Fundamental Policies for Three Gorges Resettlement

The fundamental policies refer to the policies that have long-term and stable guiding significance for resettlement, and can fundamentally regulate and promote the resettlement work. The fundamental policies of the Three Gorges Resettlement mainly include the following items.

针,并首先在三峡工程移民中系统地实施。1993年《长江三峡工程建设移民条例》第3条指出"国家在三峡工程建设中实行开发性移民方针"。

开发性移民的实质就是把移民安置工作纳入国民经济和社会发展的轨道,把移民规划和区域经济发展计划、生态与环境保护结合起来。通过移民安置促进地区经济发展和生态环境的改善,使移民和原来居民的生产生活水平达到或者超过原有水平。围绕开发性移民方针,国家对三峡工程实施了十几项扶持措施,主要有:①建立三峡库区移民后期扶持基金,按照一定比例分配给湖北省、重庆市使用;②三峡水电站投产后,缴纳税款留给地方的部分按照一定比例分配给湖北省、重庆市;③三峡水电站投产后,三峡库区用电应优先安排;④三峡库区有水电资源的市县,优先列为农村水电初级电气化县,并对其建设予以扶持;⑤三峡坝区、库区淹没的土地按应纳税额的40%征收,城镇企业搬迁等占用土地按规定纳税后全部上交再全部返还,用于农村移民安置;⑥三峡库区具备一定条件的县、区应优先列入生态农业试点示范县,予以扶持;⑦安置农村移民开发的土地和新办的企业,应依法减免农业税、农业特产税、企业所得税;⑧中央和三峡库区所在地的省、市政府在安排建设项目或分配资金时,对三峡库区应给予倾斜和照顾。

1. Development-driven resettlement policy

In 1984, when reviewing the Feasibility Report on the 150 m Construction Plan for the Three Gorges Project, the Party's Central Committee and the State Council proposed to explore new ways of carrying out development-driven resettlement. For this reason, the Chinese government put forward the development-driven resettlement policy, which was first carried out systematically during the TGP resettlement. Article 3 of the *Regulations on Residents-Resettlement for the Yangtze River Three Gorges Project Construction* in 1993 points out that "the State shall implement the development-driven resettlement policy in the construction of the TGP".

The essence of development-driven resettlement is to bring resettlement in line with the national economic and social development, and design resettlement planning with regional economic development, ecology and the environment. Promoting regional economic development and improving the ecological environment through resettlement, so that the production and living standards of resettlers and native residents can reach or exceed their original levels. Centering on development-driven resettlement policy, the State implemented more than a dozen supporting measures for the TGP, mainly including: 1) A later-stage support fund for resettlement of the Three Gorges Reservoir Area, allocating it to Hubei Province and Chongqing Municipality in a certain proportion. 2) Part of the tax paid to local governments shall be distributed to Hubei Province and Chongqing Municipality in a certain proportion after the Hydropower Station is put into operation.3) Power supply priority shall be given to the reservoir area after the Hydropower Station is put into operation. 4) The cities and counties in the reservoir area with hydropower resources are given priority to be enlisted as rural primary electrification areas, and their construction shall be supported. 5) The tax of the land flooded in the Three Gorges Dam area and the reservoir area shall be collected at 40% of the taxable amount, and the tax shall be fully returned after it is paid as required for the land occupied by the relocated urban enterprises, funding the resettlement of rural resettlers. 6) Priority shall be given to listing the counties and districts with certain conditions in the reservoir area as pilot demonstration counties and districts of ecological agriculture, and support should be given to them. 7) The land developed and new enterprises established for resettlement of rural resettlers shall have tax exemptions and reductions on agricultural tax, agricultural special product tax, and enterprise income tax according to law. and 8) The central government and the provincial and municipal governments of the places where the reservoir area is located shall give preference and favorable conditions to the area when arranging construction projects or allocating funds.

2. 高效的管理体制

"中央统一领导,分省负责,以县为基础"的管理体制,从组织上确保开发性移民方针的贯彻落实。从中央到淹没区和移民安置区所在的省、市、县及乡镇,可根据工作需要建立专门的移民工作机构,并增配领导职数,专门分管移民工作,强化各级政府责任。

3. 移民安置任务与补偿资金的"双包干"方针

为了保证移民安置任务在规划投资内完成,避免反复调整设计概算的情况出现,国务院三峡建委将移民总投资切块到湖北省、重庆市包干使用,即根据移民规划任务将移民补偿经费按淹没实物指标切块到湖北省、重庆市,再由两省市切块包干到三峡库区各区县,实行包干使用。在实施过程中,有关技术监督单位运用"匹配分析法"对这项政策进行量化分析,按移民专项内容进行分解。通过实施这项重大改革,国家明确了各级政府和移民迁建单位的权利、义务和责任,在全库区建立了移民任务和移民资金的刚性约束机制,增强了地方各级政府的工作责任,为管好用好移民经费、保证各项移民任务的顺利完成奠定了基础。

4. 全国有关省市对口支援三峡移民的方针

《长江三峡工程建设移民条例》明确提出,国家鼓励和支持国务院有关部门和各省、市、自治区采取多种形式,从教育、科技、人才、管理、信息、资金、物资等多方面,对口支援三峡移民安置。要求国务院各部委在安排计划时结合三峡移民,多安排些项目;各省、自治区、直辖市的有关部门,要在互惠互利的基础上,积极开展与三峡库区各县(市)的经济、技术合作,并且每1~2年召开一次全国性的对口支援三峡工程移民工作会议。

2. Effective management system

The management system of "unified leadership by the central government, assigned responsibilities by provinces, and county-based management" ensured the implementation of the development-driven resettlement policy in terms of organization. From the central government to the provinces, municipalities/cities, counties, and townships where the flooded areas and resettlement areas are located, special resettlement work institutions may be established according to the work needs, and the number of leadership positions may be increased. These leaders shall be in charge of resettlement work and strengthen the responsibilities of governments at all levels.

3. Policy of "double dedicated uses" for resettlement tasks and compensation

In order to complete the resettlement tasks within the planned fund and avoid repeated adjustments to the estimated budget, the TGPCC divided and allocated the total fund for resettlement to Hubei Province and Chongqing Municipality for dedicated use. The resettlement compensation fund shall be divided and allocated to Hubei Province and Chongqing Municipality according to the resettlement planning tasks and quotas of flooded tangible assets, and then Hubei Province and Chongqing Municipality shall divide and allocate the funds to districts and counties in the Three Gorges Reservoir Area for dedicated use. During implementation, the relevant technical supervision units used the "matching analysis method" to carry out a quantitative analysis of this policy and break it down according to the specific contents of the resettlement. Through this major reform, the State clarified the rights, obligations, and responsibilities of governments at all levels, and of resettlement and construction organizations. The State established a rigid constraint mechanism on the resettlement tasks and funds throughout the reservoir area, which strengthened the work responsibilities of local governments at all levels, laid a foundation for the proper management and use of resettlement funds, and ensured the successful completion of various resettlement tasks.

4. Policy of relevant provinces and municipalities for supporting the Three Gorges resettlement

It is specified in the *Regulations on Residents-Resettlement for the Yangtze River Three Gorges Project Construction* that the State encourages and supports the relevant departments of the State Council and the provinces, municipalities, and autonomous regions to take various actions to provide partner support to the Three Gorges resettlement in such aspects as education, science and technology, human resources, management, information, funds, and

表5.1 三峡移民工程对口支援单位一览表

三峡库区受援区县	重点结对支援省、自治区、直辖市
夷陵区	黑龙江省、上海市、青岛市
秭归县	江苏省、武汉市
兴山县	湖南省、大连市
巴东县	北京市
巫山县	广东省、广州市、深圳市、珠海市
巫溪县	吉林省
奉节县	辽宁省
云阳县	江苏省
万州区	上海市、天津市、福建省、南京市、宁波市、厦门市
开县	四川省
忠县	山东省、沈阳市
石柱土家族自治县	云南省、江西省
丰都县	河北省
涪陵区	浙江省
武隆县	江西省、云南省
长寿区	广西壮族自治区
渝北区	安徽省
巴南区	河南省

二、对三峡移民政策的调整和补充（1999—2001年）

自三峡移民大规模搬迁以来，国家于1999年至2001年先后提出了实施农村移民外迁安置、工矿企业迁建政策的"两个调整"政策和水污染防治、地质灾害防治"两个防治"政策，这两个政策的制定和落实是党和国家综合考虑三峡移民，以及经济、社会和生态环境的协调可持续发展做出的重大决策，对保证三峡库区生态环境安全发挥了重要作用。同时，国家对《中华人民共和国土地管理法》进行了修订，提高了征地移民补偿标准，并对后期扶持政策进行了调整。

materials. All ministries and commissions under the State Council shall arrange more projects in conjunction with the Three Gorges resettlement when making plans. The relevant departments of all provinces, autonomous regions, and municipalities shall, on the basis of mutual benefit, actively carry out economic and technical cooperation with the counties (municipalities/cities) in the area. A national conference on partner support for the TGP resettlement shall be held every 1-2 years.

Table 5.1 List of PartnerSupporters of Three Gorges Resettlement

Assisted District/County in Three Gorges Reservoir Area	Key Provinces/Autonomous Regions and Municipalities for Pairing Support
Yiling District	Heilongjiang Province, Shanghai Municipality, and Qingdao Municipality
Zigui County	Jiangsu Province and Wuhan City
Xingshan County	Hunan Province and Dalian City
Badong County	Beijing Municipality
Wushan County	Guangdong Province, Guangzhou City, Shenzhen Municipality, and Zhuhai City
Wuxi County	Jilin Province
Fengjie County	Liaoning Province
Yunyang County	Jiangsu Province
Wanzhou District	Shanghai Municipality, Tianjin Municipality, Fujian Province, Nanjing City, Ningbo City, and Xiamen City
Kaixian County	Sichuan Province
Zhongxian County	Shandong Province and Shenyang City
Shizhu Tujia Autonomous County	Yunnan Province and Jiangxi Province
Fengdu County	Hebei Province
Fuling District	Zhejiang Province
Wulong County	Jiangxi Province and Yunnan Province
Changshou District	Guangxi Zhuang Autonomous Region
Yubei District	Anhui Province
Banan District	Henan Province

II. Updates and Supplements to the Three Gorges Resettlement Policies (1999-2001)

Since the large-scale relocation, the State successively put forward from 1999 to 2001 the policy of "two adjustments" for relocating rural people and industrial and mining enterprises. It also put forward the policy of "two preventions and controls" to mitigate water pollution and geological hazards. The formulation and implementation of these two policies was a major decision made by the Communist Party of China (CPC) and the State, consid-

1."两个调整"政策

一是外迁安置政策调整。在三峡移民安置之初,主要采取的是就地后靠安置的模式。但随着移民安置工作的逐步开展,三峡库区有限的环境容量和巨大的安置任务之间的矛盾逐渐显现出来。三峡库区土地容量严重不足,库区人口资源的环境承载压力巨大,同时由于我国宏观经济环境发生了变化,原来规划的二、三产业安置和自谋职业安置渠道难以落实,部分移民需要调整为农业安置。为确保三峡库区良好的生态环境和可持续发展,保证移民安居乐业,1999 年 5 月国务院对三峡库区移民安置政策做出重大调整,鼓励和引导更多的移民外迁安置,实行以多种方式安置农村移民的方针,把本地安置与异地安置、集中安置与分散安置、政府安置与自找门路安置结合起来,努力实现移民搬得出、稳得住、逐步能致富和长治久安的目标。1999 年 6 月 6 日,国务院办公厅发布了《关于三峡工程库区移民工作若干问题的通知》,对农村移民安置去向进行了重大调整,增加了农村移民出县外迁安置比例。

二是工矿企业迁建政策。工矿企业迁建政策由结合技术改造进行迁建,

图 5.1　山东对口支援项目——山东如意集团投资 20 亿元的 50 万锭紧密纺织项目

ering the Three Gorges resettlement and the coordinated and sustainable development of the economy, society, and ecological environment. It played an important role in protecting the ecology of the Three Gorges Reservoir Area. In addition, the State revised the *Land Administration Law of the People's Republic of China*, raised the compensation standard for land acquisition and resettlement, and updated the later-stage support policy.

1. Policy of "two adjustments"

First, the relocation and resettlement policy was updated. At the beginning of the Three Gorges resettlement, people were mainly moved backwards within the local area. However, with the gradual development of resettlement work, a contradiction between the limited environmental capacity and the huge quantity of resettlers gradually emerged. Land capacity was seriously insufficient in the area, and the pressures of population, resources, and the environment were huge. The originally planned resettlement channels for secondary and tertiary industries and self-employment were difficult to implement due to changes in the macroeconomic environment in China, and some people needed to continue work in agriculture after moving. In order to maintain a good ecological environment and sustainable development in the reservoir area, and ensure that resettlers live and work in peace and contentment, the State Council made major updates to the resettlement policy for the area in May 1999. It encouraged and guided more resettlers to relocate in other places, adopted more ways of resettlement, consisting of local resettlement and relocation in other places, centralized and dispersed resettlement, and resettlement by governments and self-resettlement. It made efforts to achieve the goal of resettlement, stability, wealth, and long-term stability. On June 6, 1999, the General Office of the State Council issued the *Circular on Some Issues Concerning the Resettlement Work in the Three Gorges Project Reservoir Area*, which made major updates to the resettlement destinations of rural resettlers and increased the proportion of rural resettlers moving out of their original counties.

Second, the policy for the relocation of industrial and mining enterprises was updated. It was changed from relocation in combination with technical transformation to relocation focusing on structural updates, quality improvement, and benefit enhancement. The updates to structures of organization, ownership, and product of industrial and mining enterprises were to be intensified and they would not be simply replicated. State-owned and collectively-owned industrial and mining enterprises that had serious pollution, no market for their products, and were insolvent in their assets were to resolutely go bankrupt or be closed down.

◔ Figure 5.1　Partner Support Project of Shandong Province - 500,000 Compact Spinning Project with an Investment of RMB 2 Billion from Shandong Ruyi Group

调整为工矿企业迁建要把工作重点、着力点放在结构调整、改进质量和提高效益方面。加大工矿企业组织结构、所有制结构和产品结构的调整力度，绝不能搞原样搬迁。对污染严重、产品无市场和资不抵债的国有、集体搬迁工矿企业，坚决执行破产或关闭；对产品有市场、资产质量好、领导班子强的搬迁工矿企业，通过与对口支援名优企业合作或合资，进行组合搬迁。对于符合上市条件的已经迁建的工矿企业，有关部门要积极给予支持。三峡库区要大力引进名优产品和名优企业，努力改善投资环境，尤其要转变观念，认真改善投资软环境，切实加大对口支援工作力度。国家制定优惠政策，鼓励和调动迁建工矿企业结构调整的积极性。

2. "两个防治"政策

一是加强三峡库区水污染防治工作。国家安排专项资金用于三峡库区污水和垃圾处理项目建设。按照《三峡库区及其上游水污染防治规划》，"十五"期间建成37个城市污水处理厂和30个小城镇污水处理项目，建成26个城市垃圾处理场和30个小城镇垃圾处理项目，规划建成4个危险废弃物处理项目。

二是加强三峡库区地质灾害防治工作。成立三峡库区地质灾害防治领导小组，由国土资源部牵头，重庆市、湖北省和国务院有关部门负责同志组成。国家安排专项资金用于三峡库区地质灾害防治项目。

3. 土地补偿标准调整

《长江三峡工程水库淹没处理及移民安置规划报告》对土地补偿费及安置补助费计算的价格基期为1993年5月，主要依据为1988年12月29日第七届全国人民代表大会常务委员会第五次会议修正的《中华人民共和国土地管理法》以及1991年颁布实施的《大中型水利水电工程建设征地补偿和移民安置条例》确定的补偿补助标准。

1998年8月24日，第九届全国人大常委会第四次会议修订通过了新的《中华人民共和国土地管理法》（以下简称"新《土地管理法》"）。新《土地管理法》在进一步强调土地管理和切实保护耕地重要性的同时，扩大了土地有偿使用的范围，提高了征用土地补偿标准。为保证三峡库区移民安置工作的顺利进行，中央决定根据新《土地管理法》，先后调增"库区移民迁建征地投

Industrial and mining enterprises that had product markets, good asset quality, and strong leadership teams were to be relocated through cooperation or joint ventures with well-known or high-performing enterprises. Relevant authorities were to actively support the relocated industrial and mining enterprises that met the listing conditions. The reservoir area was to introduce high-quality products and famous enterprises, improve the investment environment, change their thinking in particular, carefully improve the soft investment environment, and effectively increase their efforts in partner support. The State was to formulate preferential policies to encourage and stimulate the enthusiasm of relocated industrial and mining enterprises for structural updates.

2. Policy of "two preventions and controls"

First, measures for mitigating water pollution in the Three Gorges Reservoir Area were enhanced. The State arranged special funds for the construction of wastewater and garbage treatment works in the reservoir area. According to the *Water Pollution Prevention Planning for the Three Gorges Reservoir Area and Its Upper Reaches*, 37 municipal wastewater treatment plants, 30 small municipal wastewater treatment works, 26 municipal garbage treatment plants, and 30 small municipal garbage treatment works were completed. Yet to be completed are 4 hazardous waste treatment works which are scheduled for completion during the Tenth Five-Year Plan period.

Second, measures for mitigating geological hazards in the area were enhanced. A Leading Group for Prevention and Control of Geological Hazards in the area was established, which was led by the Ministry of Land and Resources and composed of leading staff from Chongqing Municipality, Hubei Province, and relevant departments of the State Council. The State arranged special funds for the works of mitigating geological hazards in the area.

3. Updates to land compensation criteria

In the *Report on Treatment of Reservoir Inundation and Resettlement Planning of the Three Gorges Project on the Yangtze River*, the price base period for the calculation of the land compensation and resettlement subsidy was May 1993, mainly based on the criteria of compensation and subsidy determined in the Land Administration Law of the People's Republic of China, amended by the Fifth Session of the Standing Committee of the Seventh National People's Congress on December 29, 1988 and the *Regulations on Land Acquisition Compensation and Resettlement for Construction of Large and Medium-sized Water Conservancy and Hydropower Projects* promulgated in 1991.

On August 24, 1998, the Fourth Session of the Standing Committee of the Ninth National People's Congress amended and adopted the new Land Administration Law of the

资"和"农村淹没土地补偿投资",其主要调整内容有:

一是农村淹没土地新增补偿投资。土地征用费是农村移民生产安置的主要资金来源,与保持移民生活水平不低于原有生活水平和三峡库区的可持续发展密切相关。长江设计院根据实际工作和有关要求,于2004年5月编制完成《长江三峡工程库区淹没土地新增补偿投资测算报告》报批稿。按批准的方案,农村淹没耕园地土地补偿倍数取9倍,安置补助倍数取6倍,非耕园地补偿标准按照《中华人民共和国森林法》的有关规定确定。农村淹没土地补偿费及安置补助费新增投资,等于淹没土地补偿费、林木补偿费、安置补助费三项增加额之和,共计22.95亿元,分配给湖北省4.22亿元、重庆市18.73亿元。

二是移民征地新增补偿投资。2001年6月,长江设计院根据新《土地管理法》的有关规定和国务院三峡建委办公室的要求,完成了三峡库区移民迁建征地需增加投资的测算工作,经国务院三峡建委批准实施。其测算内容为三峡库区城集镇迁建、工矿企业迁建、专业项目复建、农村移民迁建及中央直属军工企业迁建在1999年1月1日前未征用土地新增加的土地补偿和安置补助费。在上述测算内容中,与农村移民安置有关的是农村移民迁建征用土地新增投资。农村移民迁建征用土地包括农村建房占地及居民点对外交通、库周交通、村组副业等占地。

People's Republic of China (hereinafter referred to as "the new Land Administration Law"). While further emphasizing the importance of land management and practical protection of cultivated land, the new Land Administration Law expanded the scope of compensated land use and raised the standard of compensation for land acquisition. In order to ensure smooth resettlement in the reservoir area, the central government decided to increase the "investment in land acquisition for resettlement and relocation of the Three Gorges Reservoir Area" and the "compensation investment for rural inundated land" in accordance with the new Land Administration Law. The main updates included:

First, new compensation fund was increased for inundated rural land. The land acquisition fee was the main fund source for production and resettlement of rural resettlers, which was closely related to keeping the living standard of resettlers equal to or higher than the original living standard and the sustainable development of the reservoir area. Based on the actual work and relevant requirements, the Yangtze River Survey, Planning and Design Institute prepared the draft for approval of the *Report on Calculation of New Compensation Fund for Inundated Land in the Three Gorges Project Reservoir Area on the Yangtze River* in May 2004. According to the approved plan, the compensation for inundated cultivated farmland in rural areas increases by 9 times, and the compensation for resettlement subsidy increases by 6 times. The compensation criteria for non-cultivated farmland shall be determined in accordance with the relevant provisions of the Forest Law of the People's Republic of China. The new compensation for rural inundated land and resettlement subsidy shall be equal to the sum of the three increases in inundated land compensation, forest compensation, and resettlement subsidy, totaling RMB 2.295 billion, of which RMB 422 million was allocated to Hubei Province and RMB 1.873 billion to Chongqing Municipality.

Second, new compensation fund was increased for land acquisition due to resettlement. In accordance with the relevant provisions of the new Land Administration Law and the requirements of the Office of the TGPCC, the Yangtze River Survey, Planning and Design Institute completed the calculation of the new investment required for the resettlement, relocation and land acquisition of the reservoir area in June 2001, which was approved by the TGPCC. The calculation included newly-added land compensation and a resettlement subsidy for land not acquired before January 1, 1999 for the relocation of cities and towns in the reservoir area, relocation of industrial and mining enterprises, reconstruction of specialized projects, relocation of rural resettlers, and relocation of military industrial enterprises directly under the central government. In the above calculation, what was related to the resettlement of rural resettlers was the added fund for land acquired for the relocation of rural resettlers. The land acquired for the relocation of rural resettlers included the land occupied by rural houses, by the external transportation of the settlements, the transportation around the reser-

三、后期扶持政策调整（自 2006 年以来）

2006 年 5 月 17 日，《国务院关于完善大中型水库移民后期扶持政策的意见》文件（简称"17 号文件"）明确规定，对大中型水库农村移民进一步实行后期扶持。三峡水库属大型水库，其农村移民为"17 号文件"政策适用范围，同时，在国务院三峡建委办公室争取下，国务院决定对三峡城集镇移民也进行相应的后期扶持。

这一时期扶持政策的主要内容包括：一是"17 号文件"将三峡库区农村移民纳入全国大中型水库移民后期扶持范围，每人每年补助 600 元，连续扶持 20 年。二是从三峡库区基金中每年安排部分资金用于解决移民遗留问题的处理。三是建立三峡库区移民后期扶持基金。2006 年，从三峡水电站上网售电收入中按每千瓦时 8 厘钱的标准提取库区基金，分配给湖北省和重庆市，连续扶持 20 年，比原来 4.5 厘钱的提取标准有了较大幅度提高。该政策规定能将后期扶持资金直补到移民个人的尽量直补到移民个人，也可以采取项目扶持或直补与项目扶持相结合的方式。三峡库区各级政府加大了对库区养老保险基金等的调剂力度，将资金及时划拨到位，保证了城镇居民最低生活保障金和企业离退休人员养老金的及时足额发放，城镇低保基本实现"应保尽保"，为三峡移民后期的生产生活提供了有力的保障。四是建立三峡移民专项资金。国家从三峡水电站上网售电收入中按每千瓦时 0.5 厘钱的标准提取专项资金，提取期限为 10 年，用于解决搬迁后的三峡移民生产生活困难问题。五是支持发展三峡库区四大产业，国务院三峡建委办公室重点支持三峡库区发展柑橘、畜牧、水产养殖和旅游等四大优势产业，并以此带动三峡库区的经济发展。

在扶持方式上，新政主要调整为"一个尽量，两个可以"。第一，尽量直补到人，后期扶持资金能够直接发放给移民个人的应尽量发放到移民个人；第二，可以实行项目扶持，在充分尊重移民意愿的基础上，后期扶持资金用于移民所在农业社范围内的公益事业项目或生产生活项目，解决移民生产生活中存在的突出问题；第三，可以采取两者结合的方式，一部分后期扶持资金直接发放到移民个人，其余部分由农业社用于公益项目或生产发展项目。具体方式由地方各级人民政府在充分尊重移民意愿的基础上，听取移民安置社群众意见后确定，要求做到公开透明。

voir, and the sideline businesses of the villagers' groups.

III. Updates to Later-stage Supporting Policies (Since 2006)

Issued on May 17, 2006, the document *Opinions of the State Council on Improving the Later Support Policy for Resettlement of Large and Medium-Sized Reservoirs* (Document No. 17) clearly stipulated that further support shall be provided for rural resettlers of large and medium-sized reservoirs. The rural resettlers of the Three Gorges Reservoir as a large reservoir was within the purview of "Document No. 17". In addition, with the efforts of the TGPCC, the State Council decided to provide corresponding later-stage support to the resettlers from the cities and towns in the Three Gorges.

The supporting policies during this period mainly included: First, in "Document No. 17", the rural resettlers of the reservoir area were included in the later support of resettlement of large and medium-sized reservoirs in China, with a subsidy of RMB 600 per person per year which lasted for 20 years. Second, part of the Three Gorges Reservoir Area Funds was allocated each year to solve the problems left by the resettlement. Third, a later support fund for resettlement of the reservoir area was established. In 2006, the on-grid electricity sales revenue of the TGHS was withdrawn at the rate of RMB 0.008 per kWh to be put into the Reservoir Area Fund, then the fund was allocated to Hubei Province and Chongqing Municipality for 20 years in succession, which was substantially increased from its original withdrawal rate of RMB 0.0045 per kWh. This policy stipulated that the later support funds should be directly subsidized to the resettlers as much as possible, and the project support or the combination of direct subsidy and project support could also be adopted. The governments at all levels in the reservoir area put more efforts into the updates of the endowment insurance fund and other funds in the area, and allocated the funds timely, which ensured the timely and full payment of a minimum living security fund for urban residents and a pension for retirees from enterprises. This basically accomplished "universal coverage for eligible residents", providing a strong guarantee for the production and living of the Three Gorges resettlers in the later stage. Fourth, a special fund for the resettlement was established. The State financed this special fund from the on-grid electricity revenue of the TGHS at the rate of RMB 0.0005 per kWh for 10 years, in order to solve the production and living difficulties of the resettlers. Fifth, the development of four major industries in the reservoir area was supported. The TGPCC of the State Council focused on supporting the development of four advantageous industries in the area, including citrus, animal husbandry, aquaculture, and tourism, so as to drive the economic development there.

In terms of support modes, the new policy was mainly updated as follows. First, the

第三节 三峡移民工程的实施和成效

一、三峡移民工程的实施

三峡移民搬迁安置工作总体上可分为试点阶段和正式实施阶段。三峡工程动工前的1985年至1992年，进行了8年开发性移民试点工作。在积累了丰富经验后，1993年至2009年开始长达17年的移民搬迁安置工作。按照"先淹先搬，移民进度与枢纽工程相衔接""一级开发，一次建成，分期蓄水，连续移民"的工作方案，搬迁安置工作又分四期进行，保证了移民工作连续有效的开展。

1. 移民搬迁安置试点阶段

1984年，党中央、国务院在审议三峡工程150米建设方案的可行性报告时，中央领导要求在三峡库区进行开发性移民试点工作。1985年开始进行三峡库区的移民试点工作，主要由三峡省筹备组负责。1985年6月，三峡省筹备组撤销后，中央决定成立国务院三峡地区经济开发办公室，具体负责三峡移民试点工作。1991年7月，为了加强三峡移民试点工作，国务院成立了三峡工程移民试点工作领导小组，进一步加强对三峡移民试点工作的领导，并且增拨资金，加大移民试点工作的力度。到1992年，三峡库区的开发性移民试点工作已持续不断地进行了8年。在这8年间，原三峡省筹备组和后来成立的国务院三峡地区经济开发办公室，在国务院和三峡工程移民试点工作领导小组的领导下，会同四川、湖北两省和所属有关各级政府，认真贯彻执行中央制定的开发性移民方针，在农村移民安置、城镇搬迁、工厂扩建、人才培训等方面，进行了认真的探索和实践。

funds shall be allocated to resettlers as much as possible, that is, the later support funds shall be directly subsidized to them as much as possible. Second, project support can also be adopted. On the basis of fully respecting resettlers' will, the later support funds could be used for public welfare projects or production and living projects within the agricultural associations of the resettlers to solve outstanding problems in the production and living of resettlers. Third, both modes can be combined, part of the later support funds could be directly paid to the resettlers and the rest could be used by the agricultural associations for public welfare projects or production development projects. The specific implementation shall be determined by the local people's governments at all levels on the basis of fully respecting resettlers' will and after listening to their opinions. It shall be open and transparent.

Section 3 | Implementation and Results of the Three Gorges Resettlement

I. Implementation of Three Gorges Resettlement

The relocation and resettlement work of the resettlers can generally be divided into a pilot stage and a formal implementation stage. Development-driven resettlement pilot work had lasted for eight years, from 1985 to 1992, before the TGP was commenced. After rich experience was gained, resettlement work began and lasted for 17 years, from 1993 to 2009. According to the work plan of "the affected in the flooded areas moving first and carrying out the resettlement along with the project" and "Grade-I development, completion at one go, staged water storage, and successive resettlement", the relocation and resettlement work was completed in four stages, which ensured successive and effective resettlement.

1. Relocation and resettlement pilot stage

In 1984, when the Party Central Committee and the State Council reviewed the Feasibility Report on the 150 m Construction Plan for the Three Gorges Project, the central leadership ordered to launch development resettlement pilot programs in the Three Gorges Reservoir Area. In 1985, the resettlement pilot work in the reservoir area began and was mainly charged by the Three Gorges Provincial Preparatory Group. In June 1985, after the group

移民试点工作的成功开展为后来的大规模安置移民积累了丰富的经验，主要从开垦土地、改造土地、建立果园安置移民、在企业中安置农村移民、在城镇迁建中实行基础设施先行、智力开发及对农村移民进行培训等方面进行试点工作。在8年试点工作期间，国家进行了一系列尝试，出台了包括《三峡工程移民政策和迁建补偿标准》《关于加强水库移民工作若干意见的通知》等在内的一系列政策，取得了良好的效果。

8年试点工作的主要经验包括：一是要坚持开发性移民方针，试点中通过对这一方针的探索，进一步加深了对这一方针内涵的理解；二是总结出移民安置的"六个先行"；三是农村移民安置坚持以大农业为基础，以土地为根本，实行"就地就近后靠安置"为主的路子；四是移民工作宜早不宜晚，移民工作越早进行，安置的人数越少，工作难度越小，移民生产生活恢复和发展就越早，反之结果相反；五是搞好智力开发、人才培训工作；六是探索出移民资金周转重复使用、滚动增值的路子。

2. 大规模移民实施阶段

1992年4月3日，全国人民代表大会七届五次会议通过了《关于兴建三峡工程的决议》，标志着三峡库区的移民工作开始进入有计划的组织实施阶段。1993年至2009年，配合三峡枢纽工程建设，先后开展了四期移民工作。

was revoked, the central government decided to establish the Three Gorges Regional Economic Development Office of the State Council, which was specifically responsible for the resettlement pilot work. In July 1991, in order to intensify the pilot work, the State Council established the Leading Group of Resettlement Pilot Work of the TGP to further strengthen its leadership, and allocated additional funds. By 1992, the development resettlement pilot work in the area had been carried out continuously for eight years. During that eight years, under the leadership of the State Council and the Leading Group of Resettlement Pilot Work of the Three Gorges Project, the former Three Gorges Provincial Preparatory Group and the later-established Three Gorges Regional Economic Development Office of the State Council, together with Sichuan Province and Hubei Province and their relevant governments at all levels, conscientiously implemented the development resettlement policy formulated by the central government and conducted careful exploration and practice in the aspects of resettlement of rural resettlers, urban relocation, factory expansion, and talent training.

Rich experience was gained from this successful resettlement pilot work for the subsequent large-scale pilot work. It was carried out mainly in land reclaim, land transformation, establishment of orchards for resettlers, resettlement of rural resettlers in enterprises, priority of infrastructure construction, intellectual development, and training for rural resettlers. During the eight-year pilot work, the State issued a series of policies including the *Resettlement Policy and Relocation Compensation Standard for Three Gorges Project,* and the *Circular of Several Opinions on Strengthening Reservoir Resettlement Work*, which have achieved good results.

The main experience of the eight-year pilot work is as follows: First, adhering to the development-driven resettlement policy. Through the exploration of this policy in the pilot, the understanding of this policy has been furthered; Second, "six firsts" are summarized for resettlement. Third, the resettlement of rural people shall adhere to the base of great agriculture, the foundation of land, and the principle of "moving backwards in the neighborhood". Fourth, the resettlement work should be carried out as early as possible. The earlier the resettlement work is carried out, the fewer the number of people needed to be resettled, the less difficult the work, and the earlier the production and living recovery, or vice versa. Fifth, intellectual development and talent training shall be completed well. Sixth, the question of how to re-use and add value to resettlement funds was solved.

2. Large-scale resettlement stage

On April 3, 1992, the Fifth Session of the Seventh National People's Congress passed the *Resolution on the Construction of the Three Gorges Project*, which marked the beginning of the planned organization and implementation of the resettlement work in the Three Gorges Reservoir Area. From 1993 to 2009, four phases of resettlement work were carried out along

（1）一期移民（1993—1997年）以满足实现大江截流为目标

这一阶段以大江截流为标志，移民主要是为保证三峡枢纽大江截流的需要，任务是完成坝址处90米水位（20年一遇洪水回水线为82.28米）以下的移民搬迁，共涉及移民39 077人，其中湖北27 721人，重庆11 356人；房屋迁建172.21万平方米，集镇迁建24个。

一期移民采取的主要措施包括：一是确立移民领导和管理体制，实行移民任务和移民资金包干的原则，将移民补偿资金切块安排到县，由各区县包干使用；二是建立对口支援的稳定机制，逐渐建立起全国各省市、各部门对口支援三峡库区的稳定机制，从资金、技术、信息、人才等方面全力支援三峡库区；三是加强三峡库区水利建设，加强资金扶持力度；四是国家给予三峡库区特殊扶持。1994年8月25日，国务院下发《国务院关于三峡工程库区进一步对外开放问题的批复》，同意将三峡工程库区各市县列为长江三峡经济开发区，实行沿海经济开放区的优惠政策。

（2）二期移民（1997—2003年）以满足135米蓄水为目标

二期移民施工期为6年，以实现水库蓄水135米、首批机组发电和双线五级船闸通航为目标。二期移民完成移民工程建设投资290亿元，累计完成城乡移民搬迁72.16万人，三峡库区13个全淹和半淹的城市、县城已基本完成迁建。国务院三峡工程移民验收组于2003年4月27日审议并通过了三峡库区二期移民终验报告，标志着三峡库区135米水位线以下移民工程迁建及清库能够满足三峡工程蓄水的需求。

在此阶段，国务院对农村移民安置、工矿企业迁建政策做了重大调整。一是在农村移民安置上，国家安排上海、山东、江苏、浙江、江西、安徽、湖南、湖北、四川、重庆、广东、福建等12个省市接受三峡库区外迁农村移民；二是在工矿企业迁建结构调整中，国家给予破产关闭工矿企业以特殊优惠政策，同时为实行结构调整的工矿企业提供技改专贷，对其进口设备实行减免关税，有力地推动了工矿企业结构调整和破产关闭工作的顺利完成，为三峡库区重构新的产业体系、培育新的经济增长点奠定了坚实基础。

with the construction of the TGP.

(1) First phase of resettlement (1993-1997) aimed to complete river closure

This phase was marked by closing the river and the resettlement was mainly for the river closure of the TGP too. The task was to relocate the resettlers living at the dam site below the water level of 90 m (the backwater line of the once-in-20-year flood was 82.28 m), involving the resettlement of 39,077 people, including 27,721 people in Hubei Province and 11,356 people in Chongqing Municipality. 1.7221 million m^2 of houses and 24 market towns were relocated.

Main measures for the first phase of resettlement included: First, resettlement leadership and management systems were established. Resettlement tasks and funds were assigned, and the funds were divided and allocated to districts and counties for dedicated use. Second, a stable mechanism of partner assistance was established by provinces, municipalities/cities, and government departments across China, to fully support the area in terms of funding, technology, information, talents, etc. Third, the construction of water conservancy in the area and the financial support were enhanced. Fourth, the State gave special support to the area. On August 25, 1994, the State Council issued the *Reply of the State Council on the Further Opening of the Three Gorges Project Reservoir Area*, agreeing to include the cities and counties in the TGP Reservoir Area in the Yangtze River Three Gorges Economic Development Zone and offer them the preferential policies for the coastal open economic zones.

(2) Second phase of resettlement (1997-2003) aimed for water storage at a depth of 135m

The second phase of resettlement lasted for six years and was aimed at reaching water storage at a depth of 135 m, enabling the power generation of the first units, and opening to shipping of the double-line five-level ship locks. In the second phase, RMB 29 billion was invested in the resettlement, 721,600 people were relocated, and the relocation of 13 fully inundated and semi-inundated cities and counties in the reservoir area were basically completed. On April 27, 2003, the Three Gorges Project Resettlement Acceptance Group of the State Council reviewed and approved the *Final Acceptance Report for the Second Phase of Resettlement of the Three Gorges Reservoir Area*, indicating that the relocation and cleanup of the area below the 135 m water level could meet the requirements of the TGP's water storage demand.

In this phase, the State Council made major updates to the resettlement policies for both rural resettlers and relocation of industrial and mining enterprises. First, in terms of rural resettlers, the government asked 12 provinces and municipalities to accept them from the Three Gorges Reservoir Area, including Shanghai, Shandong, Jiangsu, Zhejiang, Jiangxi, Anhui,

(3) 三期移民（2003—2006年）以满足150米蓄水为目标

三期移民搬迁安置24万人，建设各类房屋702万平方米，搬迁破产关闭工矿企业262家。10座县城、2座城市和112座集镇大部分已完成整体搬迁，三峡库区基础设施得到很大改善。2006年9月通过国家验收，与前两期移民搬迁工作相比，第三期农村移民搬迁政策从"后靠安置"为主转变为"后靠""外迁"并重，强调要加大外迁的比例，搬迁方式也由一期的"先拆房、再走人"转变为"先看房后搬家"的方式，彻底消除移民群众的后顾之忧，体现出政策向更人性化的方向转变。

(4) 四期移民（2006—2009年）以满足175米蓄水为目标

经过四期移民，累计搬迁安置131万人，淹没涉及的12座城市和144座集镇已完成整体搬迁。国务院决定三峡工程开始175米水位实验性蓄水，并于2008年8月在三峡库区召开了长江三峡四期移民工程阶段性验收会议，认为三峡库区175米坝前水位线下移民搬迁安置任务已全部完成，库底清理目标已经按规定全部实现，移民安置区基本具备生产生活条件。

Hunan, Hubei, Sichuan, Chongqing, Guangdong, and Fujian. Second, in the updates to the relocation structure of industrial and mining enterprises, the State provided special preferential policies to bankrupt and closed industrial and mining enterprises, offered special loans for technological transformation for industrial and mining enterprises with structural updates, and reduced or exempted the tariffs on their imported equipment. This effectively promoted the successful completion of structural updates and bankruptcy closure of industrial and mining enterprises and laid a solid foundation for rebuilding a new industrial system, and cultivating new areas of economic growth in the reservoir area.

(3) Third phase of resettlement (2003-2006) with the aim to reach water storage at a depth of 150m

In the third phase of resettlement, 240,000 people were relocated, 7.02 million m² houses of various types were constructed, and 262 industrial and mining enterprises were relocated or closed due to bankruptcy. Most of the 10 counties, 2 cities, and 112 market towns were relocated in a whole and the infrastructure of the reservoir area was greatly improved. In September 2006, the third phase of resettlement was accepted by the State. Compared with the previous two phases, this resettlement policy for rural resettlers in the third phase was changed from mainly "moving backwards" to both "moving backwards" and "moving out", with a larger proportion of moving out. The relocation mode was also changed from "demolition first and then relocation" in the first phase to "house checking first and then relocation", which completely released the worries of the resettlers, reflecting the policy change in a more humane manner.

(4) Fourth phase of resettlement (2006-2009) with the aim to reach water storage at a depth of 175 m

After the four phases of resettlement, 1.31 million people were relocated and resettled, and 12 inundated cities and 144 market towns were relocated in a whole. The State Council decided to start the experimental water storage of the TGP at a depth of 175 m, held a phased acceptance meeting for the fourth phase of resettlement in Yangtze River Three Gorges in August 2008, and considered that the relocation and resettlement tasks for the resettlers living below the water level of 175 m in front of the dam in the reservoir area had been completed. The goal of reservoir bottom cleanup had been fully completed as required, and production and living conditions had been basically available in the resettlement areas.

二、三峡移民工程的成效

三峡工程建设及其移民安置工作促进了三峡库区经济社会的快速发展，产业结构得到调整优化，特色产业体系逐步形成，城乡面貌焕然一新。

1. 三峡库区经济发展加速

三峡库区经济总量快速增长，地方财政实力显著增强。1992年至2013年，三峡库区生产总值由152亿元增加到5 708亿元，人均生产总值由974元增加到39 373元，三峡库区生产总值和人均生产总值的年均增长率均超过同期湖北省、重庆市和全国平均水平。公共财政预算收入由9.16亿元增加到360.23亿元，超过同期湖北省、重庆市和全国平均水平。

经济布局趋于合理，产业结构逐步优化。1992年至2013年，三峡库区第一产业、第二产业、第三产业结构由4∶3∶3调整到1∶5.5∶3.5，第一产业结构优化，第二产业大幅提升，第三产业不断发展，特别是特色农业和生态农业逐步壮大，一批环保、清洁能源等新型工业初具基础，旅游业和物流业蓬勃发展。

图5.2 就地后靠移民因地制宜种植柑橘致富

II. Results of Three Gorges Resettlement

The construction of the TGP and its resettlement work has promoted the rapid economic and social development of the reservoir area, the industrial structure has been updated and optimized, a characteristic industrial system has been gradually formed, and a new look has taken form in the urban and rural areas.

1. Accelerated economic development of the Three Gorges Reservoir Area

The GDP of the reservoir area has been growing rapidly and local financial strength has been greatly enhanced. From 1992 to 2013, the GDP of the area increased from RMB 15.2 billion to RMB 570.8 billion, the per capita GDP increased from RMB 974 to RMB 39,373, and the average annual growth rate of GDP and per capita GDP in the area exceeded that of Hubei Province, Chongqing Municipality, and the average of China at the same time. The public budget revenue increased from RMB 916 million to RMB 36.023 billion, exceeding that of Hubei Province, Chongqing Municipality, and the average of China at the same time.

The economic layout tends to be rational and the industrial structure has been gradually optimized. From 1992 to 2013, the structure of primary, secondary and tertiary industries in the reservoir area was adjusted from 4∶3∶3 to 1∶5.5∶3.5, the primary industry structure was optimized, the secondary industry was greatly improved, and the tertiary industry continuously developed. In particular, the characteristic agriculture and ecological agriculture gradually grew, a number of new industries such as environmental protection and clean energy began to take shape, and tourism and logistics industries were flourishing.

⇐ Figure 5.2   Resettlers Moved Backwards Within the Local Area Becoming Rich by Planting Citrus

2. Improved urbanization in the Three Gorges Reservoir Area

The urbanization process of the reservoir area has accelerated significantly. From 1992 to 2013, the urbanization rate of the area increased from 10.68% to 52.18%, which was close to the national urbanization rate of 53.7%, with an average annual growth rate of 1.98%, higher than the average of China in the same period.

The scale of urban areas has multiplied. By 2013, 12 counties (cities) in the area had a built-up area of 259.2 km^2 and grew its permanent population to 2,284,500, 6.52 times and 3.58 times that of 1992, which was 39.73 km^2 and a population of 788,100.

Cities and towns have seen a leap forward in promotion in functions. The large-scale resettlement brought profound changes to the appearance of the cities and towns in the area.

2. 三峡库区城镇化水平提高

三峡库区城镇化进程明显加快。1992年至2013年，三峡库区的城镇化率由10.68%提高到52.18%，已接近全国城镇化率（53.7%），以平均每年1.98个百分点的速度发展，高于全国同期平均发展速度。

城镇规模成倍增长。2013年三峡库区12座县城（城市）建成区面积259.2平方千米，常住人口282.45万人，分别是1992年39.73平方千米和78.81万人的6.52倍和3.58倍。

城集镇功能跨越式提升。大规模的移民工程建设使三峡库区城集镇面貌发生了深刻变化。城集镇迁建新区布局合理、功能齐全、配套完善、具有现代风貌。三峡库区一批小城镇迅速崛起，产业和人口向城集镇集聚进程明显加快。

3. 三峡库区基础设施建设水平提高

三峡库区形成了较为完善的交通网络体系。自三峡移民安置实施以来，三峡库区高速公路、铁路、机场从无到有，长江"黄金水道"优势进一步凸显，基本形成"公、铁、水、空"一体化的综合交通体系。截至2013年12月底，三峡库区公路总长度达到85 477千米，其中高速公路1 499千米。三峡库区农村公路乡镇通达率达100%、行政村通达率达95%以上。

三峡库区的供电能力、电网标准和等级得到大幅度提高。变电站容量和10千瓦及以上输配电线路较搬迁前大幅增加，三峡库区电网布局优化，全面提高了用电可靠性，改善了用电质量。

三峡库区城乡供水综合生产能力增强。2013年年底，城市供水综合生产能力为211.92万立方米每天，是1992年的3.72倍。农村自来水普及率达72.2%，较1992年增加了64.5个百分点。

三峡库区邮电通信、广播电视事业迅速发展。邮电通信设施更新换代，程控交换容量扩展，传输网络更加完善，覆盖面大大提高。2013年，三峡库区已形成无线、有线、卫星三位一体，互为补充、交叉服务的广播电视传输覆盖格局，从落后的"摇把子"时期一跃跨入现代化通信时代。

4. 三峡库区城乡居民生活水平提高

三峡库区城乡居民收入水平逐年提高。2013年，三峡库区农村居民人均

The new relocation areas in cities and towns had a reasonable layout, was fully-functional, and had complete supporting facilities and modern features. A number of small towns in the area rose rapidly, and the agglomeration of industries and population to those cities and towns accelerated significantly.

3. Improved infrastructure in the Three Gorges Reservoir Area

A relatively complete transportation network has been formed in the reservoir area. Since the implementation of the Three Gorges resettlement, expressways, railways, and airports in the area have come into being, the advantages of the Yangtze River's golden waterway were further highlighted, and an integrated transportation system of "highway, railway, waterway, and air" was basically formed. By the end of December 2013, 85,477 km of highways were laid, including 1,499 km of expressways. The accessibility rate of rural roads in townships and towns in the area was 100%, and that of administrative villages was at or above 95%.

The capacity of power supply and the grid standards and grades of the area have been greatly improved. The capacity of transformer substations and distribution lines of 10 kW and above have been increased significantly compared with those before the relocation. The layout of the power grid in the area has been optimized, which comprehensively improves power supply stability and quality.

The comprehensive production capacity of urban and rural water supply in the area has also been enhanced. By the end of 2013, the comprehensive production capacity of urban water supply reached 2.1192 million m^3 per day, representing 3.72 times that of 1992. The coverage of tap water supply in rural areas reached 72.2%, an increase of 64.5% compared with that in 1992.

Posts, telecommunications, radio and television in the area have developed rapidly. Post and telecommunication facilities have been upgraded, program-controlled exchange capacity has been expanded, and the transmission network has a much greater coverage. In 2013, there was radio and television coverage of wireless, cable, and satellite transmission in the area, which complement each other and provide cross services. The reservoir area took a huge leap forward, from the backward period of being limited only to old-fashioned landlines into the era of modern communication.

4. Improved living standard of urban and rural residents in the Three Gorges Reservoir Area

The income of urban and rural residents in the reservoir area has increased year-on-year. In 2013, the per capita net income of rural residents in the area was RMB 8,342, 14.48 times

纯收入8 342元，是1992年576元的14.48倍。三峡库区城镇居民人均可支配收入23 204元，是1992年1 724元的13.46倍。

图5.3　迁建后的万州新城，居民住房条件明显改善

三峡库区城乡居民住房条件明显改善。2013年，三峡库区农村居民人均住房面积43.8平方米，城镇居民人均住房面积35.9平方米，分别比1992年（农村23.8平方米，城镇19.6平方米）增加20平方米、16.3平方米。

三峡库区城乡居民社会保障体系逐步完善。2013年，三峡库区城乡居民养老保险参保人数达626万人，城乡居民合作医疗保险城镇居民参保人数达255万人，城乡居民合作医疗保险农村居民参保人数达1 189万人。移民已基本实现养老保险和合作医疗全覆盖。

三峡库区劳动力就业状况逐步好转。2013年，湖北库区城镇登记失业率为3.15%，重庆库区城镇登记失业率为2.44%，农村劳动力转移就业351.62万人。

5. 三峡库区社会事业取得长足进步

三峡库区教育水平不断提高。三峡库区"两基"人口覆盖率、适龄儿童入学率和完学率接近100%，初中毕业生升入高中阶段学校的比例已达96.5%。

三峡库区卫生事业快速发展。2013年，三峡库区卫生技术人员数6.05万人，卫生机构床位数6.56万张，分别是1992年的1.76倍和2.93倍。

三峡库区文化体育事业蓬勃发展。三峡库区各个区县都拥有一所博物馆、

that of RMB 576 in 1992. The per capita disposable income of urban residents in the area was RMB 23,204, 13.46 times that of RMB 1,724 in 1992.

The housing of urban and rural residents in the area has significantly improved. In 2013, the per capita housing area of rural residents in the area was 43.8 m^2, and that of urban residents was 35.9 m^2, having an increase of 20 m^2 and 16.3 m^2 respectively, compared with that of 1992 (23.8 m^2 in rural areas and 19.6 m^2 in urban areas).

Figure 5.3　Wanzhou New City After Relocation with Obviously Improved Housing

The social security system for urban and rural residents in the area has gradually improved. In 2013, the area had 6.26 million urban and rural residents covered by endowment insurance, 2.55 million urban residents and 11.89 million rural residents covered by urban and rural cooperative medical insurance. Most resettlers are covered by endowment insurance and cooperative medical insurance.

The employment in the area has also gradually improved. In 2013, the urban registered unemployment rate in the reservoir area in Hubei was 3.15%, the urban registered unemployment rate in the reservoir area in Chongqing was 2.44%, and 3,516,200 rural laborers had been transferred and employed.

5. Great progress in social undertakings in the Three Gorges Reservoir Area

The education in the area is also improving. The population coverage rate of the "Two Basic Education Plans", the enrollment rate of school-age children, and the completion rate are close to 100%. The proportion of junior high school graduates entering high schools has reached 96.5%.

Healthcare in the area has also developed rapidly. In 2013, the reservoir area had 60,500 health technicians and 65,600 beds in health institutions, 1.76 times and 2.93 times the numbers of 1992, respectively.

Cultural and sports activities are flourishing in the area. Each district or county in the area has a museum, a public library, and a cultural center, and the sports facilities and fitness equipment tend to be perfectly installed.

In general, the construction and resettlement of the TGP has brought about a leap forward in economic and social development of the area, allowing the local people to share the fruits of reform and development, and steadily move towards "guaranteed basic living, available employment, and poverty alleviation". They sincerely thank the CPC and the government for their care, and the area is harmonious and stable.

公共图书馆、文化馆，体育场地设施建设和健身器材配置趋于完善。

总体来看，三峡工程建设和移民安置促进了三峡库区经济社会的跨越式发展，让三峡库区群众分享到了改革发展的成果，朝着"基本生活有保障、劳动就业有着落、脱贫致富有盼头"的方向稳步发展，三峡库区群众发自内心地感谢党和政府的关心，三峡库区社会和谐稳定。

6. 外迁移民安居乐业

三峡工程外迁安置移民 19.62 万人，占整个农村移民的 35.63%，极大地缓解了三峡库区的人地矛盾，促进了移民多元化发展，探索出了一条大型水库移民安置的新路子。目前，三峡外迁移民总体稳定，生产生活条件得到极大改善，后续发展良好，主要体现在以下几个方面。

第一，外迁移民生产生活条件得到极大改善，12 个接收外迁安置的省市尽量将条件较好的地方作为安置地，比如浙江的杭嘉湖平原，广东的珠三角地区，山东的胶州半岛和烟台、威海、济南等发达地区，是外迁移民的主要安置区。据统计，政府组织的外迁移民承包地基本达到或超过人均 1 亩，且多为熟地，住房条件、面积、基础设施配套等都得到极大改善。

第二，外迁使移民思想观念发生了变化，拓展了视野，增强了市场意识和竞争意识。三峡库区民众普遍勤劳，他们外迁走出大山后不再"坐井观天"，在国家的移民政策的引导下，充分发挥他们的聪明才智和勤劳本色，不怕苦、不怕累，很快走上了勤劳致富、科技致富、经营增收的道路。

第三，外迁移民搬迁入住后得到当地政府的关怀和帮扶，基本融入了当地社会。各地出台政策、创新帮扶方式，帮助移民尽快适应。一是生产帮扶，各地相继出台税费减免政策、生产开发奖励政策、金融扶持政策等，做好劳务输出和就业工作，组织有技术的当地人手把手教移民当地的种养殖技术；二是生活帮扶，安置地组织党员干部与移民结对帮扶，帮助移民解决吃、住、行、就医等各方面的生活问题，鼓励当地居民与移民交往；三是教育帮扶，不少安置地减免了移民子女的学杂费，组织品学兼优的学生与移民子女结对子等。通过一系列的帮扶措施，移民基本融入了当地社会，移民二代已经基本对安置地居民概念产生了身份认同，淡化了移民身份。

6. Resettlers living and working in peace and contentment

In the TGP, 196,200 people have been relocated and resettled, accounting for 35.63% of total rural resettlers in China, which has greatly alleviated the contradiction between people and land in the reservoir area and has helped the individual development of the resettlers. Much has also been learned for possible future resettlements of large-scale reservoir. At present, Three Gorges resettlers are living a stable life, their production and living conditions have been greatly improved, and they develop well, which can be seen in the following aspects.

First, the production and living conditions of resettlers have been greatly improved. The 12 provinces and municipalities receiving resettlers have tried their best to place them in resettlement sites with better conditions, for example, the Hangjiahu Plain in Zhejiang, the Pearl River Delta in Guangdong, the Jiaozhou Peninsula, Yantai, Weihai, and Jinan in Shandong, and other developed areas. According to statistics, the contracted land for resettlers organized by the government basically reaches or exceeds 1 *mu* (about 666.7 m^2) per capita, and most of it is cultivated land. The housing conditions, housing area, and supporting infrastructure have all been greatly improved.

Second, the mindset of resettlers has been changed, their horizons have been broadened, and their awareness of the market and of competition has been sharpened due to relocation. The people in the Three Gorges Reservoir Area are industrious. After being relocated from the mountains, they no longer have a limited outlook. Under the guidance of the State's resettlement policies, they give full play to their ingenuity and hard-working instinct, and they are not afraid of hardship and fatigue, so they quickly embark on the road to wealth through hard work, technology application and business operations.

Third, the resettlers have received care and assistance from the local government and are generally integrated into the local community. Local governments have introduced policies and innovative ways to help resettlers adapt to the new environment as soon as possible. The first is production assistance. Local governments have successively introduced tax reduction and exemption policies, production and development incentive policies, financial support policies, etc., in order to do a good job in labor export and employment, and organize skilled local people to teach resettlers local farming techniques. The second is life assistance. Party members and officials in the resettlement sites are organized to provide pairing-assistance to the resettlers, helping them solve their living problems in terms of food, housing, transportation, and medical care, and native residents are encouraged to communicate with the resettlers. The third is education assistance. In many resettlement sites, the tuition and miscellaneous fees of resettlers' children are reduced or exempted and local students with excellent academic performance are paired up with resettlers' children. Through a series of

―― **本章小结：**――

三峡工程顺利推进的关键在于移民工作的顺利完成。在经过8年的前期试点，积累了丰富的经验后，政府于1993年至2009年正式开展了长达17年的移民搬迁安置工作。三峡移民工程坚持发展性移民重建，创新移民安置社会管理模式，推进和谐稳定新型库区建设。三峡移民工程成功完成了131万人的伟大迁徙，移民生产生活条件显著改善，搬迁城镇和基础设施建设实现跨越式发展，库区经济结构和社会结构实现重大转型，实现了在移民中发展、在发展中移民，为破解水库移民这道世界性难题树立了成功范例。

参考文献：

［1］本书编委会. 百问三峡［M］. 北京：科学普及出版社，2012.

［2］陈夕. 中国共产党与长江三峡工程［M］. 北京：中共党史出版社，2014.

［3］尹忠武，袁永源. 三峡工程移民规划设计［J］. 人民长江，2003(8).

［4］《中国三峡建设年鉴》编纂委员会. 中国三峡建设年鉴（2006）［J］. 宜昌：中国三峡建设年鉴社，2006.

［5］《中国三峡建设年鉴》编纂委员会. 中国三峡建设年鉴（2013）［J］. 宜昌：中国三峡建设年鉴社，2013.

assistance measures, the resettlers have largely integrated into the local community, the second generation of resettlers has formed an identity along with other residents there, and their status as being considered "resettlers" has been diminishing.

Chapter Summary:

The key to the success of the Three Gorges Project lies in the smooth resettlement work. After rich experiences were accumulated in the eight years of pilot work, the government officially started the 17-year resettlement work from 1993 to 2009. During the Three Gorges Resettlement, development-driven resettlement and reconstruction had been adhered to and the social management model of resettlement had been innovated, promoting the development of a new harmonious and stable reservoir area. In the Three Gorges Resettlement, 1.31 million people have been relocated, the production and living conditions of resettlers have been remarkably improved, leaping development has been realized in the relocated cities and towns, as well as in the infrastructure. The economic and social structures of the reservoir area have been greatly transformed, achieving the goal of having development and resettlement without coming into conflict, setting a successful example for solving the worldwide problem of reservoir resettlement.

References:

[1] Editorial Board of This Book. *One Hundred Questions about Three Gorges* [M]. Beijing: Popular Science Press, 2012.

[2] Chen Xi. *The Chinese Communist Party and the Three Gorges Project* [M]. Beijing: Chinese Communist Party History Publishing House, 2014.

[3] Yin Zhongwu, Yuan Yongyuan. *Resettlement Planning and Design of Three Gorges Project* [J]. Yangtze River, 2003(8).

[4] Editorial Board of Construction Yearbook of China's Three Gorges. *Construction Yearbook of China's Three Gorges* (2006) [J]. Yichang: Publishing House of Construction Yearbook of China's Three Gorges, 2006.

[5] Editorial Board of Construction Yearbook of China's Three Gorges. *Construction Yearbook of China's Three Gorges* (2013) [J]. Yichang: Publishing House of Construction Yearbook of China's Three Gorges, 2013.

> **阅读提示：**

　　三峡工程是中国改革开放确立社会主义市场经济体制后兴建的第一个大型水利水电工程。从三峡工程建设之初，国家就确定了按照市场经济原则组织工程建设和国家宏观调控相结合的工程建设管理体制，理顺了市场经济条件下政府规范管理、企业独立运营、市场配置资源之间的关系，实现了一系列建设管理机制创新。

　　作为20世纪世界上规模最庞大、技术最复杂的巨型水电工程，三峡工程在规划设计、工程建设中面临诸多挑战和难题。三峡工程建设者会同国内众多科研单位和专家学者，以全球视野谋划和推动创新，通过原始创新、集成创新和引进消化吸收再创新，在水利水电工程技术方面取得了重大突破，形成了一大批科技进步创新成果和工程质量及技术标准。

　　The Three Gorges Project (TGP) was the first large water conservancy and hydropower project implemented after the establishment of the socialist market economy during the reform and opening-up of China. From the very beginning of the TGP, China applied the project construction and management mechanism where project construction and national macroeconomic control were integrated under the principles of the market economy. China rationalized the relations between standardized government administration, independent operation of enterprises, and market dominated allocation of resources. This brought many innovations regarding construction and management of the project.

　　The TGP, the world's most technologically complex and largest hydropower project of the 20th century, was faced with many challenges and difficulties in terms of planning, design and construction. The developer of the TGP worked with many scientific research institutions and experts in the country to plan and innovate the project with a global vision. Through innovations in original technologies, integrated technologies, and imported and assimilated technologies, they achieved major breakthroughs in water conservancy and hydropower technologies. They also obtained a large number of advanced and innovative technologies, and established a set of technical standards regarding engineering quality and technologies.

Chapter 6 >>>>

第六章
三峡工程的创新
Innovations of the Three Gorges Project

第一节 三峡工程的建设管理创新

三峡工程的建设管理创新是在中国改革开放的大背景下，在社会主义市场经济体制下和充分总结国内基本建设项目经验教训的基础上，通过借鉴国内外的成功经验而形成的。实践证明，三峡工程的建设管理是科学的、符合中国国情的，有效地加快了三峡工程的建设开发，有力地控制和缩短建设工期，确保工程质量，提高三峡工程的投资效益，对国内基本建设管理体制产生了重要影响。

一、建设管理机制创新

在计划经济体制下，实行的是特殊的资源配置模式，由于大型水利工程的建设管理等完全通过政府的计划调度完成，一方面造就了中国水利工程建设的奇迹，另一方面也造成了水利工程建设管理责任不清、建管脱节等问题。伴随中国社会的转型而来的市场经济资源配置模式，则从根本上改变了原有的大型水利工程等公用事业的建设管理机制，原有的大型水利工程建设管理机制就很难与市场经济的要求相协调，需要进行变革。

从三峡工程建设伊始，国务院就决定三峡工程要按照社会主义市场经济的方式组织建设和运营。因此，三峡工程采用国家宏观调控和项目法人负责制有机结合的建设管理体制，即国家在工程建设中起宏观调控和监督作用，项目法人对设计、施工和运行管理有较大的自主权，按照市场经济的规律和三峡工程的建设特点进行管理；管理过程中实行以项目法人责任制为中心，招标承包制、工程监理制和合同管理制有机结合的建设项目管理体制，确保三峡工程建设的顺利实施。

Chapter 6 Innovations of the Three Gorges Project

Section 1 | Construction and Management Innovations of the TGP

Amid the reform and opening-up of China and establishment of the socialist market economy, innovations in the construction and management of the TGP were achieved by learning lessons from domestic capital construction projects and borrowing other successful experiences, domestically and abroad. It has been proven that the construction and management approach of the TGP was rational and met the actual needs of China at the time, effectively accelerating progress of the project, substantially shortening construction duration, ensuring high quality, enhancing efficiency of investment, and generating great influence over the domestic capital construction management system.

I. Innovations to Construction and Management of the Project

In the former planned economy system, a special resources allocation model was implemented, and construction and management of the large water conservancy projects were accomplished entirely through the overall planning of the government. This model brought about miracles to the water conservancy projects in China, but caused problems such as unclear responsibilities for construction and management of such projects, and disconnection between different responsible entities. Under the market economy dominated resources allocation model that came with the social transformation of China, the old construction and management methods used for public utility projects, such as large water conservancy projects, were fundamentally obsolete, because such mechanisms could no longer meet the needs of the market economy and thus had to be reformed.

From the beginning of the TGP, the State Council decided to implement construction and management of the project under the socialist market economy principles. Therefore, the TGP was incorporated into a construction and management system that integrates both national macroeconomic control and project legal person responsibility system, i.e. the country undertakes macroeconomic control and supervision over the project, while the project legal person is given extensive autonomy over the design, construction, operation and manage-

1. 项目法人责任制

1993年9月27日，经国务院批准，中国长江三峡工程开发总公司（以下简称"三峡总公司"，2009年更名为中国长江三峡集团公司，简称"三峡集团"）成立。三峡总公司按照国家赋予的项目法人职责，根据国家批准的三峡工程建设方案、工程概算、工期目标，全面承担三峡枢纽工程总体筹划、资金筹措、工程建设、枢纽运营、债务偿还及资产的保值增值等职责，充分运用市场规律配置资源，组织各方力量完成枢纽工程建设任务。三峡总公司具体负责组织技术设计、招标设计和施工详图设计，审定具体实施方案；主持决定建设项目年度计划与立项、分年投资与筹资的实施方案，对工程施工和设备招标采购、质量与安全、进度、造价、运营等目标进行全面控制；负责组织协调施工、监理、设计、运行等全过程的工作。

2. 招标投标制

三峡工程实行招标投标制，运用市场竞争机制，采取分项招标方式，择优选择建筑安装、物资供应、设备制造及工程监理等方面的承包商参与三峡枢纽工程的建设。项目法人成立招标委员会，严格执行"三公"的招标原则，即公开招标、公平竞争、公正评标，对招标工作实施统一管理。评标采用专家评标、定量打分、定性分析、集体决标的程序，具体步骤为：编制招标设计文件——经审查批准后公开发售标书——承包商购买标书并编制投标文件——在公证机构代表监督下公开开标——聘请有资格的外部专家评标——招标委员会决标。

ment of the project, and manages the project according to the laws of the market economy and the actual needs of the project. The construction project management system, centered on the responsibility system of a project legal person, combined the bidding and contracting, project supervision, and contract management systems, to ensure successful implementation of the TGP.

1. Responsibility system of the project legal person

On September 27, 1993, China Yangtze Three Gorges Project Development Corporation (hereinafter referred to as "the CYTGPDC", renamed China Three Gorges Corporation (TGP) in 2009) was founded under the authorization of the State Council. In line with the project legal person's responsibilities, charged by the state and in accordance with the construction plan, budget and duration objectives of the Three Gorges Project (TGP) approved by the state, the CYTGPDC undertook the responsibilities, including general planning, fund raising, construction, hydroproject operation, debt repayment and assets preservation/accretion for the TGP, allocation of resources according to market rules, and coordination between all participants to complete construction of the project. Specific responsibilities of the CYTGPDC: technical design, bidding design and shop drawing design; audit of the implementation plan; preparing and deciding on the annual plans of the construction project and the implementation plans for project approval; year-by-year investment and fund raising; total control over the objectives including construction, equipment bidding and procurement; quality; safety; schedule; costs and operation; organizing and coordinating the whole process of the project, including construction, supervision, design and operation.

2. Tendering-bidding system

The TGP was implemented by using the tendering-bidding system, i.e. an invitation to bid was issued for each subproject through the market competition mechanism and the highest-rated contractors were hired to undertake certain responsibilities such as construction, installation, materials supply, equipment manufacture and project supervision. The project legal person instated the Tender Committee and strictly adhered to the three principles of tendering, i.e. open tendering, fair competition and impartial bid evaluation, and maintained centralized management over tendering. The bid evaluation process consists of expert bid evaluation, quantitative scoring, qualitative analysis and collective awarding of bids. The procedure includes: 1) preparation of the bidding design document, 2) public offering of the approved tendering document, 3) contractor buying the tendering document and preparing the bid, 4) public bid opening under the supervision of the representative of notary authority, 5) bid evaluation by hired independent qualified experts--awarding of bid by the Tender Committee.

3. 工程建设监理制

工程建设监理制是保证工程建设达到预期的质量、进度和投资目标而制定的重要制度。三峡总公司通过招标等方式委托具有相应资质的专业机构，对建筑安装、设备制造及大宗物资生产供应等项目进行全过程监控和管理。监理分项目监理、工程监理和施工监理三个层级。项目监理由三峡工程的项目法人担任，代表三峡总公司全面负责三峡工程的建设。工程监理代表三峡总公司工程建设部直接负责三峡工程的施工管理，对各分项施工监理进行综合协调。施工监理由三峡总公司选聘有监理资格的公司、设计院等单位分项对承包合同的履约进行监督管理、分项监督施工，有权发布开工、停工命令，有权协调甲乙方和设计方合同外的补偿和索赔，有责任进行现场过程检查、安全监督，有责任审查设计施工详图。三峡总公司主要聘请6家施工监理单位，施工高峰期工程监理人数达900余人。

4. 合同管理制

三峡枢纽工程建设涉及各种类型的合同，项目法人通过控制各类合同的执行过程从而达到对整个工程建设的有效管理。参建各方通过签订商业合同，形成以项目法人为核心，参建各方分工协作的利益共同体。三峡工程合同涉及金额较大，其中最大单个合同金额达66.85亿元，合同执行期长达数年。设置合同管理制有很大益处，一旦合同出现偏差，则承包商可通过监理反馈给业主的工程项目部，对设计图纸和技术上的问题可会同现场设计代表及时处理，重大问题及时反馈到三峡总公司工程建设部，经研究决策后付诸实施，较好地处理了合同执行过程中出现的各类问题。

国有大型企业按照市场规律具体负责实施项目管理，对质量、进度和投资负责，是中国经济建设领域管理体制的重大突破。在三峡工程建设管理机制中，项目法人责任制是核心和基础，招标投标制、工程建设监理制和合同管理制是配套支撑制度。项目法人制保证了项目法人在项目建设过程中的计划、组织、领导和协调作用，以及项目完工后的经营管理地位，实现了项目建管结合的高度统一；招标投标制解决了如何按照市场规则选择对工程最有利的承包商的问题；工程建设监理制解决了如何保证项目建设中的质量、成

3. Project supervision system

The project supervision system is an important system intended to ensure the project objectives regarding quality, schedule and investment are achieved. Through invitation to bid and other means, the CYTGPDC engaged the qualified professional entities to undertake whole-process monitoring and management over the subprojects, such as construction, installation, equipment manufacture and bulk materials production and supply. Supervisors were instated at three levels: project supervisor, engineering supervisor and construction supervisor. The project supervisor was the project legal person of the TGP, who assumed full responsibility for the TGP on behalf of the CYTGPDC. The project supervisor directly undertook construction management of the TGP on behalf of the Project Construction Department of the CYTGPDC, and coordinated all the supervisors of the subprojects. For construction supervision, the CYTGPDC hired the qualified companies or design institutes to supervise the performance of contracts and construction of the subprojects. They were authorized to give the order to commence or suspend works and coordinate compensations and claims between Parties A and B and the Designer. They were obligated to conduct onsite processes inspection and safety supervision, and review shop drawings. The CYTGPDC hired six construction supervisors. The peak number of construction supervision personnel was over 900.

4. Contract management system

The TGP involved various types of contracts, so the project legal person achieved effective management of the entire project by controlling performance of the contracts. The project participants signed the commercial contracts to create a community of shared interests which was centered around the project legal person, and coordinated the efforts of the participants. The TGP involved huge contract amounts. The largest single contract amount was RMB 6.685 billion, and execution of this contract lasted several years. The contract management system was massively beneficial. In the event of any deviation from contract, the contractor could report the problems to the Project Department of the Owner via the supervisor. Design drawing problems and technical problems could be solved by working with onsite design representatives. Major problems could be quickly reported to the Project Construction Department of the CYTGPDC, and solved after investigation and consultation. In this way, problems encountered during contract performance could be properly handled.

The large state-owned enterprise undertakes project management and bears responsibility for quality, schedule, and investment, which is a major reform for a management system of Chinese economic development. The responsibility system of the project legal person is the core of the management mechanism of the TGP, while the tendering-bidding, project supervi-

本与进度三项基本要素之间协调与平衡的问题；合同管理制解决了项目建设各参建方协同参与机制的问题，四制协同配合才能管理好三峡工程建设。

二、融资机制创新

三峡工程是一个多目标开发的综合性水利枢纽工程，投资规模大，建设工期长，跨越多个宏观经济周期，在三峡工程决策中，能否筹集到稳定可靠的资金以满足三峡工程巨大的资金需求，是一个至关重要的问题。经国家正式批准的三峡工程初步设计静态总概算（1993年5月末核算的价格，不包括物价上涨及施工期贷款利息等因素的影响）为900.9亿元，三峡工程施工期长达17年，计入物价上涨及施工期贷款利息动态总投资估算约为2 039亿元。三峡总公司在充分考虑国内外筹资环境的基础上，提出"保证稳定可靠的融资来源、保持合理的资本负债结构、尽可能降低融资成本、控制项目财务风险"的筹资目标，确立"三结合、三为主"的筹资原则，即"国内融资与国外融资相结合，以国内融资为主；长期资金与短期资金相结合，以长期资金为主；债务融资与股权融资相结合，以债务融资为主"。

三峡工程融资实践遵循企业生命周期理论，在不同阶段以当时市场的环境和政策为基础，对三峡工程资金筹措方案和财务能力进行认真分析和评估，充分考虑工程建设需要、资金供求特点和国内外筹资环境，并紧密结合三峡总公司的成熟度，分三个阶段制定三峡工程总体筹资方案。

sion and contract management systems are supporting systems. The responsibility system of the project legal person ensures that he fulfills the functions of planning, organizing, leading, and coordinating, and performs operation and management of the completed project. In this way, construction and management of the project are put under unified authority. The tendering-bidding system ensures that under the market rules, the most favorable contractor among all the candidates is hired to undertake the project. The project supervision system effectuates coordination and balance between the three basic elements of the project: quality, cost and schedule. The contract management system ensures participants will work collaboratively to complete the project. Only through these four systems could the TGP proceed under sound management.

II. Innovations of the Financing Mechanism

The TGP is a multipurpose integrated hydroproject which is characterized by large investment, long duration and continuance through multiple macro-economic cycles. During the decision-making process of the TGP, the ability to raise stable and reliable funds to support the huge investment of the project was a crucial issue. The state-approved static general estimate for preliminary design of the project was RMB 90.09 billion (the price estimated by the end of May 1993, excluding factors such as price inflation and loan interest in the construction period). The construction period of the project was 17 years, and the total dynamic investment that takes account of price inflation and loan interest during the construction period was estimated to be approximately RMB 203.9 billion. After adequately examining the domestic and foreign environments for raising funds, the CYTGPDC set the fundraising objectives as follows: "find stable and reliable financing sources, maintain a rational capital liability structure, minimize financing costs, and control financial risks of the project". They also established three financing principles: "The project is to be supported by both domestic and foreign financing, but dominated by domestic financing; it is to be supported by both long-term and short-term funds, but dominated by long-term funds; it is to be supported by both debt financing and equity financing, but dominated by debt financing".

The practical financing process of the TGP was guided by the enterprise life cycle theory. On the basis of the actual market environment and policies in various stages, the fundraising plan and financial capacity were carefully analyzed and assessed. The project's needs, supply and demand of funds, and domestic and foreign fundraising environments were adequately examined, and the maturity of the CYTGPDC was realistically evaluated. Hence, the general fundraising plan of the TGP was divided into three stages.

1. 三峡工程大江截流前的施工阶段

由于当时三峡总公司刚成立，金融市场对三峡总公司的认知度低，加上三峡工程自身建设的不确定性因素，资金来源的稳定性与可靠性成为三峡总公司这一阶段融资的首要目标。该阶段以国家资本金和国家政策性贷款为主，考虑到三峡工程防洪、航运、水资源配置等社会效益和公益性功能，国家以设置三峡基金等方式累计向三峡总公司拨入资本金1 289亿元。1994年，国家开发银行在成立之初就向三峡工程承诺贷款300亿元，每年可提款30亿元。国家注入的资本金和政策性银行贷款为三峡工程后来其他各项融资发挥了"种子效应"。

2. 三峡工程大江截流后的建设阶段

该阶段以国内商业银行贷款、国际优惠信贷资金和企业债务融资为主。此阶段项目建设风险大幅度降低，金融市场对三峡工程建成后的现金流和三峡总公司发展前景普遍看好，国家资本金与政策性银行资金对三峡的支持为市场化融资奠定了坚实的基础。

在国家开发银行的政策性贷款引导下，从1994年起，参与三峡工程融资的商业银行逐年增加。三峡总公司把竞争机制引入商业银行贷款，享受了多年同期贷款基准利率下浮10%的优惠贷款利率，同时与商业银行签署长期借款协议，滚动使用借款资金，通过借新还旧、蓄短为长，增加资金调度的灵活性。1998年至2003年期间，各商业银行根据三峡总公司资金需求提供了近200亿元中期贷款和流动资金贷款。

1997年，三峡总公司进入国内资本市场发行企业债券，较早地将大型工程建设项目引入资本市场。此后三峡总公司不断进行债券品种的创新，如发行第一只附息债券、第一只浮动利率债券、第一只固定利率长期债券等。截至2003年，三峡总公司发行6期企业债券，募集190亿资金。三峡债以其合理的定价水平、符合国际惯例的发行方式、良好的流动性及较高的信用等级，深受投资者青睐。

三峡总公司在引进国外先进技术、设备的同时，以招标的方式创新性地引进了国外优惠贷款。以德国、法国、瑞士为首的政府积极支持本国企业投

第六章 | 三峡工程的创新
Chapter 6　Innovations of the Three Gorges Project

1. Construction stage before river closure of the TGP

The CYTGPDC was a newly founded corporation that received low recognition in the financial market, and there were uncertainties in the TGP, so the primary goal of the CYTGPDC in this stage was to secure stable and reliable funding sources. Back then, the funds were mostly from the state capital fund and national policy-related loan. Given the social benefits and public welfare functions of the TGP, such as flood control, shipping and water resources allocation, the country earmarked RMB 128.9 billion of capital fund for the CYTGPDC through channels such as the TGP Construction Fund. In 1994, the newly founded China Development Bank issued a loan of RMB 30 billion to the TGP, with RMB three billion allocated each year. The capital fund allocated by the state and the policy-related bank loan were "seed funds" that preceded many other financing sources for the TGP.

2. Construction stage after river closure of the TGP

In this stage, the funds were mostly from domestic commercial bank loans, international concessional credits, and corporate debt financing. Project risks were substantially reduced in this stage, and the financial market generally showed a positive view of the potential post-construction money flows and of the development prospects of the CYTGPDC. The state capital fund and national policy-related loan laid a firm foundation for market-based financing.

Encouraged by the policy-related loan from China Development Bank, more and more commercial banks started financing TGP every year since 1994. The CYTGPDC solicited commercial bank loans through the competition mechanism, and gained a preferential loan interest rate that was 10% less than the prevailing multi-year coterminous benchmark interest rate. Many of these were long-term borrowing agreements with commercial banks, so that the CYTGPDC could use borrowed funds on a rolling basis and make fund procurement more flexible by using new debts to pay old debts and converting short-term borrowings into long-term borrowings. From 1998 to 2003, commercial banks offered nearly RMB 20 billion medium-term and working capital loans to support the fund demands of the CYTGPDC.

In 1997, the CYTGPDC started issuing enterprise bonds in the domestic capital market and became one of the first enterprises to bring large construction projects into the capital market. Since then, the CYTGPDC kept innovating in terms of the variety of their bonds. For example, it issued its first coupon bond, first floating rate bond and first fixed interest rate bond. By 2003, the CYTGPDC had issued six enterprise bonds and thereby raised RMB 19 billion in funds. The CYTGPDC bonds were strongly preferred by investors, thanks to their methods of issuing being in line with international conventions and having excellent liquidity

标三峡工程设备，使得各国信贷机构积极提供宽松的出口信贷，实现了设备采购与融资的有机结合，提升了三峡总公司在全世界的融资声誉。

3. 三峡工程首批机组投产发电后的阶段

该阶段三峡工程建设风险进一步降低，三峡总公司的收益能力逐渐凸显，融资机制呈现股权融资和债权融资并重的特点。三峡总公司改变传统水电工程依靠银行贷款导致资产负债率过高的融资方式，充分利用资本市场，依靠滚动开发机制，通过将优质资产改制上市的方式进行股权融资，打造中国长江电力股份有限公司（以下简称"长江电力"）这一资本运作平台，并逐步将三峡工程投产的发电机组资产分批注入长江电力，开创了边建设、边评估、边上市的先例。

2002年，三峡总公司在原葛洲坝电厂的基础上组建长江电力，实施优质资产改制上市，并于2003年11月在上海交易所上市进行股权融资；2005年8月，长江电力实施股权分置改革，以相对合理的代价获得了所持股份的流通权，实现了国有资产的动态保值增值，提高了对长江电力的控制力；2009年9月，三峡工程发电资产全部注入长江电力，成功实现整体上市。同时，三峡集团继续运用贷款、债券"两条腿走路"的融资方式，持续获得低成本融资。在2002年至2009年期间，通过高层对话和签订战略合作协议的方式，累计获得总授信额度1 000亿元，并推出公司债、短期融资券、中期票据等融资品种，获得银行间市场重点AAA发行人资信，多次成为相关债券发行银行当年同期限品种票面利率最低的品种。

三峡工程多元化的融资策略降低了融资风险，保证了三峡工程建设和经营的顺利进行，充分实现了融资理论与实践创新的紧密结合，取得了良好的经济效益。在三峡工程进行工程决算后，这些措施又在三峡工程的经营和长江流域水力资源的滚动开发中发挥作用，并给三峡工程的未来发展带来更大的经济效益。

and high credit ratings.

While introducing and importing advanced foreign technologies and equipment, the CYTGPDC introduced foreign preferential loans. Countries such as Germany, France and Switzerland encouraged their enterprises to bid on supplying equipment for the TGP, and allowed their credit institutions to provide less heavy-handed export credits so that equipment procurement was comfortably integrated with financing. This enhanced the reputation of the CYTGPDC across the world in terms of financing.

3. The stage after commissioning of the first units of the TGP

In this stage, the risks of the TGP were further reduced, the earning power of the CYTGPDC gradually became clear, and equal importance was attached to both equity financing and debt financing. Breaking away from the traditional financing model where hydropower projects were plagued by excessively high asset-to-liability ratios caused by heavy reliance on bank loans, the CYTGPDC made full use of the capital market, implemented the rolling development mechanism, conducted equity financing through restructuring, listing high quality assets, creating a capital operation platform known as China Yangtze Power Co., Ltd. (hereinafter referred to as "Yangtze Power"), and gradually transferring the commissioned generator units of the TGP as assets to Yangtze Power. This set a precedent for the model of "construction, assessment and listing in parallel".

In 2002, the CYTGPDC founded Yangtze Power by restructuring the former Gezhouba Hydropower Plant, and implemented the restructuring and listing of high-quality assets. In November 2003, it started equity financing through listing on the Shanghai Stock Exchange. In August 2005, Yangtze Power implemented equity division reform and obtained the circulation right of its shares at a relatively reasonable cost. This enabled dynamic value preservation/accretion of the state-owned assets and strengthened control over Yangtze Power. In September 2009, all power generation assets of the TGP were transferred to Yangtze Power, which led to its successful overall listing. Besides, CTG maintained the "two-leg" financing model consisting of debts and bonds, and continued obtaining low-cost financing. From 2002 to 2009, the company obtained a total line of credit of RMB 100 billion through high-level dialogues and strategic cooperation agreements, and offered a variety of financing instruments such as corporate bonds, short-term financing bonds and medium-term notes. Obtaining a Class AAA issuer rating in the interbank market also boosted its ability to obtain funds. Several times among issuing banks, its financing had the lowest nominal interest rate compared to other organizations seeking financing of the same or similar variety and with the same duration.

The diverse financing strategy of the TGP reduced its financing risks, ensured success-

三、投资控制创新

三峡工程开创性采用"静态控制、动态管理"的投资管理模式,对影响工程投资的各种静态因素和动态因素进行分别控制和管理,并从管理职责上准确区分企业(项目法人)与国家承担的风险和责任,实现了与国际重大工程项目投资控制和管理的接轨。

1. 采用"静态控制、动态管理"的投资管理模式

项目法人负责制的一项重要内容是,投资控制要实行"静态控制、动态管理"。1997年2月,国务院三峡建委《关于三峡工程输变电系统设计概算的批复》明确指出:"三峡输变电工程按'静态控制、动态管理'办法实行投资管理。建设委员会按275.32亿元静态投资总量,以及因价格、利率、汇率因素引起的动态投资总量进行控制。"

"静态控制",由设计单位长江委以动工当年1993年5月末材料价格为基础,测算出一个工程造价概算额,经国务院三峡建委批准,作为静态投资的最高限额。在设计不发生重大变更、不扩大工程规模的情况下,法人业主不得调整,只能在概算限额内对其子项进行变更,确保静态投资控制在限额以内。

"动态管理",是在静态控制投资额的基础上,对影响工程总投资的物价等不确定因素,以静态投资额为基数,进行年度跟踪测算,参考国家物价局公布的物价指数,科学、合理确定各种材料的物价指数,以及人员工资指数,然后确定总指数,进而推算出价差。国务院三峡建委批准的年度差价,是向各参建单位按合同条款结算费用的唯一依据。"动态管理"的内容包括,测算因每年物价上涨(价格指数)而使当年投资增加的数额(价差资金),测算因国家政策变动,利率、汇率变化而增加的投资。其中最繁杂的是每年的价格指数和价差资金测算。鉴于移民工程每年的价格指数和价差资金,关系到千家万户移民群众的切身利益,国家计委价格信息中心专门在三峡库区建立了价格信息采集网,使其价格指数更加符合三峡库区实际、更加公正和公平。

"静态控制、动态管理"的投资控制模式与计划经济体制下若干年调整一次工程概算的管理模式有着本质区别,这是一个责任清晰、科学合理、公正

ful construction and operation of the project, with practical innovation adequately integrated financing theory, and created sizable economic benefits. After a final accounting of the TGP, the aforesaid measures continued to play a part in its operation and the rolling development of water resources in the Yangtze River basin, and brought greater economic benefit for future development of the TGP.

III. Innovations of Investment Control

The innovative investment management model of "static control and dynamic management" was applied to the TGP. The static and dynamic factors affecting the project's investment were individually controlled and managed, and the risks and responsibilities of the enterprise (project legal person) and the state were clearly defined in their management responsibilities, which was exactly in line with the international model of major project investment control and management.

1. Investment management model of "static control and dynamic management"

A key element of the responsibility system of the project legal person is that investment is subject to "static control and dynamic management". In February 1997, the *Official Reply to Design Estimate for the Power Transmission and Transformation System of the TGP* given by the TGPCC explicitly stated: "The Three Gorges Power Transmission and Transformation Works are subject to investment management by the means of 'static control and dynamic management'. The Construction Committee will control both the total static investment of RMB 27.532 billion and the total dynamic investment resulting from factors including price, interest rate and exchange rate."

In "static investment", the Changjiang Water Resources Committee, which is the Designer, calculated the estimated project cost on the basis of the materials prices at the end of May, 1993, the year construction commenced. After approval of the TGPCC under the State Council, the estimate was used as the maximum limit for the static investment. Without major change of design or expansion of project scale, the owner could not adjust the limit but could only change the scale and/or limits of the subprojects, in order to ensure that the static investment would not exceed the limit.

In "dynamic management", which is a step further from static control of investment, the static investment is used as the base number for calculation, and uncertain factors affecting total investment, such as commodity price, are taken into account to annually track and calculate the investment. The price indexes of the materials and the personnel wage index are

公平的投资控制机制，有助于形成业主自我约束的激励机制。

2. 编制项目法人执行概算

编制项目法人执行概算是具体实施"静态控制、动态管理"的一项重要举措。为了认真执行"静态控制"，三峡总公司建立了概算价体系和成本价体系，确定了合同价、合同实施控制价，编制单项工程执行概算和总执行概算。湖北省、重庆市政府按照淹没实物指标将静态补偿资金切块包干到县区，县区政府再切块包干到每个项目。鉴于三峡输变电工程规模大、单项工程多、时间跨度长，国务院三峡建委专门颁发了《三峡输变电工程静态控制、动态管理办法》，其中规定单项工程初步设计概算实行"总量控制、合理调整"。国家电网公司对单项工程设计、监理、施工及主要物资设备采购实行招投标，通过市场竞争有效并合理降低了工程造价；严格实行合同管理，规范合同流转监督，严格工程结算等。

rationally calculated by using the price indexes publicized by the State Bureau of Commodity Prices. Then the total index is calculated and thus the price difference is calculated. The price difference approved by the TGPCC under the State Council is the only basis for the project participants to make expense settlement according to the Conditions of the Contract. "Dynamic management" covers the following: calculate the increase in investment (price difference fund) for the year caused by price inflation (price index), and calculate the increase in investment caused by changes in national policies, interest rate changes or exchange rate changes. The most complex task is to calculate the price index and price difference fund for the year. The price index and price difference fund of the resettlement efforts are related to the vital interests of the many resettled households, so the Price Information Center of State Planning Commission established a price information collection network in the Three Gorges reservoir area. This was to ensure that price indexes more accurately reflect the actual conditions in the Three Gorges reservoir area and that they are more impartial and fair.

The investment management model of "static control and dynamic management" is fundamentally different from the management model in the planned economic system where project estimates are adjusted every several years. This is an investment control mechanism defined by clear accountability, rationale, impartiality and fairness, and can help the project owner establish a self-disciplined incentive mechanism.

2. Preparation of the execution estimate of the project legal person

The execution estimate of the project legal person is an important part of "static control and dynamic management". In order to conscientiously execute "static control", the CYTG-PDC established an estimate price system and a cost price system to determine the contract price and contract implementation control price and prepare the execution estimate and total execution estimate of each individual subproject. On the basis of the physical land and properties to be inundated, the Hubei and Chongqing governments divided the static compensation fund and allocated it to the counties and districts. The governments of the counties and districts further allocated the compensation funds to the individual subprojects. Given its large scale, large number of subprojects, and long duration of the Three Gorges Power Transmission and Transformation Works, the TGPCC under the State Council enacted the *Methods for Static Control and Dynamic Management of the Three Gorges Power Transmission and Transformation Works*, which stipulates that the preliminary design estimate of each subproject is subject to "total quantity control and reasonable adjustment". The State Grid Corporation of China organized tendering and bidding for the design, supervision and construction and for main materials and equipment procurement in each subproject, so that the project's cost would be effectively reduced through market competition. Contract management, stand-

3. 推行限额计划

推行限额计划指的是由设计单位按照总执行概算投资，控制技术设计和施工图设计投资，从设计源头上严格控制不合理变更，以控制静态投资。国务院三峡建委于 1996 年 5 月颁发的《三峡工程概算、计划和资金管理暂行办法》第五条规定："经国家审定的枢纽工程初步设计总概算、水库淹没处理及移民安置补偿投资总概算、输变电工程系统设计总概算，是国家控制工程总投资的依据，不得随意调整。凡对设计作重大变更或增加项目而引起工程总概算增加，要报建设委员会批准。"

四、电能消纳机制创新

在政府的统一规划安排下，"三峡电"跨省远距离输电，促进了全国电网互联，实现了西电东送、南北互供，促进了中国统一电力市场的形成，有利于资源使用效率的提高和区域间电力资源的优化配置。

1. 三峡工程电能消纳的特点

三峡水电站是国家重点能源工程和"西电东送"骨干电源，电站装机 2 250 万千瓦，承担着特殊的国家使命和责任，具有不同于一般电站的显著特点。一是承载着国家能源战略和区域发展战略，对落实国家"西电东送"能源战略以及"西部大开发""长江经济带"战略意义重大；二是承担着防洪、航运、抗旱、供水、生态环保等众多重要的社会责任，综合效益显著；三是工程本身及其配套输变电工程投资巨大，对周边区域基础设施建设和人民生活水平提升具有很好的带动作用，对地方经济发展贡献巨大；四是大范围资源优化配置，有利于受电省市能源结构调整、实现节能减排目标，符合中国资源与负荷分布不均衡的国情及推进能源供给侧改革的需求。

三峡水电站电能通过跨省跨区专用配套线路点对网外送，纳入受电省（市）电力电量平衡。输电通道一旦建成便无法更改，消纳区域相对固定，且大水电消纳过程中涉及 10 个省（市）的电网公司和诸多电力企业，必须通过科学合理的机制实现电能的充分消纳和资源的优化配置。

ard contract transfer supervision and project settlement were strictly enforced.

3. Limit plan

In the limit plan, the Designer controls the investment based on technical designs and construction drawing designs according to the total execution estimate so that unreasonable changes are excluded during design, and static investment is under control. Article 5 of the *Interim Procedure for Estimate, Planning and Fund Management of the TGP,* enacted by the TGPCC under the State Council in May, 1996, stipulates: "In a state audited hydroproject, the general estimate of preliminary design, general estimates of investment for treatment of reservoir inundation and resettlement compensation, and for system design of power transmission and transformation works, are the basis for the state to control the total investment of the project, and that they shall not be adjusted without authorization. Any major design change or increased works that will lead to any increase to the project's general estimate must be reported to the Construction Committee for approval."

IV. Innovations of Electricity Consumption Mechanisms

Under centralized planning and arrangement of the government, "TGP power" is transmitted over long distances to other provinces, which enhances interconnection between the country's power grids, achieves west-to-east power transmission and north-south mutual supply, facilitates creation of the unified electricity market in China, and helps improve resources utilization efficiency and optimize the allocation of electricity resources between different regions.

1. Electricity consumption in the TGP

The Three Gorges Hydropower Station (TGHS) is the backbone power source for the national key energy development projects and the West-To-East Power Transmission project. Its installed capacity is 22,500,000 kW. It bears a special national mission and responsibility and is clearly distinct from ordinary power stations. First, it is a part of both the national energy strategy and regional development strategy, and has great strategic significance for the West-To-East Power Transmission strategy, the development of the western region in China and the development of the Yangtze River Economic Belt. Second, it bears much responsibility for flood control, shipping, drought control, water supply and ecological conservation, and brings significant comprehensive benefits. Third, the project itself and the supporting power transmission and transformation works required huge amounts of investment, so they would positively promote infrastructure development, improve the livelihoods of those in the

2. 三峡水电站电能消纳机制

在国家发展改革委和国家能源局的统筹协调下，三峡集团积极开展工作，从推动国家计委下发关于三峡水电站电能消纳方案的通知，到推动国家能源局下达溪洛渡、向家坝水电站电能消纳方案的通知，逐渐形成较完善的大水电跨省跨区消纳的科学机制，具体为：电站规划阶段，三峡集团寻找市场，与相关方协商达成初步消纳意向并联系落实送电通道；在此基础上，国家主导制定电能消纳方案，同步规划建设配套外送线路，电站纳入受电地区电力电量平衡，电价采取市场化倒推方式形成等。十多年来，该机制保障了国家重大能源工程的成功建设与可持续运营，大大减少了弃水窝电，也改善了受电省（市）的电源结构，确保国家层面的水资源综合利用和电力资源优化配置，获得各方认可。实践证明，该机制设计科学、合理，可操作性强。目前这一机制已成功推广到"西气东输""南水北调"和其他"西电东送"重点项目中。

surrounding regions, and contribute immensely to the local economic development. Fourth, optimized allocation of these massive resources would help the provinces and cities receiving the electricity adjust their energy structure and accomplish the objectives of energy conservation and emission reduction. It would also help China solve the unbalanced distribution of resources, and would support the need for energy supply side reform.

TGHS-generated electricity is transmitted via dedicated trans-provincial and trans-regional transmission lines to achieve electric power and energy balance for the receiving provinces (cities). Once completed, these power transmission lines will not change, and the areas consuming its power have become relatively stable. The hydropower consumption process involves power grid companies and numerous electric power enterprises in 10 provinces (cities), so a rational mechanism had to be established to achieve adequate electricity consumption and optimum allocation of resources.

2. Electricity consumption mechanism of the TGHS

Under the centralized planning of the National Development and Reform Commission and the National Energy Administration, the CTG made strong efforts in assisting and disseminating the notice on the electricity consumption plan for the TGHS issued by the State Planning Commission and the notice on the electricity consumption plan for the Xiluodu and Xiangjiaba hydropower stations issued by the National Energy Administration. It gradually established a sophisticated and rational mechanism for trans-provincial and trans-regional consumption of major hydropower. The mechanism: in the planning stage of the power station, the CTG seeks the market, consults with the concerned parties to reach the preliminary intent for consumption, and works to secure power transmission channels. On this basis, the state takes the lead in preparing the electricity consumption plan and simultaneously prepares the plan for the construction of transmission lines. The power station is included into the electric power and energy balance plans of the consuming regions. The price of electricity is determined by the market. In over a decade, this mechanism has ensured successful construction and sustainable operation of the nation's key energy development projects, substantially alleviating electricity overstock, improved the power source structures of the consuming provinces (cities), and ensured national-level comprehensive utilization of water resources and optimization of electric power resources. It has therefore been positively regarded by the concerned parties. It has been proven that this mechanism is rationally designed and highly practicable, and it has so far been applied to other key projects, including West-East Natural Gas Transmission, the South-to-North Water Diversion Project and for West-to-East Power Transmission.

The electricity consumption mechanism of the TGP is established based on the con-

三峡工程电能消纳机制以消纳方案为基础确定，主要包括分电方式和电价机制两部分。分电方式确定三峡水电站电力电量分电比例，按比例纳入各省市基础电力电量平衡，应分尽分。三峡左右岸26台机组按照汛期分电力、枯期分电量的原则进行分配。5月至9月汛期，按照先保障广东300万千瓦、华东720万千瓦，剩余电力留华中的原则分电；10月至次年4月枯期，按照华中52%、华东32%、广东16%的比例分电。后来为支持三峡库区发展，每年送电重庆40亿千瓦时，其余电力电量仍按消纳方案比例应分尽分。三峡地下电站投产后，电能全部留湖北，通过汛期送华东、枯期再置换给湖北的方式消纳，并因此新建一条华中至华东输电容量300万千瓦的直流输电线路。在电价机制方面，明确三峡水电站采用按照受电省市平均购电价倒推上网电价的电价机制，据此核定各受电省市上网电价，加权平均水平约每千瓦时0.25元，并率先建立落地电价随受电省市平均上网电价浮动的动态电价调整机制。数据表明，从2003年至2020年10月底，三峡输变电工程累计输送三峡上网电量超1.36万亿千瓦时，相当于节约标准煤4.3亿吨，减少二氧化碳排放量11.69亿吨。

五、工程管理创新

三峡工程投资量大，施工环境复杂，各种机械交错林立，交叉作业不可避免，给质量安全和信息管理带来较大难度。三峡工程从项目管理的需要出发，按照矩阵式管理方式设置工程建设管理机构，提出"质量、安全（双零）"的管理目标及配套举措，开发了大型集成化管理信息系统——三峡工程管理信息系统。在参考世界各国水电工程最高标准和我国已有国家标准的基础上，三峡工程还制定了严格的质量标准，相关标准和条文经过实践检验成为新的行业标准和国际标准。

1. 质量、安全（双零）管理

2001年，三峡总公司提出了"零质量事故、零安全事故"的管理目标。为了达到"零安全事故"的目标，三峡总公司制定了《安全生产管理办法》《工程建设施工安全管理规定》等操作性很强的安全管理制度。同时，三峡总

sumption plan and consists of two parts, including a power distribution method and an electricity price mechanism. The power distribution method dictates the proportions of electric power distributed, and energy of the TGHS. The electricity is included into the basic electric power and energy balance of the provinces and cities according to such proportions, with electricity going to whoever needs it. The 26 units on the left and right banks of the TGP are allocated under the principle that electricity is distributed in flood season and electricity is distributed in dry season. During the flood season from May to September, power is distributed under the principle that three million kW goes to Guangdong, 7.2 million kW goes to East China and the rest stays in Central China; during the dry season, from October to April, power is so distributed that Central China receives 52%, East China 32% and Guangdong 16%. Later, in order to support development of the TGP reservoir area, four billion kWh was transmitted to Chongqing every year, while the rest of the electricity was distributed in certain proportions according to the consumption plan. After commissioning of the TGP underground power plant, all the electricity stayed in Hubei. The consumption plan was that electricity would be transmitted to East China during the flood season and returned to Hubei in the dry season. Therefore, a DC transmission line with three million kW transmission capacity was built from Central China to East China. In the electricity price mechanism, it was explicitly stipulated that the feed-in tariff of TGHS will be calculated according to the average electricity purchase price of the provinces and cities receiving the electricity, and on this basis, the feed-in tariff of each receiving province/city is determined. The weighted average price is approximately RMB 0.25/kWh. This innovative, dynamic electricity price adjustment mechanism has been established, where the actual tariff fluctuates with the feed-in tariff of the receiving provinces/cities. From 2003 to the end of October 2020, according to statistics, the Three Gorges Power Transmission and Transformation Works transmitted a total of over 1.36 TWh of TGP grid power. That represents the equivalent of 430 million tons of standard coal and therefore a reduction of 1.169 billion tons of carbon dioxide emission.

V. Project Management Innovations

The TGP is characterized by large investment, a complex construction environment, and a great amount of machinery in complex placement and unavoidable cross operations, which makes quality, safety, and information management very difficult to maintain. Given the need for management of the TGP, the project's management organization was instated in a matrix management model, management objectives regarding "quality and safety (zero accident)" and the supporting measures were proposed, and a large integrated management information system was developed: the Three Gorges Project Management System. After examining best

公司针对不同文化层次的工程参建人员，采取讲授、夜校、培训和现场观摩等形式开展全员安全教育，提高参建人员的安全素质，为落实安全生产制度打下了坚实基础，营造了良好的安全生产氛围。

三峡总公司还健全组织机构，成立了三峡工程安全生产委员会和三峡总公司安全总监办公室，明确三峡总公司安全总监办公室负责对工程建设各项安全管理工作进行组织、协调、监督与指导。三峡总公司安全总监办公室于2001年聘请了2名日本安全总监充实安全管理队伍，确定爆破施工、高边坡坍塌、排架垮塌、地下洞室开挖工程坍塌等8大危险源，并相应制定了重大危险源防范安全措施和事故应急救援预案。

2. 三峡工程管理信息系统

三峡工程管理信息系统（TGPMS）是一个面向三峡工程建设和管理全过程的具有辅助决策和预测功能的综合信息服务系统，它包含成本管理、计划与进度管理等13个功能子系统。TGPMS主要依靠现代信息技术、办公自动化技术、多媒体技术，是一个为设计、承包商、监理、业主共同完成项目目标而搭建的集成式协同工作平台，形成从项目实施层、管理层到决策层以及各层级对外联系的信息体系，实现对三峡工程全过程、全方位的信息控制与管理。TGPMS是中国水电界、中国工程界首次建设开发的大型集成化管理信息系统，通过融合国际先进管理理念、方法、模型并结合三峡工程建设的实际情况开发而形成，是一套既包含国际先进管理理念又符合中国国情的工程管理系统。

3. 三峡标准

三峡工程在集成创新、引进消化吸收再创新、原始创新、工程管理等方面取得了一系列重大成果，形成了具有自主知识产权的"三峡品牌"技术和管理经验，许多技术和管理经验已在国内外大型水电工程中得到应用和推广。通过十多年的建设管理，三峡工程已经成为水电行业的标杆，形成了一系列的"三峡标准"，引领中国水电向国内纵深发展，并扬帆远航走向海外。

practices in terms of standards for hydropower projects in other countries and also the existing standards in China, the company established rigorous quality standards for the TGP. The successful application of these standards and provisions made them new industrial and international standards.

1. Quality and safety (zero accident) management

In 2001, the CYTGPDC proposed the management objectives of "zero quality issues and zero accidents". To accomplish the "zero accidents" objective, the CYTGPDC established highly practicable safety management systems, including *Measures for Work Safety Management* and *Regulations for Construction Safety Management in Construction Project*. On top of this, the CYTGPDC raised safety awareness by organizing various types of safety classes by means of lecture, night school, training and site visits for project personnel of various educational backgrounds. This laid a foundation for implementing the work safety system and fostered an environment conducive to work safety in general.

The CYTGPDC also improved its organization by instating the TGP Work Safety Committee and the CYTGPDC Safety Director Office, and explicitly stipulated that the Safety Director Office is responsible for organizing, coordinating, supervising and guiding the work safety management efforts of the project. In 2001, the CYTGPDC Safety Director Office hired two Japanese safety directors to supplement the safety management team. They identified eight hazards including blasting, risks of high slope collapse, bent frame collapse and collapse during underground cavern excavation. So the company formulated safety measures and emergency rescue plans to counteract these major hazards.

2. Three Gorges Project Management System

The Three Gorges Project Management System (TGPMS) is an integrated information service system with decision aiding and prediction functions dedicated to the construction and management of the TGP. It consists of 13 functional subsystems, including cost management, planning, and schedule management. The TGPMS, which is built upon modern information technology, office automation technology and multimedia technology, is an integrated, collaborative work platform used by the Designer, Contractor, Supervisor, and Owner to accomplish the project's objectives. It is an information system for communication between the project's implementation level, management level, and decision-making level, and also between these levels and external parties. Its goal is to implement whole-process, comprehensive information control and management over the TGP. The TGPMS was the first large integrated management information system developed for China's hydropower and engineering industries. It was created by integrating a combination of internationally advanced

图 6.1　三峡水电站发电机组安装图（摄影：黄正平）

（1）大型水电工程建设的"三峡标准"

三峡工程在建设方面面临许多世界级难题，广大建设者通过加大科研力度，开展自主创新，在土建开挖、截导流、大坝混凝土浇筑、金结安装等方面取得了一大批科研成果和专利技术，形成了一大批技术标准和规范，编印了《中国长江三峡工程标准》（TGPS）等有关手册。此外，在工程质量、安全管理等方面也形成了三峡工程自己的标准。上述种种都已成为国内外水电行业具有标杆意义的"三峡标准"。

（2）大型水力发电机组的"三峡标准"

在三峡 70 万千瓦机组安装前，已有的国家标准及其他相关现行安装标准只涵盖 55 万千瓦容量的机组（二滩机组）。通过三峡工程建设，形成了 70 万千瓦以上的机组设计、制造、安装、运行、管理的三峡技术标准和规范，其关键性指标均高于国内外同行业其他标准。在此基础上，又制定了三峡"精品机组"评价标准，其主要指标大大高于国家标准和国际标准。中国制造的水电机组在国际和国内市场占有越来越重的份额，在三峡机组上确立的中

◐ Figure 6.1　Installation Drawing for Generator Unit of TGHS (Photographed by Huang Zhengping)

management concepts, methods and models, tailored for the TGP, making it a system which encompasses internationally advanced management concepts and which serves the needs of China.

3. TGP standard

The TGP realized a range of major achievements in integrated technologies innovation, re-innovation of imported and assimilated technologies, of original technologies and of project management, and created the "Three Gorges brand" technologies which are proprietary intellectual properties. Many of the technologies and management approaches have been applied and popularized in large domestic and foreign hydropower projects. After over a decade of construction management, the TGP has now become a benchmark in the hydropower industry. It has established a set of "TGP standards", and leads China's hydropower industry in developing domestically and expanding internationally.

(1) "TGP standards" for large hydropower projects

The TGP was faced with many world-class challenges, so project participants intensified their scientific research efforts, sought independent innovations, had a large number of scientific achievements and obtained patents in excavation technologies, in closure/diversion, and in dam concrete placing and metal structure installation. It established a large set of technical standards and specifications, and published the manuals such as *Yangtze River Three Gorges Project Standard* (TGPS). In addition, its work quality and safety management standards specifically designed for the TGP were also established. These standards are called the "TGP standards" which are considered the benchmark for the domestic and foreign hydropower industries.

(2) "TGP standards" for large hydropower generator unit

Before installation of the 700,000 kW units of the TGP, the existing national standards and other applicable installation standards only covered 550,000 kW units (Ertan units). Through their implementation, the TGP technical standards and specifications have been established regarding design, manufacture, installation, operation and management of > 700,000 kW units, and its key indexes are higher than those of other national and international standards in the industry. On this basis, the TGP's "quality unit" evaluation criteria has

国标准已经成为国际标准和"中国制造"的亮点,成为国际机设备电厂商标榜自己产品质量的标准。

(3) 大型船闸建设管理的"三峡标准"

三峡船闸为双线连续五级船闸,是目前世界上已建成船闸中连续级数最多、总水头和级间输水水头最高、总体规模最大的船闸,一系列技术难题在世界上尚无实践经验可循。通过大量的设计研究工作,三峡船闸在复杂条件下的总体设计、特高水头船闸输水技术、高陡边坡稳定与变形控制技术、"全衬砌"式船闸结构技术、超大型人字门和启闭机技术、复杂运行工况下的船闸监控系统等方面取得了多项技术创新,将世界船闸建设提升到新的水平。

(4) 大型升船机建设管理的"三峡标准"

三峡升船机是目前世界上规模最大的升船机,提升总重量 15 500 吨,过船规模 3 000 吨级,没有一个国家有如此大规模的齿轮齿条爬升式升船机工程建设经历,在建设中无可供参考的标准。三峡升船机是不折不扣的"中国制造"产品,工程总成设计、设备制造、现场安装调试和土建施工全部由国内

图 6.2 轮船夜间通过船闸(摄影:刘华)

been established, and its main indexes are much higher than those of national and international standards. The hydropower units manufactured by China are gaining more share in the international and domestic markets. The Chinese standards, developed during the TGP, are now international standards. They are a highlight of the "Made in China" brand and are also the standards that the international equipment manufacturers use to advertise their products.

(3) "TGP standards" for construction and management of large ship lock

The TGP's ship lock is a continuous double-line five-stage lock and has the largest number of stages, the highest total head and interstage head, and the largest overall scale in the world among completed ship locks. There was no practical experience to learn from in solving technical challenges. Through extensive design research, many technological innovations have been achieved with the Three Gorges ship lock, including overall design under complex conditions, ultra-high head ship lock water conveyance technology, high steep slope stabilization and deformation control technology, "full lining" ship lock structure technology, ultra-large miter gate and hoist technology, and ship lock monitoring systems under complex operating conditions. This took the world's ship lock construction technology to the next level.

(4) "TGP standards" for construction management of large ship lifts

The TGP ship lift is the largest one in the world. Its total lifting capacity is 15,500 t and supportable ship weight is 3,000 t. No country had any experience in building such a large gear-and-rack climbing ship lift and there were no prior standards to guide its construction. It is a genuine "Made in China" product. Its assembly design, equipment manufacture, onsite installation and commissioning and civil works were all accomplished by domestic design institutes and enterprises, and imported equipment accounted for no more than 10% of total investment. To complete the project, Chinese enterprises innovated their design concepts, manufacture technologies, construction processes and management methods, and successfully solved a range of world-class technical challenges. Through independent innovation and also the import, assimilation and application of foreign technologies, the CTG established a whole set of technical standards and specifications regarding design, manufacture, construction and installation of ship lifts, and has set a benchmark for large navigation structures of hydropower projects in the world. This benchmark genuinely bears the "Made in China" brand.

◌ Figure 6.2　Ships Passing Through the Ship Lock at Night (Photographed by Liu Hua)

设计院和企业完成,进口设备不超过总投资的一成。围绕工程建设,中国企业创新设计理念、制造技术、施工工艺和管理方法,成功地解决了一系列世界级技术难题。三峡集团通过自主创新和引进、消化、吸收,形成了一整套关于升船机设计、制造、施工、安装等的技术标准和规范,在全球水电工程大型通航建筑物中树立了一个标杆,在标杆上深深地打上"中国制造"烙印。

第二节 三峡工程的科技创新

三峡工程从勘测、规划、论证、设计到施工面临着一系列世界级难题,当时既无成功的经验可供借鉴,又无成熟技术可利用,必须进行大量的自主创新,克难攻坚,才能完成这一宏伟的工程。经过几代人的不懈努力,上万名科技人员参加了三峡工程的科技创新工作,取得了一系列重大突破。

一、总体设计创新

1. 成功解决了特大型枢纽建筑物的布置难题

三峡枢纽工程总体布置受坝址流量大、地形地质条件和弯曲河道等限制,综合考虑坝址自然条件和有利于泄洪、排沙、通航、发电,以及便于导流、截流和提前发挥通航发电效益等因素,创新性地提出了河床中部布置泄洪坝段、两侧布置厂房坝段、两岸山体布置通航建筑物和地下电站,在主要建筑物之间布设排沙和排漂设施的布置方案,成功解决了三峡水利枢纽泄洪流量大、机组台数多、运行条件复杂的布置难题,实现了分期施工、两次导流截流、提前发挥通航和发电效益的要求。

Section 2 | Technological Innovations of the TGP

The TGP was faced with a series of world-class challenges in surveying, planning, demonstration, design and construction. There was no successful preceding experience to borrow from and no proven technology to utilize, so extensive independent innovation had to be achieved to solve the challenges and accomplish this great project. Thanks to the unremitting efforts of several generations of people and the 10,000 talents in technology, who contributed to technological innovations for the TGP, many major breakthroughs were achieved.

I. Innovations of Overall Design

1. Successfully solving the problems in layout of the extra-large hydroproject structure

The general layout of the TGP was constrained by factors such as the large discharge at the dam site, topographical and geological conditions and the meandering river. With comprehensive consideration of the natural conditions at the dam site, and factors including flood discharge, desilting, navigation, power generation, diversion, closure and earlier accomplishment of the navigation and power generation functions, an innovative solution was designed, where the flood discharge dam monolith was built in the middle of the riverbed, the powerhouse monoliths on both banks, the navigation structure and underground power plant in the mountains on both banks, and the desilting and floating debris discharge systems between the main structures. This solved the challenges brought by large flood discharges, large number of units and complex operating conditions at the Three Gorges hydroproject. It fulfilled the requirements for phase-by-phase construction, two-stage diversion and closure and the earlier accomplishment of navigation and power generation functions.

2. 成功解决了高水头超大泄量泄洪消能技术的难题

三峡枢纽泄洪量大、水头高、水位变幅大，为满足导流截流、放空、泄洪、排沙、排漂等多目标的运行要求，创新性地提出了在泄洪坝段布置三层泄洪孔，采用"平面相间、高低重叠"型式，解决了主河槽段坝体集中布置多种大孔口的难题。而且提出泄洪深孔采用有压短管、跌坎掺气和导流底孔采用有压长管、跨缝布置的方案，研究确定适应大水位变幅的深孔跌坎体型和底孔采用不同的进口高程、鼻坎高程及挑角的优化组合方式，采用钢筋混凝土非线性配筋设计方法和横缝止水后移、横缝灌浆等工程措施，解决了泄洪坝段深孔与底孔水头高、运行时间长、空化空蚀、单独及联合泄洪消能防冲以及坝体集中开孔的结构设计等难题。

3. 成功解决了坝基高连通率结构面的抗滑稳定难题

大坝左岸厂房 1~5 号坝段和右岸厂房 24~26 号坝段位于临江岸坡上，坝基存在倾向下游的缓倾角结构面，最大裂隙连通率达到 83.2%，构成了岸坡厂房坝段沿缓倾角结构面的深层滑动稳定问题。通过采用特殊的勘测技术，查明了长大缓倾角结构面的空间位置、产状、分布范围和组合方式，建立了确定性地质概化模型，运用"改进等 K 法"、有限元法和地质力学模型试验进

图 6.3　三峡大坝深孔和表孔同时泄洪实景

2. Successfully solving the problems of high head ultra-large-discharge flood discharges and energy dissipation structure

The TGP has a large flood discharge, a high head and wide variation in water level. In order to achieve operational objectives, including diversion, closure, venting, flood discharge, desilting and floating debris discharge, an innovative solution was designed, where the flood discharge dam monolith contained three levels of flood discharge outlets and the "planar alternation and vertical overlap" structural form was built. This solved the problem of concentrating multiple types of large holes in the main channel of the dam's body. One other solution was also designed, where the low-level outlets for flood discharge were fitted with pressurized short pipes and step-down aeration structures, while the bottom diversion outlets were fitted with pressurized long pipes and built across the joints. After investigation, it was decided that the low-level outlet step-down structure and the bottom outlet that were designed to adapt to large water level variation would be optimized to have different inlet elevations, bucket elevations and jet angles. The design with non-linear reinforcement in reinforced concrete and engineering measures such as transverse joint water stops moved back and grouting in transverse joint were also used. This solved the challenging problems such as the high head of low level outlets and bottom outlets in the flood discharge dam monolith, long operation time, cavitation, cavitation erosion, scour prevention in separate and combined flood discharges, and energy dissipation and structural design of concentrated holes in the dam body.

3. Successfully solving the problem in stability against sliding at the discontinuity with a high connectivity in dam's foundation

The left-bank and right-bank powerhouse monoliths 1~5 and 24~26 were built on bank slopes. The dam foundation has a gently dipping discontinuity that is inclined towards the downstream, and the maximum fissure connectivity rate is 83.2%. This creates a problem of low-level sliding stability of the slope powerhouse monolith along the gently dipping discontinuity. By using special survey technology, project personnel ascertained the spatial position, attitude, distribution range and composition of the large and long discontinuity and built a deterministic, geological generalized model. By using the "improved equivalent K method", the finite element method and geomechanical model testing, they conducted comprehensive research and made a comprehensive assessment of the foundation's stability against sliding, proposed innovative solutions such as gear slot in dam heel, curtain moved forward, low

◅ Figure 6.3　Real Flood Discharge of Both the Low Level Outlet and the Crest Outlet of Three Gorges Dam

行综合研究，综合判定基础的抗滑稳定性，创新性地提出"坝踵设齿槽、帷幕前移、基础深层排水和厂坝联合受力、预应力锚索加固"等综合措施，成功解决了岸坡坝段的深层抗滑稳定难题。

二、导截流技术创新

1. 大流量深水河道截流技术

与国内外大型水电工程相比，三峡工程大江截流具有截流流量大、截流水深、截流时抛投强度大等特点，均居世界之最，戗堤下的河床带新淤沙覆盖层厚，截流中有严格的通航要求。

针对大江截流水深的特点，戗堤进占出现堤头坍塌的难题，因此探讨了深水截流堤头坍塌的机理，提出并创新性地采用深水平抛垫底措施，有效防止了堤头坍塌事故的发生，1997年11月8日，龙口顺利合龙。实测截流流量 8 480～11 600 米每秒，落差 0.66 米，最大流速 4.22 米每秒，截流最高日抛投量 12.09 万立方米。

三峡工程创造性地采用"预平抛垫底、上游单戗立堵、双向进占、下游尾随进占"的施工方案，解决了深水截流的一系列技术难题。

2. 深水土石围堰关键技术

二期上游土石围堰最大高度 82.5 米，堰体施工最大水深 60 米，为深水土石围堰。围堰挡水水头高，基础地形地质条件复杂。围堰形式为两侧石渣及块石体、中间风化砂及砂砾石堰体，塑性混凝土防渗墙上接土工合成材料防渗心墙。围堰填筑方量达 1 032 万立方米，且 80% 堰体为水下抛填，防渗墙面积达 8.4 万平方米，施工强度及难度在国内外已建成的水利水电工程中实属罕见。围堰于 1998 年 6 月抢至度汛高程，先后经受 8 次长江洪峰考验，在洪水流量 61 000 米每秒、最高水位 77.8 米时，围堰运行正常。

围堰工程的技术难题主要有：断面的结构和防渗形式的选择，60 米水深下抛填风化砂密度的确定，深槽、陡坡、硬岩防渗墙的施工技术，新型柔性墙体材料研制及其质量控制方法，新淤沙的动力稳定性及其处理。为此，在

level dewatering at the foundation, stress sharing between plant and dam, and reinforcement with prestressed anchor cables. This successfully solved the challenging problem of low level stability against sliding with the slope dam monolith.

II. Innovations in Diversion and Closure Technologies

1. Closures in large-discharge deep rivers

Compared with the other large domestic and foreign hydropower projects, the river closure of the TGP is characterized by a large closure discharge, deep water and high dumping intensity during closure, which all ranked as the greatest in the world. The riverbed beneath the berm bears a thick silt overburden and strict navigation requirements are imposed during closure.

Given the deep water in the closure area and the problem of head collapsing during berm advancement, the mechanism behind berm head collapsing during deep water closures was investigated, and a solution of bottom filling through horizontal dumping in deep water was devised and innovatively used. This effectively precluded accidents with the berm head collapsing from happening. On November 8, 1997, the closure gap was successfully closed. The measured closure discharge was 84,80~116,00 m/s, fall was 0.66 m, maximum flow rate was 4.22 m/s, and maximum daily dumping during closure was 120,900 m^3.

The creative solutions of bottom filling by horizontal dumping, blocking by single upstream berm, bidirectional advancing and downstream follow-up advancing were used to solve many technical problems with deep water closure of the TGP.

2. Key technologies for deep water earth-rock cofferdam

The maximum height of the second-stage upstream earth-rock cofferdam was 82.5 m and the maximum water depth during its construction was 60 m, so it is classified as a deep water earth-rock cofferdam. The cofferdam has a high water retaining head and the foundation has complex topographical and geological conditions. Structurally, both sides of the cofferdam were built of rock ballasts and rock blocks, and its middle was built of weathered sand and also sand and gravel. The top of the plastic concrete cut-off wall was connected to the cut-off core built of geosynthetics. The filling volume of the cofferdam was 10.32 million m^3, and 80% of the cofferdam's mass was built by underwater dumping. The area of the cut-off wall is 84,000 m^2. The construction intensity and difficulty are really rare among domestic and foreign water conservancy and hydropower projects. The cofferdams were built to flood control elevation in June 1998, and they have withstood eight flood peaks of the Yang-

三峡工程技术设计和施工阶段，从围堰的设计、施工、运行到拆除安排了一系列研究课题，2003年5月顺利完成使命，并安全拆除。

图6.4　三峡二期工程土石围堰拆除爆破实景

三、大体积混凝土优质高效施工技术创新

1. 研究提出以耐久性为主导的高性能混凝土配合比设计新理念

首次提出了"低用水量、低坍落度、高粉煤灰掺量并降低水胶比"大坝混凝土设计理念，形成了"高内含氧化镁中热水泥加Ⅰ级粉煤灰，联掺高效减水剂和引气剂"大坝混凝土配制新技术，破解了花岗岩人工骨料混凝土用水量高的技术难题，研制出具有高耐久、高抗裂、施工性能优良的高性能混凝土。首次提出的"限制原材料碱含量和混凝土总碱量"抑制大坝混凝土碱—骨料反应综合技术措施纳入到行业标准。

2. 建立了大坝混凝土优质快速浇筑方案及温度控制防裂新技术

三峡工程混凝土工程量巨大，总混凝土量达2 800万立方米，其中大坝混凝土量达1 600万立方米。为满足混凝土特高施工强度，三峡工程采用以架空

tze River. When the flood discharge was 61,000 m/s and maximum water level 77.8 m, the cofferdams were functioning normally.

Main technical challenges in the cofferdam's engineering are: selection of cross section structure and cut-off structure; determination of the density of weathered sand dumped in 60 m-deep water; construction of cut-off walls at deep grooves, steep slopes and hard rocks; manufacture of new flexible wall material and its quality control; kinetic stability of fresh silt and its treatment. Therefore, during technical design and construction of the TGP, a range of research programs were conducted regarding design, construction, operation and demolition of cofferdams. In May 2003, the cofferdams were safely demolished after they served their purpose.

Figure 6.4   Demolition Blasting of the Earth-Rock Cofferdam in Phase II of the TGP

III. Innovations of Technologies for High-quality, Efficient Construction with Mass Concrete

1. The new design concept of durability-oriented, high-performance concrete mix proposed during research

It was the first time the design concept for dam concrete of low water consumption, low slump, high fly ash content and reduced water-binder ratio" was proposed, and the new concrete preparation technology of adding grade I fly ash in hot cement with high magnesium oxide content and mixing in a high-efficiency, water-reducing agent and an air-entraining agent was implemented. This solved the problem of high water consumption in granite-based artificial aggregate concrete. The high-performance concrete with high durability, high crack resistance and excellent workability was manufactured. The inventive technical solution of inhibiting alkali-aggregate reactions in dams by limiting alkali content in raw material and total alkali in concrete was included into the industrial standard.

2. Invention of a fast, high-quality concrete placing method, and new technology for crack prevention through temperature control

The TGP has a huge quantity of concrete works, with a total of 28 million m^3 of concrete, including 16 million m^3 for the dam. To support ultra-high construction intensity, the TGP used an overhead belt conveyor and a tower belt conveyor that was connected to a silo to achieve continuous placing, and also used a whole set of concrete production equipment and lifting equipment. This was a new large concrete construction system. It was the first sys-

的皮带输送机加塔式皮带机入仓连续浇筑为主的方案，再辅以成套的混凝土生产设备、吊运等专用设备，形成了一个崭新的大规模混凝土施工系统，首次实现塔带机、门塔机、缆机三类浇筑机械联合作业，创造了年浇筑混凝土548万立方米、月浇筑混凝土55.35万立方米、日浇筑混凝土2.2万立方米的世界最高纪录，远远超过了由古比雪夫工程保持的年浇筑313万立方米的前世界纪录。三峡工程采用了个性化和精细化的温控措施，提出了大仓面、厚层浇筑施工技术，研究应用混凝土预冷二次风冷骨料新技术、软冷却水管铺设、分期通水和个性化通水、新型保温材料适时跟进温控等整套技术，确保了混凝土浇筑质量，显著提升了施工效率，攻克了三峡大坝孔洞多、结构复杂、坝块尺寸大、混凝土温度控制严格、防裂难度大等一系列技术难题。右岸大坝浇筑混凝土416万立方米，经现场检查未发现一条温度裂缝，创造了世界混凝土重力坝筑坝史上的奇迹。

3. 建立了三期碾压混凝土围堰施工技术

三峡三期碾压混凝土围堰是三峡工程二、三期衔接的关键性控制工程，混凝土工程量约167万立方米。三期碾压混凝土围堰必须在明渠截流后不到半年的一个枯水期内完建并挡水，施工质量要求很高，施工难度极大。为适应碾压混凝土快速施工，通过结构优化研究，有效解决了堰体结构与快速施工之间的矛盾。设计研究采用不分纵缝，加大横缝间距，全断面采用碾压混凝土施工；防渗采用掺防水剂的薄层变态混凝土，既简化了堰体结构，又保证了碾压混凝土的施工质量。通过周密分析影响连续高强度施工的各种因素，提出应对措施，保证了施工方案和施工技术的先进性。紧紧围绕碾压混凝土快速施工进行研究，提出了合理的质量控制指标和质量保证措施，解决了如何在确保施工质量的情况下实现快速施工的技术难题。

tem that used a tower belt conveyor, a portal crane and a cable crane for one concrete placing operation. It achieved an annual concrete placing rate of 548 million m^3, a monthly concrete placing rate of 553,500 m^3 and a daily concrete placing rate of 22,000 m^3, which are world records and far exceed the previous record of 3.13 million m^3 accomplished by the Kuibyshev Project. The TGP uses customized and delicate temperature control. Concrete placing technology with a large pouring area and a thick layer was designed. Other technologies were also developed and applied, including concrete precooling with secondary air-cooled aggregate, flexible cooling water pipes, stage-by-stage water injection, customized water injection and timely temperature control on new thermal insulation material. These measures ensure high quality of concrete placing, significantly enhance construction efficiency, and solve many technical problems in the Three Gorges Dam such as a large number of holes, a complex structure, a large dam block, delicate temperature control of concrete and difficult crack control. A total of 4.16 million m^3 of concrete was placed in the dam's right-bank. On-site inspection detected no temperature cracks, which is a miracle and is unprecedented in the history of concrete gravity dam construction.

3. Invention of the third-stage roller compacted concrete (RCC) cofferdam construction technology

The third-stage RCC cofferdam of the TGP was a key controlling project that linked Phase II and Phase III, and its concrete quantity was approximately 1.67 million m^3. The third-stage RCC cofferdam had to be completed with a retention wall built within a dry season of less than half a year after closure of the open channel. This requires high quality construction and the construction itself is very difficult. To support fast construction of roller compacted concrete, a structure optimization method was developed, which effectively resolved the conflict between cofferdam structure and fast construction. In the design, no longitudinal joint was built, the spacing between transverse joints was increased, and roller compacted concrete was used in the entire cross section. For the cut-off structure, thin metamorphic concrete mixed with a waterproofing agent was used, which simplified the cofferdam's structure and ensured high construction quality of roller compacted concrete. After close analysis of the factors affecting continuous high-intensity construction, proper solutions were designed to ensure a superior construction scheme and superior construction technology were used. So intensive research into fast construction of roller compacted concrete was conducted, and rational construction quality control criteria and quality assurance measures were designed, which have solved the technical problem of achieving fast construction while maintaining high construction quality.

四、巨型水电机组研发、设计和制造技术创新

1. 巨型水电机组研制核心技术的突破

突破了巨型水电机组研制的核心技术，攻克了三峡巨型水电机组研发、设计和制造难题，实现了巨型水电机组的自主化。在三峡工程之前，中国自主研制的最大水电机组仅为32万千瓦，没有70万千瓦水电机组，而三峡机组比世界上已有的70万千瓦机组的运行条件更复杂，是国际公认的设计、制造技术难度最大的机组。在总体设计方面，于世界上首次提出了三峡水电机组的设计准则和技术规范，通过优化比选，确定水电机组总体方案、设计参数、工程尺寸等，解决了水轮机运行稳定性及向高水头区要效率、向低水头区要出力的技术难题，还解决了发电机电磁、冷却、推力轴承等技术难题，仅用7年的时间，中国就实现了水电重大装备研制核心技术30年的跨越。

2. 巨型水轮机设计理念的突破

突破了巨型水轮机传统设计理念，在世界上首创"将水轮机运行稳定性放在首位"的设计准则。三峡水电站水轮机的运行水头为61～113米，水头变幅远远超出业界统计的经验范围，水轮机更易发生有害振动。当时全球大型水电机组多数受到强烈振动、转轮裂纹等运行故障的困扰。为此，突破了以效率和出力等能量指标为主导的传统设计理念，三峡工程首次提出了"将水轮机运行稳定性放在首位"的设计准则，以及巨型水电机组运行稳定性控制指标；创新性地开展了提高水轮机设计水头、改善水轮机高水头运行稳定性的研究，取得重大研究成果和重大设计技术突破。32台机组已运行十几年，并经历了各种水头考验，水轮机运行安全、稳定、可靠。

3. 巨型水轮发电机组技术难题的突破

首创了蒸发冷却技术的巨型水轮发电机，攻克了巨型水轮发电机组电磁振动的世界技术难题。在世界上首次自主研制了70万千瓦"蒸发冷却"水轮发电机，其定子绕组温升低，且温度分布均匀，优于电力行业相关标准，是巨型水轮发电机冷却技术领域的重大突破，技术内涵原创，为世界首创全新冷却概念的巨型水轮发电机组，具有完全自主知识产权。首创了发电机振动

IV. Innovations in R&D, Design, and Manufacture of Large Hydropower Generator Units

1. Breakthrough in core technology of large hydropower generator unit

Breakthrough in the core technology of large hydropower generator units was achieved, challenges in R&D, design and manufacture overcome, and their independent manufacture made possible. Before the TGP, the largest hydropower generator unit independently developed by China had only 320,000 kW capacity. No 700,000 kW units had ever been built by China. The operating conditions of the TGP generator unit were more complex than those of the existing 700,000 kW units in the world, so it was universally recognized as being the most difficult to design and manufacture. In overall design, the design criteria and technical specifications for the TGP hydropower unit were the first of their kind. Through comparison of the available options, the optimum overall scheme, design parameters and engineering size of the unit were identified. This solved technical problems in the operational stability of turbines, high efficiency of high head areas, and high output of low head areas. This also solved technical problems in electromagnetism, cooling, and thrust bearings of generators. In only seven years, China achieved breakthroughs in the core technologies of major hydropower equipment that would have otherwise taken 30 years.

2. Breakthrough in the design concept of large turbines

The traditional design concept of large turbines was outmoded, and was replaced by the innovative design criterion that operational stability is the priority. The operating head of TGHS turbines is 61~113 m. Their head variation was far beyond the empirical range known to the industry, and they were more susceptible to harmful vibrations. Back then, most large hydropower units in the world were troubled by strong vibration and runner cracking. Therefore, the traditional design concept, which was dominated by energy indicators such as efficiency and output, had to be discarded. With the TGP, the innovative design criterion that operational stability is the priority was proposed, and the criteria for stability control of large hydropower units were established. Innovative research into the techniques to increase design heads of turbines and improve high head stability of turbine operation was conducted, and major achievements and design breakthroughs were obtained. The 32 units have been in service for a dozen years, have withstood various heads and have been operating safely, stably and reliably.

图 6.5 三峡水轮机制作现场图

源定量分析方法,利用电磁振动激振源特性计算方法,攻克了巨型水轮发电机电磁振动世界性技术难题,成功解决了三峡左岸部分进口发电机和三峡右岸部分发电机的电磁振动问题,将 100 赫兹电磁振动幅值削弱了 87%,从根源上消除了有害高频电磁振动和噪声。该成果推广至向家坝、溪洛渡等机组,运行情况表明定子铁心水平通频振动优良。

4. 国内外大型水轮发电机组安装技术和标准的创新

首创了国内外大型水轮发电机组安装技术和标准,创新了机组启动调试和在线监测技术。创新了主要部件的安装和吊装技术,缩短了安装周期,创新了机组启动调试和在线监测技术。创造了单个电站年投产 500 万千瓦装机容量的世界纪录。制定了《三峡水轮发电机组安装标准》和《三峡"精品机组"评价标准》,填补了 70 万千瓦水轮发电机组安装标准的空白,其关键性指标均高于国内外同行业标准。按照压力脉动、振动、摆度综合判定原则,在世界上首创了机组运行区域的划分标准,确保机组安全、高效、稳定运行。

⊃ Figure 6.5　Manufacturing Site of a TGP Turbine

3. Solving the technical problems of large turbine generator units

The turbine generator units based on inventive evaporative cooling technology were built, and the world-class technical problem of electromagnetic vibration of large turbine generator units was solved. The world's first 700,000 kW evaporation cooled turbine generator was independently developed. This generator had low stator winding temperature rise and uniform temperature distribution, and was superior to the applicable standards in the electric power industry. It represented a major breakthrough in turbine generator cooling technology and was based on completely original technology. The turbine generator units were built upon a completely original cooling concept and were protected by completely proprietary intellectual property. The method for quantitative analysis of generator vibration sources was invented. By using the electromagnetic vibration excitation characteristic calculation, they have solved the world-class technical problem of electromagnetic vibration in large turbine generators, solved the problem of electromagnetic vibration in some of the imported generators on the left bank and some on the right bank of TGP, reduced 100 Hz electromagnetic vibration amplitude by 87%, and fundamentally eliminated harmful high-frequency electromagnetic vibration and noise. The technology was applied in the generator units of Xiangjiaba and Xiluodu hydropower stations, whose operation indicates that the stator core has excellent horizontal direct frequency vibration condition.

4. Innovations in installation technology and standards for domestic and foreign large turbine generator units

The innovative installation technology and standards for domestic and foreign large turbine generator units were established, and unit startup commissioning and online monitoring technology were created. The installation and lifting technology for the main components were created, which shortened the installation period, and unit startup commissioning and online monitoring technologies were created. The project set the world record for having five million kW of installed capacity commissioned in one year for a single power station. The *Standard for the Installation of Turbine Generator Units of the TGP* and the *Evaluation Criteria for Quality Units of TGP* were established, which filled the gaps in installation standards for 700,000 kW turbine generator units. Their key indicators are superior to those of the other domestic and foreign standards in the industry. On the basis of the evaluation, criteria of pressure pulsation, vibration and swing, the world's first criteria for dividing unit operation area were established. It would ensure safe, efficient and stable operation of the units.

五、船闸高边坡及完建施工技术创新

1. 深切高陡边坡的稳定、变形控制与大型衬砌结构研究

船闸高边坡集高、陡、长于一体，不仅规模大、形态复杂，而且岩石开挖后存在深切开挖卸荷变形的问题，船舶过闸对边坡稳定的要求高，如此复杂的船闸高边坡问题在国内外尚无先例。不仅要保持高边坡岩体在施工期和运行期的稳定，要求岩体作为船闸结构的一个组成部分与衬砌结构协同工作，还要考虑边坡岩体变形对船闸设备正常运行，特别是对人字形闸门正常运行的影响。通过应用大量高新技术进行地质勘测和多种现场科研试验，用不同模型进行计算分析，采用开挖、加固、防渗、排水等综合技术，可靠地解决了高边坡的稳定与变形问题。在此基础上，通过合理采用岩槽的开挖形式（保留两线船闸间岩体隔墩）和船闸的结构型式，大量节省了工程量和相关投资，保证了船闸的建设工期，形成了一整套大规模岩石开挖及高陡人工边坡稳定性预测和控制的岩石工程方法技术；它在建设完工后的整体综合质量评价表明工程质量优良，长期安全性能可靠，是一项成功的世界级岩石工程。在工程兴建过程中形成了一整套岩石高陡边坡工程的安全和质量的技术保障体系，包括工程超前预报、多种模型综合分析和变形预测、先进高效的监测系统、大型岩石开挖成型技术和岩体稳固的综合锚固排水技术，以及新型安全和质量管理模式和监控评价体系等。

2. 高难度的船闸施工技术

三峡船闸施工工程量大、工期紧、技术难度高。175米深切岩坡开挖，其下部直立开挖部分需作为船闸结构的组成部分，要求保持岩坡的强度和完整性，高薄衬砌墙混凝土浇筑、高大闸阀门设备的安装等施工难度均非一般船闸施工可比。针对复杂地质条件下高达68.5米直立岩坡的开挖、300吨级长达60米的水平锚索施工，对施工工序、直立坡成型、爆破控制、锚固的设备和器材，提出了成套工艺和技术要求，并分别提出了多种控制岩体质量的新技术和水平锚固工程的高精度施工工艺及技术标准。混凝土浇筑首次采用了已获国家专利的先进立模施工新技术。针对金属结构和设备安装提出了大型

V. Innovations to High Slope Construction and Construction Technologies for Ship Locks

1. Research into stability, deformation control and large lining structure of deep-cut, high, and steep slopes

Slope of the ship lock was high, steep and long. It was characterized by a large scale and complex form, and also by deep-cut excavation and unloading deformation after rock excavation took place, so the slope had to be stable enough for ships to pass through. This complex problem of a high-sloped ship lock was unprecedented domestically and internationally. It was necessary to maintain stability of the high slope rock mass during the construction and operation periods, and also to use the rock mass as an integral part of the ship lock's structure to support the function of lining structure. It was also necessary to investigate how sloped rock mass deformation would affect normal operation of ship lock equipment, particularly the miter gate. By using many advanced technologies for geological survey, conducting many types of field tests, using various models for calculation and analysis, and using the technologies including excavation, reinforcement, seepage control and dewatering, they successfully solved the problems related to stability and deformation of high slopes. On this basis, through a rationalized rock groove excavation method (retain the rock mass separating pier between two lanes of the ship lock) and ship lock structure, work quantities and investment were reduced by a large margin, while the ship lock was completed on schedule. A whole set of rock engineering methods and techniques were invented that could be used for large-scale rock excavation and prediction of slope stability of high and steep man-made slopes. According to a comprehensive quality assessment after completion, the project was deemed to be of excellent quality, reliably safe in the long-term and a successful world-class rock engineering work. During construction, a complete technical support system for safety and quality of high, steep-sloped rock engineering was invented, which encompassed advanced engineering prediction, comprehensive analysis and deformation prediction, an advanced efficient monitoring system, large rock excavation forming technology, integrated anchoring and dewatering technology for rock mass stabilization, and an innovative safety and quality management model, monitoring, and evaluation system.

2. Very difficult ship lock construction technology

The TGP ship lock construction involved large work quantities, tight schedule and a high level of technical difficulty. In 175 m of deep-cut rock slope excavation, the vertically excavated part in the lower section needed to be an integral part of the ship lock's structure,

人字门、阀门、设备安装的专用标准和安装工艺等,保证了船闸施工的质量和工期。

3. 高精度巨形人字门安装技术

高精度特大型船闸人字门安装技术：三峡船闸人字门的最大高度为38.5米,最大单扇门重850吨,人字门高度和单扇门重居当今世界之最,具有安装精度要求高、吊装难度大、安装部件复杂、测控难点多、焊接变形控制要求严等特点。经过对人字门的测量控制、焊接变形控制、背拉杆调试等方面进行专题研究,采取改装和加固吊耳、卸扣,吊机滑轮重组、平衡承重和两节连续吊装的新方案,提高了吊装安全系数和工效;研究了综合布置测量网点的办法,保证了测量放样精度;采用新的焊接工艺,保证了门体焊接质量。这些施工技术的综合应用,成功地解决了高精度、特大型人字门安装技术难题。

4. 船闸完建施工

三峡双线五级船闸为了适应三峡工程围堰发电期、初期和后期不同运行水位的需要,在设计上采用了第一、二级船闸底槛及相应闸门和启闭设备分两次建设的方案。船闸第一次建设只能适应水库水位在135.0～156.0米之间运行,为适应最终设计规模水库水位在145.0～175.0米之间运行,在2006年9月进行船闸完建工程施工。将一、二闸首及闸室底板抬高8米,拆除并重新定位安装顶底枢、承压条,以及浇筑闸首及底板混凝土。为满足施工,需将重850吨的人字门抬升悬空约80天,研究提出液压顶升、计算机控制钢丝绳悬吊、辅助支撑整体框架的方案。研究解决了老混凝土约束区的浇筑方法和工艺,选用低热水泥、取消宽槽等措施,成功解决了船闸完建施工过程中的一系列重大技术问题,确保了工程安全和质量、进度,为大型船闸设计、建设技术积累了丰富的经验,具有极强的推广应用价值。

so the rock slope had to retain its strength and integrity. Concrete placing in the wall with high thin lining, and installation of high and large gates and valves are much more difficult than with ordinary ship locks. For excavation of a 68.5 m-high vertical rock slope and installation of 60 m of 300 t horizontal anchor cable under complex geological conditions, a set of technological and technical requirements were imposed. Vertical slope forming, blasting control and anchoring equipment and materials were used, and multiple new technologies for rock mass quality control and high-precision construction and technical standards for horizontal anchoring works were established. For the first time, the new and advanced formwork process which was granted a national patent was used for concrete placing. For metal structure and equipment installation, specialized standards and installation processes were established regarding large miter gates, valve and equipment installation, which ensured ship lock construction was completed with high quality and on schedule.

3. High-precision installation technology for large meter gates

Installation of miter gates for the high-precision extra-large ship lock: the maximum height of the TGP ship lock miter gates was 38.5 m and the maximum weight of a single gate was 850 t, which makes them the tallest and heaviest single meter gates in the world. The ship lock's construction was characterized by high installation precision, highly difficult lifting, complex components, difficulties in measurement and control and strict requirements for limiting welding deformation. After special research into the issues including measurement and control of the miter gates, limiting welding deformation, and testing of back tie bars, a new solution was designed. In this solution, lifting lugs and shackles were modified and reinforced, crane pulleys were rearranged, load was balanced and two sections were lifted continuously, which increased the safety and efficiency of installation. A method for the complex arrangement of measurement points was developed to ensure the accuracy of surveying and setting-out. A new welding procedure was used to ensure quality gate body welding. The comprehensive application of these construction technologies successfully solved the technical problems with installation of high-precision, extra-large miter gates.

4. Ship lock completion

To adapt to various operating water levels during the cofferdam power generation period, in the initial stage and later stage of the TGP, the two-lane, five-stage ship lock was designed so that the embedded sills of the first and second ship locks and the corresponding gates and hoisting equipment were built in two stages. The ship lock completed in the first stage was only suitable for a 135.0~156.0 m water level in the reservoir. In order to support a 145.0~175.0 m water level in the final design, ship lock completion works were conducted

六、超大型升船机建造技术创新

1. 引进、消化吸收再创新、再完善

为进一步提高三峡升船机的安全性和可靠性,在充分借鉴和吸收国外在升船机建设方面的成功经验,三峡总公司委托国外相关研究机构对三峡升船机采用齿轮齿条爬升式方案进行了可行性研究。并组织众多单位对设计单位提交的设计成果的设计深度、完整性、正确性及技术经济合理性进行评价,对涉及升船机安全的综合性、系统性问题进行复核、分析和论证;开展了近30项专项科研和试验,进一步验证设计成果的安全可靠性与可实施性。经专家审查,2003年9月,国务院三峡建委第十三次全体会议批准,三峡升船机由"钢丝绳卷扬全平衡垂直提升式"升船机修改为"齿轮齿条爬升式"升船机。通过引进、消化吸收再创新,不断优化和完善设计方案,解决了升船机设计方案的安全性和可靠性问题。

2. 攻克齿条、螺母柱等关键部件制造技术难题

经过历时两年多时间的研制攻关,解决了驱动机构齿条、安全机构螺母柱等关键部件铸造工艺(ME级最高质量等级)和制造技术,攻克齿条表面淬火开裂这一世界级难题;掌握了132米×23米×10米(长×宽×高)巨型承船厢结构焊接变形控制及复杂机构装配制造技术;通过科学的项目组织管理,对驱动系统机械、液压、电控、检测等设备厂内整合组装,进行机电液联合调试,极大地缓解了现场安装质量和进度控制压力,保证工程建设顺利实施。三峡升船机大型部件制造达到国际领先水平,而且相关成果在向家坝升船机成功应用。

3. 解决特大型升船机施工系列技术难题

不断提升施工工艺和管理方法,攻克了140米高空混凝土梁系结构国内首次使用贝雷架方案浇筑工艺、齿条螺母柱等大型超高设备高精度安装、承船厢及其复杂设备安装,以及升船机系统集成调试等一系列技术难题;创造了168米高混凝土塔柱结构施工无裂缝、125米超高设备安装垂直度小于3毫

in September 2006. The lock head and lock chamber floor of the first and second ship locks were raised by eight meters, the top/bottom pintle and bearing plate were removed, repositioned and reinstalled, and a concrete placing lock head and floor was completed. During construction, the 850 t miter gate needed to be lifted and suspended for approximately 80 days, so a solution was formulated where hydraulic jacking was used, suspension steel wire rope was controlled by computer and the whole framework was given auxiliary supports. The concrete placing method was developed to solve the problems of old concrete confinement. Low heat cement was used and wide grooves were removed. These measures successfully solved a range of major technical problems in ship lock completion works, ensuring safety, high quality and timely completion of the works, bringing ample experience in design and construction of large ship locks, which are very worth popularizing and applying.

VI. Innovation of Construction Technology for Ultra-Large Ship Lifts

1. Import, assimilation, re-innovation and further improvement

To further enhance safety and reliability of the TGP ship lift, the CYTGPDC adequately learned from the successes in ship lift construction of foreign countries, and commissioned relevant foreign design institutes to conduct a feasibility study on the TGP gear-and-rack climbing ship lift. Many entities were engaged to assess the design depth, integrity and correctness and the technical/economic rationality of the design deliverables submitted by the designer, and to review, analyze and investigate the comprehensive and systematic issues concerning safety of the ship lift. Nearly 30 research programs and tests were conducted to further demonstrate the safety and implementability of the design deliverables. In September 2003, after the expert audit, the 13th plenary meeting of TGPCC gave the authorization to change the TGP ship lift from a steel wire, rope winched fully balanced vertical ship lift to a gear-and-rack climbing ship lift. Through import, assimilation and re-innovation of foreign technologies, they kept optimizing and improving the design scheme, and solved safety and reliability related problems in the ship lift's design scheme.

2. Solving technical manufacturing problems of the key components such as racks, nuts and studs

Through over two years of R&D efforts, they developed the casting process (ME-grade highest quality grade) and manufacturing technology for key components such as gear racks of the driving mechanism and nuts/studs of the safety mechanism, and solved world-class problems of quenching cracks on rack surfaces. They mastered the technologies for welding

米、承船厢全行程运行同步偏差小于 2 毫米等工程奇迹；解决了船厢设备与塔柱结构变形协调性这一影响三峡升船机能否正常运行的核心问题。

4. 形成齿爬式升船机设计、制造与安装调试全产业链技术标准

三峡升船机建设过程中，先后制定和颁布 30 余个技术标准，全面覆盖设备制造、安装调试及土建施工各主要工序，对检验项目、质量指标、检验状态、取样方法、试样制备、抽样比例、环境条件等都做出了详细规定；根据三峡升船机建设经验，国内陆续制定升船机设计规范，形成了升船机设计、制造、安装调试施工系列技术标准，填补了中国超大型齿爬式垂直升船机建造技术标准空白。三峡升船机的成功建成，标志着中国已掌握超大型升船机建设技术。

while controlling deformation of a 132 m×23 m×10 m (L×W×H) large ship lifting chamber structure and assembly, and manufacture of complex mechanisms. Through rational project organization and management, they completed in-factory assembly of the mechanical equipment, hydraulic equipment, electric control equipment and testing equipment of the driving system. They conducted mechanical-electrical-hydraulic testing, considerably relieved the pressure of onsite installation quality and progress control, and guaranteed successful implementation of the project. Manufacture of the TGP's ship lift is now at an advanced international level, and the technological achievements have been successfully applied with the Xiangjiaba ship lift.

3. Solving the technical problems in construction of extra-large ship lifts

They continued improving their construction process and management methods, and solved a range of challenging technical problems, including fist use of a Bailey truss in the concrete placing of a 140 m overhead concrete beam for the first time in China, high-precision installation of large extra-high equipment such as racks, nuts and studs, installation of ship lift chambers and their complex equipment, and integration and testing of ship lift systems. A 168 m high concrete pillar structure was built without cracking, 125 m extra-high equipment installed with verticality of <3 mm, and a ship lift chamber completed with total travel synchronization of <2 mm. These are nothing less than engineering miracles. They solved the problems in coordination between the ship lift chamber and pillar structure, which is a key issue that affects the ability of the TGP ship lift to work normally.

4. Establishment of technical standards for design, manufacture, installation and commissioning gear climbing ship lifts

During construction of the TGP ship lift, over 30 technical standards were developed and enacted. They covered the main processes, including equipment manufacture, installation and commissioning and civil works, and encompassed detailed stipulations regarding inspection items, quality indices, inspection status, sampling methods, sample preparations, sampling proportions and environmental conditions. Based on this experience, design specifications for ship lifts and the technical standards for designs, manufacture, and installation and commissioning have been established, which fills the gaps in technical standards for ultra-large gear climbing vertical ship lifts in China. Successful completion of the TGP ship lift signifies that China has mastered large ship lift manufacture and its accompanying technology.

── **本章小结：** ──

三峡工程在建设运行过程中，形成的科技成果获国家科技进步奖20多项、省部级科技进步奖200多项、专利数百项，创造了100多项"世界之最"。"长江三峡枢纽工程"项目获得了2019年度国家科学技术进步奖特等奖。三峡工程的创新成果和创新经验得到了实践的检验，使我国水利水电工程建设整体科技实力达到国际先进水平。三峡工程的建设标准、管理方法在国内外得到普遍的借鉴和推广，成为引领中国水电跨出国门、走向世界的标准和品牌。

参考文献：

［1］本书编委会. 百问三峡［M］. 北京：科学普及出版社，2012.

［2］李永安. 三峡工程资本运营探索与实践［M］. 北京：中信出版社，2014.

［3］《中国三峡建设年鉴》编纂委员会. 中国三峡建设年鉴（2001）［J］. 宜昌：中国三峡建设年鉴社，2001.

［4］《中国三峡建设年鉴》编纂委员会. 中国三峡建设年鉴（2006）［J］. 宜昌：中国三峡建设年鉴社，2006.

［5］国家电网公司. 中国三峡输变电工程·综合卷［M］. 北京：中国电力出版社，2008.

Chapter Summary:

The scientific and technological achievements obtained during construction and operation of the TGP have been granted over 20 prizes from the State Science and Technology Advancement Award, over 200 prizes from provincial/ministerial Science and Technology Advancement Award and hundreds of patents, and have set over 100 world's greatest records. The TGP was granted the Grand Prize of the 2019 State Science and Technology Advancement Award. The innovation achievements and experience of the TGP have withstood the test of time and pushed China's technological capacity of water conservancy and hydropower engineering to an internationally advanced level. The construction standards and management methods of the TGP have been popularized and applied domestically and internationally and have become the standards and brands that push the hydropower industry of China into the international arena.

References:

[1] Editorial committee of this book. *Questions about the Three Gorges* [M]. Beijing: Popular Science Press, 2012.

[2] Li Yongan. *Exploration and Practice of the Capital Operation on China Yangtze River TGP* [M]. Beijing: Citic Press, 2014.

[3] Editorial Board of *China Three Gorges Construction Yearbook. China Three Gorges Construction Yearbook* (2001) [J]. Yichang: China Three Gorges Construction Yearbook Press, 2001.

[4] Editorial Board of *China Three Gorges Construction Yearbook. China Three Gorges Construction Yearbook* (2006) [J]. Yichang: China Three Gorges Construction Yearbook Press, 2006.

[5] State Grid Corporation of China. *China Three Gorges Power Transmission and Transformation Works · General Volume* [M]. Beijing: China Electric Power Press, 2008.

阅读提示：

　　根据长江流域综合利用规划，三峡工程是治理开发长江的关键性骨干工程。它的主要开发任务是解决长江中下游，特别是荆江河段的防洪问题；向华中、华东和川东地区提供大量电力；改善宜昌至重庆河段以及中游的通航条件。

　　截至2021年7月，三峡工程已连续12年实现175米满蓄目标，发挥了防洪、发电、航运、水资源利用等巨大综合效益，有效保障了长江流域防洪安全、能源安全、航运安全、供水安全和生态环境安全，为推动长江经济带发展发挥了基础保障作用，在推动区域经济社会可持续发展、促进清洁能源产业升级和创新发展中发挥着不可替代的作用。

　　According to the plan for comprehensive utilization of the Yangtze River basin, the TGP is a backbone project for the control and development of the Yangtze River. Its primary tasks are: solve the flood control problem of the middle and lower reaches of the Yangtze River, particularly the Jingjiang River reach; supply a large quantity of electricity to Central China, East China and East Sichuan; improve navigation conditions in the reach between Yichang and Chongqing, and the middle reaches of the Yangtze River.

　　By July 2021, the TGP had accomplished its full water storage objective of 175 m for 12 consecutive years, brought massive comprehensive benefits in terms of flood control, power generation, shipping and water resources utilization, and effectively safeguarded the safety in flood control, energy, shipping, and water supply in the Yangtze River Basin. It also provided basic support for the development of the Yangtze River Economic Belt, and played a vital role in promoting sustainable regional socioeconomic development, and facilitating the upgrading and innovation of the clean energy industry.

Chapter 7 >>>>

第七章
三峡工程的运行效益
Operational Benefits of the TGP

第一节 三峡工程的防洪效益

三峡工程的首要效益是防洪。三峡工程初步设计中，防洪主要考虑对荆江河段进行补偿调度，提高荆江河段的防洪标准，避免荆江河段发生毁灭性洪灾。在这种调度方式下，三峡水库只拦蓄上游来水超过 55 000 立方米每秒的较大洪水。自 2003 年三峡水库蓄水运用以来，一方面，由于工程建设进展顺利，较初步设计提前实施了 156 米、175 米分期蓄水，防洪效益得以提前发挥；另一方面，在不降低对初步设计规定的大洪水防洪作用的前提下，三峡水库也充分发挥了对一般性洪水的防洪作用。可以说，三峡水库自蓄水运用以来，实现并拓展了初步设计的防洪效益。

一、三峡工程建设前的防洪情势和目标

1. 三峡工程建设前的防洪情势

三峡水库建设前，长江中下游防洪工程体系主要包括支流水库、蓄滞洪区以及堤防等。由于长江洪水量巨大，仅靠支流水库难以有效发挥滞洪错峰作用，而当时堤防尚未进行大规模加固，难以抵御大洪水；蓄滞洪区的启用，需要以局部付出巨大的淹没损失为代价，且人员转移、灾后重建等工作任务重，社会影响较大。因此一旦遭遇大洪水，长江中下游地区将损失惨重。1998 年长江大洪水以后，国家加大投入力度，对长江干堤进行加固、修缮，3 900 千米的长江干堤已经全线达标，干流堤防工程的抗洪能力从中华人民共和国成立初期的 3～5 年一遇提高到 10～30 年一遇，河道行洪能力大大增强。即便如此，如果不修建三峡工程，防洪形势依然紧张，具体有以下几方面。

第一，长江洪峰流量远大于河道安全泄量。根据有关资料，自 1153 年以

Section 1 | Benefits of Flood Control of the TGP

The primary benefit of the TGP is flood control. In the preliminary design of the TGP, the priority of flood control was to compensate the reach of the Jingjiang River, enhance flood control standards of the reach, and prevent any destructive floods in this reach. Under this flood dispatching strategy, the Three Gorges Reservoir (TGR) only stores water from large floods with an impoundment of >55,000 m^3/s from the upstream. Since the TGR was put into service in 2003, due to the smooth progress of the project construction, the 156-meter and 175-meter stages of water storage were implemented earlier than the preliminary design, and the flood control benefits were brought into play sooner than the preliminary design projected. In the meantime, the TGR also fulfilled its role in controlling general floods without compromising the major flood control function specified in the preliminary design. Undoubtedly, the TGR has realized and expanded the flood control benefits of the preliminary design since it was put into use.

I. Flood Control Situation and Objectives before Construction of the TGP

1. Flood control situation before construction of the TGP

Before construction of the TGR, the flood control system in the middle and lower reaches of the Yangtze River mainly consisted of the tributary reservoirs, flood storage and detention areas, and protective embankment. The Yangtze River has a huge amount of flood discharge, so the tributary reservoirs alone could not achieve flood detention and peak shifting. The protective embankments were not extensively reinforced and thus were not sufficient protection against major floods. If merely flood storage and detention areas were used, this would lead to disastrous inundation of local areas and necessitate the arduous tasks of relocation and post-disaster reconstruction, along with massive social impact. In this case, the middle and lower reaches of the Yangtze River would suffer great losses in the event of

来，宜昌流量超过 80 000 立方米每秒的有 8 次，城陵矶以上干流和洞庭湖的汇合洪峰流量在 1931 年、1935 年和 1954 年均超过 100 000 立方米每秒，而目前上荆江的安全泄量为 60 000～68 000 立方米每秒、城陵矶附近约 60 000 立方米每秒、汉口约 70 000 立方米每秒、湖口约 80 000 立方米每秒，洪水来量大与河道泄洪能力不足的矛盾十分突出。

第二，遭遇大洪水仍需启用蓄滞洪区。长江中下游安排了 40 处可蓄滞洪水 500 亿立方米的分蓄洪区。其中荆江地区 4 处；城陵矶附近区 25 处（洞庭湖区 24 处，洪湖区 1 处）；武汉附近区 6 处；湖口附近区 5 处（鄱阳湖区 4 处，华阳河区 1 处）。相对比较完善的分蓄洪区有荆江分洪工程和汉江杜家台分洪工程，这些工程可以发挥削减洪峰、蓄纳超额洪水、降低江河洪水位的作用。当时，长江流域蓄滞洪区仍存在分蓄洪设施不完善、区内安全建设滞后、政策法规不完善，以及缺少相应的控制手段、管理机构不健全、工程管理落后等诸多问题，要频繁使用蓄滞洪区是不现实的。

第三，荆江河段遇特大洪水可能发生毁灭性灾害。如遇 1860 年或 1870 年枝江洪峰流量达到 110 000 立方米每秒的特大洪水，荆江河段按允许泄量计算，超额洪峰流量约 50 000 立方米每秒，即使采用可能实施的各项分洪措施，超额洪峰流量还有 30 000～35 000 立方米每秒，会产生难以控制的破坏力量，在荆江南岸或北岸漫堤溃坑。不论向南漫溃入洞庭湖区，还是向北溃决进入江汉平原（其地面高程比南岸低 5～7 米），由于溃堤的洪水来势猛，水头高（北岸有十几米），洪水量大，对地区经济和社会安定产生难以估量的严重后果，发生毁灭性灾害。

2. 三峡工程建设前的防洪目标

根据长江中下游平原地区社会经济发展水平和经济发展布局的要求，本地区防洪目标是：

第一，荆江河段的防洪标准应不低于百年一遇，并能在遭遇类似 1870 年特大洪水时，配合分蓄洪措施，保证荆江河段行洪安全，防止南北两岸堤防漫溢，发生毁灭性灾害。

第二，长江中下游（荆江河段以下，包括洞庭湖区）以防御 1954 年型洪水为目标，在各类防洪工程设施配合运用下，确保武汉等大中城市和重要平

major flood. Since the major flood of the Yangtze River in 1998, China invested more in reinforcement and renovation of the main dikes along the river. Its main dikes with a length of 3,900 km all met the standards, and the flood control capacity of the embankments along the trunk stream increased from 3 to 5 years to 10 to 30 years, and the flood discharge capacity of the river increased significantly. Even so, the flood control situation was still worrying without the TGP, as can be seen in the following facts.

First, the peak discharge of the Yangtze River is far greater than the safe discharge of the river reach. Since 1153, according to statistics, floods with > 80,000 m^3/s happened eight times at Yichang, and the combined peak flow of the main trunk system above Chenglingji and Dongting Lake exceeded 100,000 m^3/s in 1931, 1935 and 1954. Back then, the safe discharge was 60,000~68,000 m^3/s in the upper reach of Jingjiang River, approximately 60,000 m^3/s near Chenglingji, approximately 70,000 m^3/s at Hankou, and approximately 80,000 m^3/s at Hukou. There was a sharp disparity between the flood volume and flood discharge capacity of the river.

Second, the flood storage and detention areas still had to be used in the event of major flooding. There were 40 flood diversion and storage areas with 50 billion m^3 of flood storage and detention capacity along the middle and lower reaches of the river. Four of them were in the Jingjiang River reach, 25 near Chenglingji (24 at Dongting Lake and 1 at Honghu Lake), six near Wuhan, and five near Hukou (four at Poyang Lake, one at Huayang River). The relatively well-functioning flood diversion and storage areas included the Jingjiang River Flood Diversion and Hanjiang River Dujiatai Flood Diversion Projects. These projects could shave flood peaks, impound excess flood waters and lower flood levels of the rivers. At that time, the flood storage and detention areas of the Yangtze River basin were still troubled by problems such as inadequate flood diversion and storage facilities, outdated security systems, deficient policies and laws, lack of control methods, deficient management organization and an outdated project management model. Thus, frequent use of the flood storage and detention areas was impractical.

Third, a catastrophic flood in the Jingjiang River reach would wreak destructive consequences. If a catastrophic flood with 110,000 m^3/s peak discharge happened in Zhijiang River, like what happened in 1860 or 1870, there would be an excess peak discharge of 50,000 m^3/s after the permissible discharge capacity of the Jingjiang River reach was used for calculation and still an excess peak discharge of 30,000~35,000 m^3/s after all possible flood diversion measures were taken. This would be an unmitigated disaster which would overwhelm the embankments on the south or north of the Jingjiang River. The flood would flow either southwest into the Dongting Lake or northward into the Jianghan Plain (ground elevation is 5~7 m lower than the south bank). Because of its high head (a dozen meters on the

原区的安全。

第三，随着干支流水库和堤防的建设，逐步减少分蓄洪区的使用机会。

二、三峡工程的防洪效益

1. 蓄水以来的防洪效益

截至2021年，三峡水库累计拦洪63次，总蓄洪量1 900多亿立方米，保障了长江中下游安澜。2010年、2012年、2020年入库最大洪峰均超过70 000立方米每秒，经过水库拦蓄，削减洪峰约40%，极大地减轻了长江中下游地区的防洪压力。特别是2020年75 000立方米每秒峰值的洪水，更是刷新了三峡工程建库以来最大洪峰的纪录。

2010年7月20日，三峡水库入库洪峰流量70 000立方米每秒，最大控泄流量40 000立方米每秒，最大削减洪峰流量30 000立方米每秒，削减洪峰达到40%多，拦蓄洪水水量76亿立方米，约占本次入库洪水总量的24.5%。

图7.1　三峡大坝泄洪气势如万马奔腾

north bank) and large volume, the flood breaching the dam would be devastating and cause unquantifiable damage to regional economic and social stability, no matter which direction it flows.

2. Flood control objectives before construction of the TGP

Based on the level of socioeconomic development and the economic development strategy for the middle and lower reaches of the Yangtze River, the flood control objectives of this region were as follows:

First, the flood control standard of the Jingjiang River reach should be no less than for a once-in-a-century flood; in the event of a flood similar to the catastrophic flood that happened in 1870, flood diversion and storage measures should be taken to help ensure flood discharge safety of the Jingjiang River reach, and to prevent a destructive disaster caused by dam breach and inundation on the south and north banks.

Second, the middle and lower reaches of the Yangtze River (downstream the Jingjiang River reach, including the Dongting Lake area) should be readily protected from floods similar to the flood that happened in 1954, and the flood control facilities should be used to protect the safety of large and medium-sized cities and the important plain areas.

Third, construction of the trunk stream/tributary reservoirs and embankments will gradually reduce the use of flood diversion and storage areas.

II. Benefits of Flood Control of the TGP

1. Benefits of flood control since the TGR was put into use

By 2021, the TGR had accomplished flood retention 63 times and its total flood storage volume was over 190 billion m^3, thus protecting safety of the middle and lower reaches of the Yangtze River. In 2010, 2012 and 2020, the maximum peak inflow discharge exceeded 70,000 m^3/s. Impoundment by the reservoirs shaved the flood peak by approximately 40%, and immensely relieved flood control pressure of the middle and lower reaches of the Yangtze River. Particularly, the 75,000 m^3/s flood peak in 2020 renewed the record of maximum flood peak since completion of the TGP.

On July 20, 2010, the peak inflow discharge of the TGR was 70,000 m^3/s, maximum controlled discharge was 40,000 m^3/s, maximum shaved peak discharge 30,000 m^3/s, with its flood peak shaved by over 40%, and impounded flood volume was 7.6 billion m^3, which

◒ Figure 7.1   Magnificent View of a Flood Discharge at the Three Gorges Dam

三峡水库在此次洪水拦蓄过程中，降低荆江河段沙市站洪水位约 2.5 米，洪湖江段 1 米，使得长江中下游河段特别是沙市水位站和武汉汉口水位站没有超过警戒水位，为下游防洪节约了大量的人力和物力。据估算，2010 年汛期三峡工程产生的防洪经济效益为 266.3 亿元。

2012 年 7 月 24 日，三峡水库遭遇成库以来的最大洪峰流量 71 200 立方米每秒。在本次洪水调度过程中，三峡最大控泄流量 45 000 立方米每秒，最大削减洪峰 28 200 立方米每秒，削峰率 41%。避免了荆南四河超过保证水位，控制下游沙市站水位未超过警戒水位，城陵矶站水位未超过保证水位，保证了长江中下游的防洪安全，同时也大大减少了下游江段上堤查险的时间和频次，节约了中下游防洪成本。

2020 年 8 月，长江上游发生一次复式洪水过程，三峡水库相继出现编号为第 4 号、第 5 号的连续洪水，洪峰流量分别为 62 000 立方米每秒（8 月 15 日 8 时）、75 000 立方米每秒（8 月 20 日 8 时）。4 号、5 号洪水期间，三峡水库最大控泄流量分别为 41 500 立方米每秒、49 400 立方米每秒，削峰率分别达 33%、34%。三峡水库坝前水位最高蓄至 167.65 米，拦蓄洪水 108 亿立方米，分别降低宜昌至莲花塘江段、汉口至大通江段水位 2~4 米、1.1~2 米，长江干流沙市水位低于保证水位 1.76 米，避免了荆江分洪区运用和转移 60 余万人、淹没耕地 49.3 万亩。2010 年、2012 年和 2020 年防洪效益巨大。

2. 中小洪水调度

2009 年 8 月 6 日 8 时，三峡水库遭遇洪峰流量为 55 000 立方米每秒的入库洪水过程。若不进行拦洪调度，荆江干流河段将超过警戒水位。应湖北省防汛抗旱指挥部办公室要求，三峡水库首次对中小洪水进行了滞洪调度尝试。本次拦洪调度，三峡水库最大出库流量 40 000 立方米每秒，取得了显著的防洪效益：一是避免了荆江河段高洪水位。如三峡水库不拦蓄洪水，湖北长江宜昌至监利段和荆南四河水位将全线超过警戒水位，共有 1 500 多千米堤段要按照警戒水位布防巡查。三峡水库控泄后，仅长江干流监利段 20.7 千米和荆南四河 649.8 千米水位超过设防水位。二是降低了响应级别。按照防汛抗旱应急预案要求，如三峡水库不拦蓄，湖北长江和荆南四河出现超过警戒水位的洪水，应启动防汛三级应急响应，经三峡水库控泄后，长江监利段和荆南四

was approximately 24.5% of the reservoir's inflow flood that day. During flood retention by the TGR, the flood level at the Shashi Station of the Jingjiang River reach was reduced by approximately 2.5 m and the Honghu Lake reach by one meter, the water level at the middle lower reaches of Yangtze River, particularly at the Shashi Water Stage Gauging and Hankou Water Stage Gauging Stations, did not exceed the warning level, and large amount of manpower and materials for flood control in the lower reach were conserved. During the flood season of 2010, as estimated, the related economic savings generated by the TGP's flood controls was RMB 26.63 billion.

On July 24, 2012, the TGR experienced peak discharge of 71,200 m^3/s, the greatest since its completion. During dispatching of this flood, the maximum controlled discharge of TGP was 45,000 m^3/s and maximum shaved flood peak was 28,200 m^3/s, with a peak shaving rate of 41%. This ensured that the four rivers in the Jingnan region did not exceed its maximum safe water level, and the water at the downstream Shashi Station did not exceed its warning water level. Also not exceeding its maximum safe water level was the Chenglingji Station. Therefore, the flood safety of the middle and lower reaches of the Yangtze River was protected, the time and frequency of embankment inspection of the downstream reaches was substantially reduced, and thus the cost of flood control of the middle and lower reaches was reduced.

In August 2020, a compound flood occurred in the upper reach of Yangtze River, and floods No. 4 and No. 5 appeared at the TGR, with peak discharges of 62,000 m^3/s (8:00, August 15) and 75,000 m^3/s, (8:00, August 20) respectively. During floods No. 4 and No. 5, the maximum controlled discharges of the TGR was 41,500 m^3/s and 49,400 m^3/s respectively, with 33% and 34% peak shaving rates, respectively. The TGR dam's front water level was at a maximum of 167.65 m, with an inflow of 10.8 billion m^3, the water level was reduced by 2-4 m in the Yichang-Lianhuatang reach and 1.1-2 m in the Hankou-Datong reach, and the water level at the Shashi Station in the trunk stream of the river was below the maximum safe water level of 1.76 m. Therefore, use of the Jingjiang River flood diversion area and relocation of over 600,000 people were avoided, and 493,000 *mu* (1 hm^2 = 15 *mu*) of cultivated land was protected from inundation. Massive benefits from flood controls were realized in 2010, 2012 and 2020.

2. Minor-moderate flood dispatching

At 8:00 A.M., August 6, 2009, the TGR experienced an inflow flood with 55,000 m^3/s of peak discharge. Without flood retention and dispatching, the water level in the trunk stream reach of the Jingjiang River would have exceeded the warning water level. As requested by the Office of Flood-Control and Drought-Fight of Hubei Province, the first flood detention of

河实际仅超出现设防水位洪水,只需启动防汛四级应急响应。三是避免了紧张态势。如果出现超过警戒水位的洪水,尚未达标的荆江大堤难免出险,尤其是堤基差、标准低的荆南四河更难免险情多发,抗洪抢险态势的紧张在所难免。三峡水库实施控泄调度后,湖北长江和荆南四河在设防水位期间,没有发生一处险情。

根据2009年中小洪水调度经验以及相关研究成果,2010年至2021年,三峡水库多次对中小洪水进行拦蓄,有效减轻了长江中下游干支流的防洪压力。

3. 城陵矶防洪补偿调度

对城陵矶进行防洪补偿调度,目的是在保证荆江河段遇特大洪水时防洪安全前提下,尽可能提高三峡水库对一般洪水的防洪作用,这种方式能获得较大的多年平均防洪效益,但也可能使荆江地区防洪标准有所降低。为稳妥起见,初设审查的结论意见为三峡工程主要采用对荆江河段的补偿调度方式作为三峡水库的设计调度方式计算防洪效益,并指出要继续研究城陵矶补偿调度方式。

the TGR was attempted for small and medium-sized floods. During this flood detention, the TGR's maximum outflow discharge was 40,000 m^3/s, and significant flood control benefits were realized. First, high flood level was averted in the Jingjiang River reach. If the TGR had not impounded the flood, the water levels at the Yichang-Jianli reach of the Yangtze River in Hubei and the four rivers in Jingnan region would have exceeded the warning water level, and over 1,500 km of embankments would have been patrolled according to the warning water level. After the controlled discharge of the TGR, only the water levels at the 20.7 km point of the Jianli reach of the Yangtze River's trunk stream and at the 649.8 km point of the four rivers in the Jingnan region exceeded the protection water level. Second, the response level was lowered. According to the emergency response plan for flood control and drought control, if the TGR had not impounded the flood water, the Yangtze River reach in Hubei and the four rivers in the Jingnan region would have seen water levels above the protection water level and level three emergency response for flood control would have been activated. Thanks to controlled discharge of the TGR, the water levels in the Jianli reach of the Yangtze River and the four rivers in the Jingnan region only exceeded the existing protection water level and only level four emergency response for flood control was activated. Third, a stressful situation was averted. If the flood had exceeded the warning water level, the subpar Jingjiang levee would have been at risk, particularly the four rivers in the Jingnan region, which had weak foundations and low performance, would have been exposed to more risks. Thanks to controlled discharge by the TGR, the water levels of the Yangtze River reach in Hubei and the four rivers in the Jingnan region remained within the protection water levels and there were no hazards.

Given the experience in dispatching small and medium-sized floods in 2009 and the related research results, the TGR has intercepted small and medium-sized floods many times in 2010 and 2021, effectively reducing flood control pressure on the trunk stream and tributaries in the middle and lower reaches of the Yangtze River.

3. Flood control compensation at Chenglingji

Flood control compensation at Chenglingji was intended to maximize the ability of the TGR to control general floods, under the premise that flood safety was maintained in the event of severe flooding in the Jingjiang River reach. This approach generated major multi-year benefits from flood controlling, but it also lowered flood control standards of the Jingjiang region. Out of great prudence, the review on preliminary design concluded that, in terms of flood dispatching design of the TGR, the flood control benefit of the TGP should be calculated by examining its function of compensation for the Jingjiang River reach, and the review highlighted that the compensation design for Chenglingji should be further

三峡工程蓄水运用以后，围绕防洪任务，充分考虑长江中下游江湖关系的变化，并结合水库泥沙淤积和水库回水等的分析，对三峡工程在上游溪洛渡、向家坝等枢纽建成发挥防洪作用前的防洪调度方式做了进一步的优化研究。其中，对城陵矶防洪补偿调度方式，将三峡水库防洪库容进行了重新划分，自下而上划分的三部分库容及运用方式。其中分别为：第一部分库容56.5亿立方米用于对荆江和城陵矶地区同时防洪补偿，相应库容蓄满的库水位即"对城陵矶防洪补偿控制水位"为155.0米；第二部分库容125.8亿立方米用于对荆江地区防洪补偿，相应库容蓄满的库水位即"对荆江防洪补偿控制水位"为171.0米；第三部分库容39.2亿立方米用于防御上游特大洪水。在此基础上拟定了主要适用于三峡水库试验性蓄水期的《三峡水库优化调度方案》，并于2009年获得国务院批准。随着2012年、2013年向家坝、溪洛渡水库相继建成投运，又研究提出了溪洛渡、向家坝配合三峡水库联合防洪调度后，三峡水库对城陵矶补偿水位可进一步提高。

2016年汛期，三峡水库最大入库洪峰为50 000立方米每秒，出现在7月1日，为长江1号洪峰。三峡控制出库流量31 000立方米每秒，削减洪峰19 000立方米每秒，削峰率38%。7月3日，长江2号洪峰在中下游形成，长江干流监利以下全线超过警戒水位，城陵矶水位直逼保证水位。为避免城陵矶超保证水位，减轻长江中下游的防洪压力，三峡水库在上游水库的配合下首次实施了典型的城陵矶防洪补偿调度，出库流量从31 000立方米每秒进一步减少至25 000立方米每秒、20 000立方米每秒，分别降低荆江河段、城陵矶附近地区、武汉以下河段水位0.8～1.7米、0.7～1.3米、0.2～0.4米，减少超过警戒水位堤段长度250千米。如果没有以三峡为核心的水库群联合调度，长江1号洪峰将与长江中下游形成的2号洪峰遭遇叠加，长江中下游干流枝城以下江段水位将全线超过警戒水位并延长超过警戒水位时间，城陵矶河段水位将两次超过保证水位，最高水位达到约35米，城陵矶地区将不可避免地分洪。

2016年汛期调度实践表明，三峡水库在保证遇特大洪水时荆江河段防洪安全前提下，实施对城陵矶防洪补偿调度，提高了对一般洪水的防洪作用，对减轻长江中下游防洪压力，尤其是减少城陵矶附近区的分洪量，提高防洪

investigated.

After the TGP started impoundment, the changes in the relation between the rivers and lakes in the middle and lower reaches of the Yangtze River were adequately examined, with the flood control function being the priority. Besides, with consideration of the analysis on sediment accumulation in the reservoir and reservoir backwater, the flood control dispatching method of the TGP before completion of the hydroprojects at the upstream Xiluodu and Xiangjiaba were further investigated and optimized. Given the flood control compensation method for Chenglingji, the TGR's flood control capacity was redivided and its storage capacity broken down into three parts. The first part of storage capacity was 5.65 billion m^3, which was used for flood control compensation for both the Jingjiang and Chenglingji regions. The full water level of the corresponding storage capacity was the control water level for flood control compensation for Chenglingji, that is, 155.0 m. The second part of storage capacity was 12.58 billion m^3, which was used for flood control compensation of the Jingjiang region. The full water level of the corresponding storage capacity was the control water level for flood control compensation for Jingjiang, that is, 171.0 m. The third part of storage capacity was 3.92 billion m^3, which was used for protection against upstream catastrophic flooding. On this basis, the *Optimized Regulation Plan of TGR* was formulated. It was mainly applicable to the trial storage period of TGR and was approved by the State Council in 2009. As the Xiangjiaba and Xiluodu reservoirs were completed and commissioned in 2012 and 2013, a new solution was proposed, where the Xiluodu and Xiangjiaba reservoirs would be used to assist the TGR in flood control dispatching. So the compensation water level of the TGR used for Chenglingji was further increased.

During the flood season in 2016, the maximum peak inflow discharge of the TGR was 50,000 m^3/s, which occurred on July 1 and was designated Changjiang Peak 1. The controlled outflow discharge of TGP was 31,000 m^3/s, with a shaved peak 19,000 m^3/s, and a peak shaving rate of 38%. On July 3, Changjiang's Peak 2 occurred in the middle and lower reaches of the Yangtze River. The trunk stream reaches of the Yangtze River downstream from Jianli all exceeded the warning water level, and the water level at Chenglingji was dangerously close to the maximum safe water level. In order to ensure the water level at Chenglingji would not exceed the maximum safe water level and to relieve pressure on flood control of the middle and lower reaches of the Yangtze River, the TGR, with the help of the upstream reservoirs, executed its first typical Chenglingji flood control compensation. The outflow discharge decreased from 31,000 m^3/s to 25,000 m^3/s and then to 20,000 m^3/s. The water levels in the Jingjiang River reach, the area near Chenglingji and the reach downstream of Wuhan were reduced by 0.8~1.7 m, 0.7~1.3 m and 0.2~0.4 m respectively. The embankment where the warning water level was exceeded was reduced by 250 km. Without the coordinated dis-

经济效益,确保人民群众生命财产及长江干堤和重要基础设施的安全大有益处。

第二节 三峡工程的发电效益

一、发电效益

1. 电力生产地位突出

2009年是三峡工程从建设阶段全面转入运行阶段的第一年。三峡工程初步设计建设任务(除批准缓建的升船机外)如期完成,汛后三峡水库试验性蓄水至高程171.4米,全面发挥防洪、发电、航运、生态补水等综合效益。根据中国电力企业联合会发布的2010年中国电力工业统计快报,截至2010年年底,全国发电装机容量9.62亿千瓦,其中水电装机容量2.13亿千瓦;2010年全国发电量42 280亿千瓦时,其中水电发电量6 863亿千瓦时,占总发电量的16.23%。

自2009年三峡水电站左、右岸26台机组1 820万千瓦全部投入运行,三峡水电站装机容量占2010年全国发电装机容量的1.89%,占当年全国水电装机容量的8.54%。在发电量方面,2010年三峡水电站全年发电843.7亿千瓦时,占当年全国发电量的2%,占当年全国水电发电量的12.29%。

patching of the reservoir group centered around the TGP, Changjiang's Peak 1 would have joined its Peak 2, which occurred in the middle and lower reaches of the Yangtze River. The river reach downstream Zhicheng, which was at the trunk stream of middle/lower reaches of Yangtze River, would have completely exceeded the warning water level and the duration of such a situation would have been prolonged. The water level of Chenglingji reach would have exceeded the maximum safe water level twice, the maximum water level would have been approximately 35 m, and consequently flood diversion in the Chenglingji region would have been unavoidable.

As proven by practical flood dispatching in the flood season of 2016, the TGR, under the premise that flood safety of the Jingjiang river reach is ensured in the event of severe floods, the Three Gorges Reservoir implemented the flood control compensation dispatching in Chenglingji. This measure improved the flood diversion volume on general floods and reduced the flood control pressure in the middle and lower reaches of the Yangtze River, especially the flood control, protecting the safety of people's lives and property, and important infrastructure.

Section 2 | Power Generation Benefits of the TGP

I. Power Generation Benefits

1. Prominence in power production

In 2009, the TGP was put in full operation after completion of construction. The construction tasks specified in the preliminary design of the TGP (except that the ship lift was delayed as approved) were completed on schedule. In post-flood trial impoundment, the TGR's water impoundment reached a level of 171.4 m, and the reservoir fully accomplished its objectives, including flood control, power generation, shipping and ecological water compensation. As of the end of 2010, according to the China electric power industry statistics bulletin published by the China Electricity Council in 2010, the total installed capacity of the country was 962 million kW, including 213 million kW installed capacity of hydropower. In 2010, the national total energy output was 4.228 trillion kWh, including 686.3 billion kWh of hydropower, which accounted for 16.23% of total energy output.

由此可见，三峡水电站在我国电力生产中地位突出，2010年三峡水电站发电量占全国水电发电量的比例高达12.29%，体现了三峡工程在我国水电资源开发中的突出地位。

自2003年首批机组投产发电以来，三峡水电站发出的强大清洁电能源源不断地输送至华中、华东和南方10省市，为半个中国提供了强大动能，为我国经济社会发展注入了强劲动力。

2. 梯级电站补偿效益明显

三峡水库具有不完全年调节能力，葛洲坝作为其反调节水库，具有日调节能力。三峡水电站和葛洲坝水电站共同组成了三峡梯级电站，这是当时我国最大的梯级水电站群。三峡水电站和葛洲坝水电站皆由长江电力统一经营管理，具有实现梯级联合调度运行的优越条件，可充分利用水能资源，增加梯级发电量，保证下游航运。

葛洲坝水电站为径流式水电站，1989年开始投产，总装机21台，装机容量271.5万千瓦，保证出力76.8万千瓦，是华中电网的骨干电源。三峡工程建成以后，保证出力提高到158万～194万千瓦，年发电量可提高到161亿千瓦时，梯级补偿效益明显。2019年年底，《三峡（正常运行期）—葛洲坝水利枢纽梯级调度规程》（2019年修订版）顺利获批，首次运用成效显著，流域梯级电站全年节水增发98.56亿千瓦时。

2014年，位于金沙江下游的溪洛渡、向家坝两座大型电站全部机组投产发电，与三峡水电站、葛洲坝水电站形成"四库联调"，进一步扩大了三峡梯级电站的规模。2020年，三峡、葛洲坝、溪洛渡、向家坝四座梯级电站全年累计发电量达2 269.30亿千瓦时，创历史新高。

Since 2009 when the 26 units with 18.2 million kW capacity on the left and right banks were commissioned, the installed capacity of TGHS accounted for 1.89% of the 2010 total national installed capacity, and 8.54% of the installed national hydropower capacity. In terms of energy output, the TGHS generated 84.37 billion kWh in 2010, which accounted for 2% of the total national energy output for the year and 12.29% of the national hydropower energy output.

As can be seen, the TGHS is very prominent in power production for the country. In 2010, the energy output of the TGHS accounted for 12.29% of the national hydropower energy output, which illustrates the prominence of the TGP in hydropower resources development of China.

Since 2003 when the first set of units were put into operation, the abundant clean energy generated by the TGHS was steadily transmitted to the 10 provinces and cities in Central China, East China and South China, which afforded strong energy support for half of China and injected strong impetus for the socio-economic development of China in general.

2. Significant compensation benefits of the Cascade Hydropower Station

The TGR has incomplete annual regulation capacity, and the Gezhou Dam, as a re-regulating reservoir for the former, has daily regulation capacity. The TGHS and Gezhouba Hydropower Station constitute the Three Gorges Cascade Hydropower Station, which was the largest cascade hydropower station group in China. The TGHS and Gezhouba Hydropower Station, which are centrally operated and managed by Yangtze Power, are able to achieve cascade combined dispatching and operation. By adequately utilizing hydropower resources, cascade energy output can be increased and downstream shipping can be ensured.

The Gezhouba Hydropower Station is a runoff hydropower station commissioned in 1989. It has installed capacity of 21 generator units, 2,715,000 kW, and 768,000 kW of firm output. It is the backbone power source for the Central China Grid. After completion of the TGP, the firm output increased to 1.58~1.94 million kW, and annual energy output increased to 16.1 billion kW, which indicates significant compensation benefit. By the end of 2019, the *Three Gorges (Normal Operation Period)-Gezhouba Hydroproject Cascade Dispatching Procedure* (2019 revision) was approved and its first application brought remarkable results. The Cascade Hydropower Station in the basin increased energy output by 9.856 billion kWh and increased water conservation.

In 2014, the large power stations at Xiluodu and Xiangjiaba in the lower reach of the Jinsha River had all of their generator units commissioned. Therefore, together with the TGHS and Gezhouba Hydropower Station, the four-reservoir combined dispatching system

二、联网效益

1. 输电效益明显

三峡水电站送电目标区域为华中、华东和南方电网。自三峡工程2003年发电以来至2009年年底，通过三峡输变电工程的输送，华中电网累计消纳三峡水电站电量1 292.32亿千瓦时，华东电网累计消纳三峡水电站电量1 551.72亿千瓦时，南方电网累计消纳三峡水电站电量813.91亿千瓦时。

通过输电半径覆盖上千千米的三峡输变电工程，源源不断的三峡电力，就像奔腾不息的江水，输送到华中、华东、华南和西南地区，惠及湖北、河南、江苏、广东、重庆等9省2市。三峡水电站为这些地区的经济发展日夜输送着能量，大大缓解了这些地区电力供应紧张局面，为国民经济发展提供了强有力的能源支撑。

由此可见，三峡水电站巨大的发电量为三峡输变电工程实现其经济效益创造了前提条件，依托三峡水电站，三峡输变电工程输电收入保证性高，输变电工程取得了良好的经济效益。

2. 联网效益突出

三峡水电站地处我国中部，它所供电的华中、华东和川东地区，供电距离都在400～1 000千米的经济输电范围以内。三峡水电站具有快速启停机组、迅速自动调整负荷的良好调节性能，可提供最高达1 200万千瓦的调峰容量，为电力系统的安全稳定运行提供可靠保障。

三峡工程和三峡输变电工程的建成，提高了水电在全国电力能源中的比重，提高了电网运行的安全可靠性，基本形成了全国联网的格局。按计划，三峡工程全部建成后总装机容量达到2 250万千瓦。得天独厚的地理位置和巨大的发电装机容量配置，使三峡水电站工程在全国大联网格局中发挥着电网支撑、输电枢纽等重大作用。

除直接的发电效益外，三峡工程通过大的跨区域联网项目取得较好的利用季节性电能、跨流域补偿调节、水火电互补调节、降低备用容量、送电量等联网效益和补偿调节效益。有关研究表明，仅华中、华东两大电网通过三

was established, which further enlarged the Three Gorges Cascade Hydropower Station. In 2020, the four cascade hydropower stations at Three Gorges, Gezhou Dam, Xiluodu and Xiangjiaba achieved annual energy output of 226.93 billion kWh, hitting a record high.

II. Benefits of Grid Connection

1. Significant power transmission benefits

The electricity from the TGHS is transmitted to the Central China Grid, East China Power System and China Southern Power Grid. From 2003 when the TGP started power generation to the end of 2009, thanks to power transmission from the Three Gorges Power Transmission and Transformation Works, the Central China Grid consumed a total of 129.232 billion kWh of electricity from the TGHS, the East China Power System 155.172 billion kWh and China Southern Power Grid 81.391 billion kWh.

Through the Three Gorges Power Transmission and Transformation Works, which have a transmission radius up to a thousand kilometers, abundant electricity is steadily transmitted to Central China, East China, South China and Southwest China, benefiting nine provinces and two municipalities including Hubei, Henan, Jiangsu, Guangdong and Chongqing. The TGHS keeps providing energy 24 h a day to support economic development of these regions, substantially relieving power shortages in these regions, and affording strong energy support for national economic development.

As can be seen, the massive energy output of the TGHS is the prerequisite for the economic benefit achieved by the Three Gorges Power Transmission and Transformation Works. Backed by the TGHS, the Three Gorges Power Transmission and Transformation Works can steadily generate revenue from power transmission and bring about valuable economic benefits.

2. Significant benefits of grid connection

Located in the central region of China, the TGHS supplies power to Central China, East China and East Sichuan, which are all within a 400~1,000 km economic transmission radius. The TGHS has excellent regulation performance, supported by its fast startup-shutdown units and its capability of quickly adjusting load. Its maximum peak regulation capacity is 12 million kW, which affords reliable support for safe and stable operation of the power system.

The TGP and Three Gorges Power Transmission and Transformation Works increase the proportion of hydropower in the national electrical energy system and enhance safety and reliability of power grid operation. A nation-wide interconnected grid has largely taken shape.

峡工程联网每年可取得300万~400万千瓦的错峰效益。

图7.2　三峡水电站工程在全国大联网格局中作用巨大

丰水期将三峡和华中富余水电输送到华东和华中电网，促进了清洁能源消纳，枯水期将华东电力输送到华中电网，实现了丰枯互济。截至2010年年底，跨区累计输送三峡和华中富余水电4 903.69亿千瓦时，大电网跨区资源优化配置作用得以充分体现。通过电网互联，实现了网间备用容量共享、错峰送电和紧急支援，仅2011年迎峰度夏期间，在华中、华东区域电网内机组故障跳闸、紧急停运等情况下，华中、华东跨区直流事故支援11次，最大支援电力280万千瓦，确保了电网的安全稳定运行。通过三峡至常州、三峡至上海及三峡至广东直流工程可分别向华东和南方电网输送电力600万千瓦时和300万千瓦时，提高了网络安全裕度，有效缓解了该地区电力供应紧张局面，特别在上海世博会、广州亚运会等重大活动期间发挥了重要作用。

2020年冬季，多轮寒潮来袭，全国多地遭遇罕见的低温雨雪天气。长江电力受电区域用电需求量持续走高，局部地区电力负荷甚至超过夏季峰值，

According to the plan, the total installed capacity of the TGP would be 22.5 million kW once completed. Due to its uniquely favorable geographical location and its massive installed capacity, the TGHS functions as the grid support and transmission hub in the nation-wide interconnected grid.

In addition to the direct benefit of power generation, the TGP, by means of large trans-regional grid connection projects, generates grid connection and compensation regulation benefits such as efficient utilization of seasonal electrical energy, cross-basin compensation dispatching, mutual complementation between thermal power and hydropower, reduced spare capacity and power transmission. According to relevant research, the Central China Grid and East China Power System alone can achieve three to four million kW of peak shifting each year through grid connection to the TGP.

 Figure 7.2 The TGHS plays a critical role in the Nationwide Interconnected Grid.

In the wet season, the surplus hydropower of the TGP and Central China is transmitted to the East China Power System and Central China Grid, which helps consume clean energy. In the dry season, the electrical energy from East China is transmitted to the Central China Grid. In this way, the grids support each other in wet season and in dry season. By the end of 2010, the trans-regional transmission of surplus hydropower between the TGP and Central China Grid was 490.369 billion kWh, which adequately illustrates how the large power grid achieves trans-regional optimized allocation of resources. Through grid interconnection, the power grids achieve spare capacity sharing, off-peak transmission and emergency support. In the peak period of summer 2011 alone, the regional power grids in Central China and East China executed DC emergency support 11 times and the maximum transmitted power was 2.8 million kW even when their own generators, which help maintain safe and stable operation of the grids units, experienced fault-induced tripping and emergency shutdowns. Through the DC projects from the TGP to Changzhou, from the TGP to Shanghai and from the TGP to Guangdong, the six million kWh and three million kWh of electricity is transmitted to East China Power System and China Southern Power Grid respectively. This increases the safety margin of the grids and effectively relieves power shortages in those regions. The transmitted power also plays a key role in important events such as the Shanghai World Expo and the Guangzhou Asian Games.

In the winter of 2020, the country was hit by rare cold, rainy and snowy weather induced by multiple cold waves. The regions receiving electricity from Yangtze Power saw a continued increase in power demand. The power load in certain regions even exceeded the summer peak, and the energy supply was put under great pressure. To maintain the safety of the grids, afford sufficient heat supply for the people and meet the power demands of

能源保供的压力和挑战巨大。为全力保障电网安全、确保居民温暖度冬、满足居民生活和工业用电需求，长江电力与国家电网、南方电网密切协作，科学统筹长江干流梯级电站水库消落和发电检修工作，加大梯级电站出力，全力支援电网高峰用电。梯级电站高峰运行机组最高达 70 台，单日发电量最高达 6.7 亿千瓦时，高峰出力最大达 3 873 万千瓦，最大日调峰量达 1 990 万千瓦，均创梯级电站冬季历史新高，为抵御寒潮做出了积极贡献。

第三节 三峡工程的通航效益

长江是我国东西交通的"大动脉"，也是联络东、中、西部的经济纽带，素有"黄金水道"之称。随着长江沿江经济带的发展，对发挥长江水运优势也提出了更高的要求。然而，历史上从重庆至宜昌间的河道狭窄、水流汹涌、礁石林立、夜不能航，长约 660 千米的长江航道流经丘陵和高山峡谷地区，落差 120 米，水流湍急，通航条件差，沿程主要有 139 处碍航滩险、46 处单行控制段、25 处需绞滩通行航段，仅能通行 3 000 吨级的船队，运输成本较中下游高 2~4 倍。因此，巨大的航运潜力得不到发挥。

三峡工程作为改善长江航运的战略措施，它的建成改善了宜昌至重庆河段的航道条件，而且随着滩险淹没、航深增大、坡降变缓、流速减小、航道加宽，万吨级船队可直达重庆九龙坡港，结合港口建设和船舶现代化、大型化，年单向下水通过能力可达 5 000 万吨，运输成本降低可 35%~37%。同时，由于水库调节，宜昌以下枯季流量可增加 1 000~2 000 立方米每秒，显著改善了中游航道枯季航运条件。

2010 年 10 月 26 日，三峡工程首次蓄水至 175 米运行，三峡库区长江干流回水可至江津猫儿沱，"高峡出平湖"盛景呈现，长江航运面貌焕然一新，航运效益进一步显现。

livelihoods and of industry, Yangtze Power closely worked with the State Grid and China Southern Power Grid, rationally arranged reservoir drawdown and generator overhaul of the cascade hydropower stations at the trunk stream of the Yangtze River, increased output of the cascade hydropower stations, and made all possible efforts to support the peak power demand. During that peak period, the cascade hydropower stations ran 70 units, the maximum daily energy output was 670 million kWh, the maximum peak output 38.73 million kW, and maximum daily peak regulation capacity was 19.9 million kW, which were all record highs for winter operation of the cascade hydropower stations. This contributed much to protection against the cold waves.

Section 3　Shipping Benefits of the TGP

The Yangtze River is the "main artery" of shipping from the east to the west in China, and is the economic tie between East, Central and West China, which is why it is traditionally reputed as the "golden waterway". Given the development of the Yangtze River Economic Belt, the water transport function of the Yangtze River is faced with higher requirements. In history, however, the river between Chongqing and Yichang was troubled by a narrow surface, turbulent flows and numerous rocks. Nighttime navigation was impossible. The Yangtze River waterway with a length of approximately 660 km passes through hills, high mountains and gorges. Its fall is 120 m high, and water flow is rapid, which are adverse conditions for navigation. Along the waterway, there are 139 obstructive shoals, 46 one-way controlled segments, and 25 segments that need warping tug, so only 3,000 t fleets can navigate there. The transportation cost is 2~4 times higher than the middle and lower reaches of the river. Therefore, the huge potential for shipping could not be fully tapped.

Completion of the TGP, a strategic step to improving shipping in the Yangtze River, has improved shipping conditions in the reach between Yichang and Chongqing. Thanks to the submerged shoals, increased navigation depth, decreased slope gradients, reduced flow rate, and widened waterway, a 10,000 t fleet can now sail all the way to the Jiulongpo Port of Chongqing. On top of that, the port development and deployment of large modernized vessels increase the annual one-way transit capacity to up to 50 million tons, reducing 35%~37% transportation cost. Furthermore, as a result of reservoir regulation, the discharge

一、改善了三峡库区及下游航运条件

三峡工程渠化重庆以下川江航道 600 多千米，结合实施三峡库区碍航礁石炸除工程，三峡库区干流航道等级由建库前的 III 级航道提高为 I 级航道，重庆至宜昌航道维护水深由 2.9 米提高到 3.5~4.5 米，三峡库区航道单向年通过能力由建库前的 1 000 万吨提高到 5 000 万吨，实现了全年全线昼夜通航。其中，重庆朝天门至坝址河段，在一年中有半年以上时间具备行驶万吨级船队和 5 000 吨级单船的通航条件。同时，位于三峡工程下游仅 38 千米的葛洲坝工程枯水期出库最小通航流量由 2 700 立方米每秒提高到 6 000 立方米每秒以上，比天然情况下增加 2 500~3 000 立方米每秒，葛洲坝下游最低通航水位由 38 米提高到 39 米。枯水期航道维护水深达到 3.2 米，比蓄水前提高了 0.3 米。

二、提高了船舶航行和作业安全度

三峡水库成库前，川江重庆段是全国水运安全的重灾区之一，据统计，1995 年至 2003 年重庆水运海损事故年均死亡 100 人左右，约占全国同类事故死亡总数的 20%。三峡水库成库以来，重庆水上安全管理"治本、严管和依靠科技进步"三管齐下，通过加大安全投入、实施科技兴安、长江分道航法、客渡船标准化和渡口改造等措施，强化安全管理，安全形势逐年好转并呈现总体稳定的趋势。三峡工程蓄水后（2003 年 6 月至 2013 年 12 月）与蓄水前（1999 年 1 月至 2003 年 5 月）相比，年均事故件数、死亡人数、沉船数和直接经济损失与建库前相比分别下降了 72%、81%、65% 和 20%。

of the reaches downstream from Yichang in the dry season can increase by 1,000~2,000 m³/s. This has significantly improved shipping conditions of the middle reach in dry season.

On October 26, 2010, the TGP operated with a 175 m water level for the first time. The backwater from the Yangtze River trunk stream at the TGP reservoir area flowed to Maoertuo, creating the miraculous view of a great lake in a gorge. The shipping business in Yangtze River took on a new look and became even more lucrative.

I. Improvement of Shipping Conditions in the TGP Reservoir Area and in the Lower Reaches

The TGP canalized the Chuanjiang waterway of over 600 km downstream from Chongqing, and the obstructive rocks in the TGP reservoir area were removed by blasting. Therefore, the trunk stream waterway in the TGP reservoir area was upgraded from Class III (before reservoir construction) to Class I; the dredging depth of Chongqing-Yichang waterway increased from 2.9 m to 3.5-4.5 m; the annual one-way transit capacity of the waterway in the TGP reservoir area increased from 10 million tons (before reservoir construction) to 50 million tons; day-and-night navigation became available. At the reach from Chaotianmen, Chongqing to the dam site, 10,000 t fleets and 5,000 t single vessels could sail there for more than half a year. At the Gezhouba Project, which is only 38 km downstream from the TGP, the minimum navigable flow rate increased from 2,700 m³/s to more than 6,000 m³/s during the dry season, an increase of 2,500-3,000 m³/s compared with natural conditions. The minimum navigation water level downstream from the Gezhou Dam increased from 38 m to 39 m. The dredging depth of the waterway during the dry season increased to 3.2 m, which is 0.3 m more than the water level before water storage.

II. Enhancements to Ship Navigation and Operation

Before completion of the TGR, the Chongqing reach of the Chuanjiang River was one of the less safe areas in China in terms of water transport. From 1995 to 2003, according to statistics, the transport industry's marine accidents on the river averaged about 100 deaths per year, which accounted for 20% of total deaths caused by the marine accidents nationwide. Since completion of the reservoir, Chongqing managed to bolster water safety through the "solve the root problems, impose strict control and promote technological advancement" policy. Measures such as increased safety efforts, technology-based safety protection, the Yangtze River traffic separation scheme, passenger ferry standardization, and ferry improvement were taken to strengthened safety management. As a result, the waterway traffic safety in

图7.3 峡江夜航图（摄影：黄正平）

三、降低了航运成本

由于三峡库区水流流速减缓、流态稳定、比降减小，船舶载运能力明显提高，油耗明显下降。2002年，重庆货运船舶单位载重吨装机功率平均为0.63千瓦，货运船舶单位能耗为每千吨千米7.6千克。三峡水库成库后，随着航道条件改善和船舶大型化进程加快，重庆货运船舶单位载重吨装机功率逐步下降到0.2千瓦左右，船舶平均单位能耗逐年下降，目前货运船舶单位能耗为每千吨千米1.9千克。按2002年燃油耗能水平计算，2013年重庆水运行业仅货运船舶就节约燃油约110万吨，节油效益约91亿元，减少二氧化碳排放量约330万吨。自三峡水库蓄水以来，由于船舶单位能耗下降，运输船舶空气污染物的单位排放量明显减少。船舶油耗的降低有利于节能减排和降低运输成本。

Chongqing reach was improved year by year, displaying a generally stable trend. According to statistics before (from January 1999 to May 2003) and after (from June 2003 to December 2013) impoundment of the TGP, the annual average number of accidents, number of deaths, number of sunken ships and direct economic loss decreased by 72%, 81%, 65% and 20% respectively, as compared with before reservoir construction.

⮢ Figure 7.3 Nighttime Navigation at the TGP (Photographed by Huang Zhengping)

III. Reduction of Shipping Costs

In the TGP reservoir area, the flow rate slowed, flow state stabilized and gradient decreased, so the carrying capacity of vessels significantly increased and fuel consumption markedly decreased. In 2002, the average installed power per unit of dead weight tonnage of cargo vessels of Chongqing was 0.63 kW and the unit energy consumption of cargo vessels was 7.6 kg/(kt · km). After completion of the TGR, the waterway's conditions improved and upsizing of vessels accelerated, so the installed power per unit deadweight tonnage of cargo vessels of Chongqing gradually decreased to around 0.2 kW and average unit energy consumption of vessels decreased year by year. Now the average unit energy consumption of the cargo vessels is 1.9 kg/(kt · km). Based on fuel consumption in 2002, the cargo vessels alone conserved approximately 1,100,000 t of fuel for the water transport industry of Chongqing in 2013. Fuel conservation was approximately RMB 9.1 billion, and carbon dioxide emission was reduced by about 3.3 million tons. Since the TGR started impoundment, the unit energy consumption of vessels decreased, so the unit emission of vessel-generated air pollutants was clearly reduced. Decrease in fuel consumption of vessels can help with energy conservation and emission reduction and reduce transportation costs.

IV. Promotion of the Standardization of Vessels

Before completion of the TGR, the vessels in the upper reach of the Yangtze River were mostly fleet units and 1,000 t individual vessels. After completion of the TGR, the waterway's conditions improved, so the proportion of fleets decreased year by year, self-propelled vessels developed rapidly, the upsizing, professionalization and standardization of vessels clearly accelerated, and the tonnage of dry bulk carriers increased from about 1,000 t in 2003 to 5,000-6,000 t. The size of container ships increased from 80-100 TEU in 2003 to 300-325 TEU. The tonnage of petroleum products or chemical vessels increased from 1,000-1,500 t in 2003 to 3,000-4,500 t. The main commercial model Ro-Ro ship can load 800 vehicles. Lux-

四、推进了船舶标准化进程

三峡水库成库前，长江上游地区的船舶主要以船队运输和千吨级单船为主。三峡水库成库后，随着航道条件改善，船队比重逐年下降，自航船快速发展，船舶大型化、专业化、标准化进程明显加快，干散货船从 2003 年的 1 000 吨左右发展到 5 000～6 000 吨；集装箱船从 2003 年的 80～100TEU 发展到 300～325TEU；油品或化学品船从 2003 年的 1 000～1 500 吨发展到 3 000～4 500 吨；商品汽车滚装船主力船型为 800 车位；船长 130 米以上、客位数 350 以上的邮轮已成为三峡豪华邮轮的主力船型。目前，长江上游地区船队运输方式已基本消失，3 000 吨级以上船舶艘次已达 74.08%，其中 5 000 吨级以上船舶艘次已达 46.25%。

五、促进三峡库区航运相关产业发展

数据显示，三峡水库蓄水后，重庆地区水运直接从业人员达 15 万人，其中近 8 万人来自三峡库区，依赖水运业的三峡库区煤炭、旅游、公路货运等产业的从业人员达 50 万人以上，水运业及其关联产业吸纳了三峡库区 200 多万剩余劳动力，为三峡库区经济社会发展发挥了重要的支撑作用。三峡库区大部分码头作业条件得到根本改变，一批现代化的新码头陆续兴建，改善了三峡库区港口货物运转环境，为构建现代化的三峡库区水运体系创造了基础条件。

三峡船闸自 2003 年 6 月投入运行以来，通过三峡河段的货运量持续高速增长。2019 年三峡断面通过航运量已达 1.54 亿吨，是三峡工程蓄水前该河段最大年货运量的 7 倍多。枢纽的运行有力促进了长江航运的快速发展和沿江经济的协调发展。据中国工程院关于三峡工程建设第三方独立评估的初步估算，船闸试运行后十年期间，三峡工程累计产生约 85.92 亿元（含区间运量）的航运效益。

2016 年 9 月 18 日，三峡升船机正式试通航，这进一步增强了三峡工程的通航调度灵活性和通航保障能力。2021 年，三峡船闸运行 1.01 万闸次，过船 4.04 万艘次，通过旅客 8 038 人次，过闸货运量约 1.46 亿吨（比 2020 年增加 6.83%），主要设备完好率 100%；三峡升船机累计安全运行 4 725 厢次，通过旅客 10.02 万人次，过船 4 803 艘次，过机货运量 365.51 万吨。

ury cruise ships on the TGP are now dominated by those over 130 m long with capacity for over 350 passengers. By now, fleet transportation has virtually disappeared from the upper reach of the Yangtze River. 3,000 t or larger ships account for 74.08% of all ships and 5,000 t or larger ships account for 46.25%.

V. Development of Shipping-Related Industries in the TGP Reservoir Area

Since impoundment in the TGR, according to statistics, the number of direct employees in the water transportation industry in Chongqing has increased to 150,000, including 80,000 people who are from the TGP reservoir area. The number of employees in the coal industry, tourism and road freight that rely on the water transportation industry has increased to over 500,000, and the water transportation industry and the related industries absorbed over two million surplus laborers. These industries are now a major part of the socio-economic development of the area. Most wharfs in the area have fundamentally improved their operation conditions. A number of new modern wharfs have been successively built. This has improved the transshipment environment of port goods in the area, and laid the foundation for creating a modern water transportation system in the area

Since the TGP ship lock was commissioned in June 2003, the freight volume borne by the TGP's reaches has been increasing rapidly. The shipping volume passing through the cross section of the TGP increased to 154 million tons in 2019, more than seven times the maximum annual freight volume of that reach before impoundment. The hydroproject has significantly contributed to rapid development of the shipping industry in the Yangtze River and coordinated development of the economy along the river. According to the preliminary estimation of a third-party independent assessment of the TGP conducted by the Chinese Academy of Engineering, the TGP generated about RMB 8.592 billion in overall shipping benefits (including inter-region transportation volume) in the decade since trial operation of the ship lock.

On September 18, 2016, the TGP ship lift was officially opened for trial navigation. This further enhanced the flexibility of navigation dispatching and of navigation support capacity. In 2021, its ship lock operated 10,100 times, served 40,400 ships and 8,038 passengers, the freight volume through the ship lock was approximately 146 million tons (increased by 6.83% compared with 2020), and the serviceability rate of the main equipment was 100%. The TGP ship lift operated 4,725 times, served 100,200 passengers, 4,803 ships, and had a freight volume of 3,655,100 t passing through the lift.

第四节 三峡工程的水资源利用（配置）效益

长江上游来水年内分配不均，6月至10月多年平均径流量占全年的70%以上，12月至次年4月多年平均来水量仅4 000~6 000立方米每秒。三峡水库建成后，具有调节库容165亿立方米、防洪库容221.5亿立方米，凭借良好的"拦洪补枯"的季调节性能，可有效利用洪水资源、增加枯水期长江中下游下泄流量，是我国重要的淡水资源库和生态环境调节器。

一、补水效益

初步设计三峡水库枯水期下泄流量应满足不低于电站保证出力及葛洲坝下游庙嘴最低通航水位39米对应的流量（约5 000立方米每秒）。2009年以来，随着下游沿江经济社会的发展，为满足越来越高的供水需求，三峡水库将枯水期1月至4月份水库最小下泄流量提高至6 000立方米每秒。与2003年至2016年年最小入库流量仅2 990立方米每秒相比，现状调度方式下，三峡水库最小下泄流量提高至6 000立方米每秒，可有效满足长江中下游沿江地区生产生活和生态用水需求。截至2020年年底，三峡水库累计为长江中下游补水2 267天，补水总量2 894亿立方米，相当于近10个鄱阳湖的蓄水量，平均增加下游航道水深近1米，较好满足了下游航道畅通及沿江两岸生产生活等用水需求。

Section 4 | Water Resources Utilization (Allocation) Benefits of the TGP

The annual distribution of water from the upper reach of the Yangtze River was not even, where the average annual runoff from June to October accounted for over 70% of annual runoff, while the average annual inflow water from December to the next April was only 4,000-6,000 m^3/s. After completion, the TGR had 16.5 billion m^3 regulation capacity and 22.15 billion m^3 flood control capacity. It performs well with seasonal regulation because it can retain flood water to compensate its capacity in the dry season. This helps efficiently utilize flood water resources and increase the discharge flow of the middle and lower reaches of the river in the dry season. It is an important freshwater resources reservoir and ecological environment regulator for China.

I. Water Compensation Benefit

According to the preliminary design, the discharge flow of the TGR in the dry season should be enough to sustain a firm output of the power station and to sustain the discharge (approx. 5,000 m^3/s) required for the 39 m minimum navigable water level. Since 2009, the socioeconomic development of the region in the lower reach brought increasingly high water demand, so the TGR increased the minimum discharge flow from January to April during the dry season to 6,000 m^3/s. Compared with the annual minimum inflow discharge of only 2,990 m^3/s from 2003 to 2016, the current dispatching mode has increased its minimum discharge flow to 6,000 m^3/s, which will effectively meet the water demands of production, livelihood and ecological conservation of the regions along the middle and lower reaches of the Yangtze River. As at the end of 2020, the TGR had replenished water for the middle and lower reaches of the Yangtze River for 2,267 days. The total replenished water volume was 289.4 billion m^3, which was 10 times the impoundage of Poyang Lake, and the water depth in the lower reach's waterway increased by nearly one meter on average. This effectively

二、生态效益

　　青鱼、草鱼、鲢鱼、鳙鱼（简称"四大家鱼"）作为长江中下游江湖复合生态系统的典型物种，是衡量长江水生态系统健康的重要指标。20世纪60年代以来，过度捕捞、水环境恶化、河道采砂等改变了四大家鱼繁殖需要的水温和水力学条件，导致四大家鱼产卵规模呈下降趋势。为促进四大家鱼繁殖，2011年以来，三峡水库采取了持续加大下泄流量的调度方式，创造促进四大家鱼繁殖的水力学条件。从2011年起，连续10年开展了14次生态调度试验。监测情况表明，在水温条件满足四大家鱼产卵的情况下，三峡水库实施生态调度期间，宜都断面均监测到四大家鱼产卵现象，宜都江段年均繁殖量达8.2亿粒（尾），凸显了三峡工程的生态效益。

　　三峡水库促进四大家鱼繁殖的生态调度取得了良好效益，其经验推广至了金沙江流域生态环境保护中，2017年，溪洛渡水库和向家坝水库首次实施了促进四大家鱼繁殖的生态调度试验。

　　2021年，三峡水库生态调度工作取得了显著成效：连续3次开展针对三峡库区产粘沉性卵鱼类自然繁殖的生态调度试验，总产卵规模约3亿粒；开展2次针对葛洲坝下游四大家鱼自然繁殖的生态调度试验，调度期间四大家鱼产卵规模超84亿颗，创历年之最。

三、应急调度

　　除了每年枯水期进行常规补水调度外，当长江中下游发生较重干旱或出现供水困难需要实施水资源应急调度时，三峡水库凭借巨大的库容和灵活的调节性能，实施了船舶应急救援调度、抗旱补水调度、压咸潮调度等，成功应对了多起突发事件。

maintained trafficability of the lower reach waterway and met the water demands of production and livelihood of the regions along the river.

II. Ecological Benefits

Black carp, grass carp, silver carp and bighead carp (known as "four Chinese carps") are typical species of the river-lake complex ecosystem in the middle and lower reaches of the Yangtze River, and are important indicators for keeping the aquatic ecosystem of the Yangtze River healthy. Since the 1960s, overfishing, water environment deterioration and sand mining changed the water temperature and hydraulic conditions required for the four Chinese carps to breed and thus populations declined. The TGR has been increasing its discharge flow since 2011 to create the hydraulic conditions required for breeding of the four Chinese carps. Fourteen ecological operation tests were conducted for 10 consecutive years, starting in 2011. According to the data, when the water temperature is conducive to spawning of the four Chinese carps, spawning was observed at the Yidu cross section during ecological dispatching of the TGR and the annual average reproduction in the Yidu reach of Yangtze River was 820 million eggs, a strong proof of the ecological benefit of the TGP.

The ecological dispatching of the TGR intended to simulate spawning of the four Chinese carps has been very successful. This practice has been applied in the ecological and environmental protection efforts along the Jinsha River basin. In 2017, the Xiluodu Reservoir and Xiangjiaba Reservoir conducted their first ecological dispatching test intended to stimulate breeding of the four Chinese carps.

In 2021, the ecological dispatching of the TGR rendered impressive results: three consecutive ecological dispatching tests were conducted to stimulate natural breeding of the fish species capable of laying viscid demersal eggs in the TGP reservoir area, and 300 million or so eggs were laid in total; two ecological dispatching tests were conducted to stimulate natural breeding of the four Chinese carps in the reach downstream from the Gezhou Dam. During the dispatching period, the eggs of the four Chinese carps exceeded 8.4 billion, the largest number in history.

III. Emergency Dispatching

In addition to regular water compensation during the dry season each year, the TGR, with a massive storage capacity and flexible regulation performance, has handled multiple emergencies by conducting the emergency rescue of ships, drought control, water compensation, and salt tide suppression when the middle and lower reaches of Yangtze River sustain severe droughts

1. 船舶应急救援

2011年2月12日，一艘载油990吨的船舶在葛洲坝下游枝江市水陆洲尾水域搁浅，因搁浅地为鹅卵石河床，实施常规拖带或过驳脱险操作困难，三峡水库应通航部门要求及时实施了应急抢险调度，先后两次增加下泄流量1 800立方米每秒和2 000立方米每秒，补水1.64亿立方米，有效抬升了遇险船舶所在水域水位，确保了施救工作顺利完成。

2020年7月以来，长江三次洪峰接踵而至，三峡水库入库流量持续居高不下，最高达6.1万立方米每秒；出库下泄流量长时间维持在3.5万立方米每秒以上。根据通航安全相关规定，大量载运危险品船舶、小功率船舶因不具备安全航行条件而在三峡大坝上下游滞留，最长待闸时间已达22天。船载生产生活物资无法到达目的地，沿江省市特别是重庆地区开始出现航空煤油等告急情况，对新冠肺炎疫情之后的复工复产产生不利影响。社会各方高度关注三峡河段船舶滞留情况，为助力复工复产、缓解通航压力，7月31日下午，根据长江委下达的调度令，8月1日5时至20时，三峡水库减小下泄流量至3.45万立方米每秒，为疏散滞留船舶创造了有利条件。在长江三峡通航管理局的指导下，此次紧急安全疏散因汛滞航船舶90艘（其中危险品船舶61艘），运送大量事关国计民生的物资过坝，助力沿江省市复工复产。

2. 抗旱补水

2011年，北半球多个国家和地区发生罕见旱情，我国长江中下游部分地区遭遇了百年一遇的大面积干旱，三峡水库水位在已经接近枯季消落水位155米且入库流量持续偏小的情况下，以满足生态、航运、电网供电为目标的运行方式调整为以全力抗旱为目标的应急抗旱调度方式。5月7日10时，三峡水库开始加大下泄流量，库水位从155.35米下降至6月10日24时的145.82米，抗旱补水总量54.7亿立方米，日均向下游补水1 500立方米每秒，有效改善了中下游生活、生产、生态用水和通航条件，为缓解特大旱情发挥了重要作用。

2021年，三峡水库累计来水4 536.38亿立方米，较建库以来均值偏多约一成，消落期充分发挥淡水资源库作用，累计为下游补水138天，补水总量约221亿立方米。

or power supply shortages and emergency dispatching of water resources is required.

1. Emergency rescue of ships

On February 12, 2011, a 990 t oil tanker was stranded in the tailwater area of Shuilu Island, Zhijiang, downstream from Gezhou Dam. Due to the cobblestone riverbed at the site of stranding, it was difficult to conduct regular towing or lightering, so the TGR, as requested by the navigation authority, immediately conducted an emergency rescue. The discharge flow was increased to 1,800 m^3/s and 2,000 m^3/s, and 164 million m^3 water was replenished. This effectively raised the water level in the water area where the ship was stranded and ensured the rescue mission was successfully accomplished.

Since July 2020, the Yangtze River saw three consecutive flood peaks, and consequently the inflow discharge of the TGR was kept at high level, which maxed at 61,000 m^3/s. The discharge flow was kept above 35,000 m^3/s for a long period. In accordance with navigational safety related regulations, many vessels carrying hazardous goods and low-power vessels were held up upstream and downstream from the Three Gorges Dam due to unsafe navigation conditions. The maximum gate holding duration was 22 days. The supplies related to work and livelihood carried by the vessels could not reach their destination and consequently the provinces and cities along the river, particularly Chongqing, reported shortage in aviation kerosene, adversely affecting the resumption of work and production after the Covid-19 pandemic. This situation with vessels held up drew much attention from the society. In order to promote economic reopening and relieve the navigational pressure, the Changjiang Water Resources Commission issued a dispatching order in the afternoon of July 31, and from 5:00 to 20:00, August 1, the TGR decreased its discharge flow to 34,500 m^3/s. This allowed the vessels to be cleared. Under the guidance of the Three Gorges Navigation Authority, 90 vessels (including 61 vessels carrying hazardous goods) were safely cleared, which carried large quantities of supplies important to the national economy and to people's livelihood through the dam, contributing to work resumption in the provinces and cities along the river.

2. Drought control and water compensation

In 2011, many countries and regions in the Northern Hemisphere were hit by a rare drought. Some regions in the middle and lower reaches of the Yangtze River were hit by a once-in-a-century, large-area drought. Even when the water level in the TGR was close to the 155 m drawdown water level in the dry season, and the inflow discharge was kept at an excessively low level, the operation model of the reservoir intended to support ecological conservation, shipping and grid power supply was shifted to the emergency drought control model for the purpose of protecting against the drought. At 10:00, May 7, the TGR started in-

3. 压咸潮

2014年2月，上海长江口水源地遭遇历史上持续时间最长的咸潮入侵，长江口青草沙、陈行等水源地的正常运行和群众生产生活用水受到较大影响。应上海市政府要求，三峡水库启动了建成以来的首个"压咸潮"调度。2月21日至3月3日"压咸潮"调度期间，三峡水库向下游累计补水17.3亿立方米，平均出库流量7 060立方米每秒。与正常消落按6 000立方米每秒控泄相比，多补水约9.6亿立方米，缓解了咸潮入侵的不利影响。

第五节 三峡工程的运行管理与效益发挥

三峡工程高效的运行管理，促进了三峡工程初步设计的综合效益的提升和拓展。三峡工程建设期间，采取了建设与运行相结合的措施，使工程提前发挥了综合效益。三峡水库蓄水运用以后，在水库运行环境较初步设计发生较大变化，各方面对三峡水库提出更高调度需求的情况下，三峡水库采取了优化调度措施，使初步设计的防洪、发电、航运、水资源利用效益得到了提升和拓展。

creasing discharge flow. The water level dropped from 155.35 m to 145.82 m at 24:00, June 10. The total compensated water volume was 5.47 billion m^3, and the average daily compensated water to the lower reach was 1,500 m^3/s. This effectively improved the conditions for people's livelihood, production, ecological water utilization, and navigation in the lower reach regions, and made major contributions to relieving the severe drought.

In 2021, the cumulative inflow of the TGR was 453.638 billion m^3, which was approximately 10% more than the average value since reservoir completion. During the drawdown period, the reservoir adequately fulfilled its function as a fresh water resources pool and provided compensation water for the lower reach for 138 days, with approximately 22.1 billion m^3 of water compensated.

3. Salt tide suppression

In February 2014, the water source locations at the Yangtze River estuary in Shanghai were hit by the longest salt tide intrusion in history. Normal operation of the water source locations at the Yangtze River estuary such as Qingcaosha and Chenhang and water supply for the people were severely impacted. As requested by the Shanghai Municipal People's Government, the TGR activated its first salt tide suppression operation since its completion. During the operation from February 21 to March 3, the TGR compensated 1.73 billion m^3 of water in total to the lower reach, and the average outflow discharge was 7,060 m^3/s. Compared with the 6,000 m^3/s controlled discharge during normal drawdowns, 960 million m^3 of additional water was compensated, which mitigated the adverse impact of the salt tide intrusion.

Section 5 Operation, Management and Benefits of the TGP

Efficient operation and management of the TGP has enhanced and expanded the comprehensive benefits specified in the preliminary design of the project. During its construction, the "construction and operation in parallel" model was implemented, which allowed the comprehensive benefits of the project to be materialized sooner. After the TGR started impoundment, the operating environment of the reservoir changed significantly from its preliminary design, and all parties concerned presented higher demand for dispatching of the reservoir.

一、运行与建设结合,提前发挥综合效益

按照初步设计规划,三峡水库2003年开始蓄水至135米,进入围堰发电期,围堰发电期运行水位为135米;2007年蓄水位升至156米,进入初期运行期;2013年蓄水位上升至175米,进入正常运行期。在对实施分期蓄水所涉及的工程建设、移民工程、地质灾害治理、泥沙等条件进行充分论证后,三峡工程实行运行与建设紧密结合,在初步设计的基础上,实现了围堰发电期蓄水至139米、2006年提前一年实现了156米蓄水和2008年提前五年实施了175米试验性蓄水,使三峡工程的综合效益提前发挥。以139米蓄水为例,初步设计三峡水库围堰发电期运行水位为135米,水库无调节能力。考虑葛洲坝下游宜昌河道冲刷下切实际情况,结合围堰发电期葛洲坝下游河床可能出现的下切预估值,经水库补偿流量调度计算,需汛后水库蓄水8.1亿~21.7亿立方米,对应最高水位为138.7米。在对139米蓄水涉及的三峡库区淹没问题和枢纽工程安全复核等逐一采取措施之后,三峡水库于2003年11月蓄水至139米。枯水期,通过实施航运补偿调度,三峡水库日均最小下泄流量由天然状态下的2 800立方米每秒提高至3 590立方米每秒,改善了下游的通航条件。同时,水库135~139米之间18亿立方米的防洪库容为汛期应急调度创造了条件,减轻了长江中下游的防洪压力。2004年汛期,三峡水库最大入库流量达60 500立方米每秒,控制最大出库流量56 800立方米每秒,削峰约4 000立方米每秒,拦蓄洪水4.95亿立方米。

Therefore, the TGR implemented optimized dispatching, so that the benefits encompassed in the preliminary design including flood control, power generation, shipping and water resources utilization were enhanced and expanded.

I. Construction and Operation in Parallel to Bring into Play Comprehensive Benefits in Advance

According to the preliminary design, the impounded water level of the TGR was to be 135 m in 2003. In the cofferdam power generation period, the operating water level at cofferdam was to be 135 m. In 2007, the storage level increased to 156 m and the preliminary operation period began. In 2013, the storage level increased to 175 m and its normal operation period began. After adequate appraisal of the conditions involved in stage-by-stage impoundment, such as construction works, resettlement efforts, geological hazard control and sediment, the TGP implemented operation and construction in a way that allowed the two to be closely integrated. On the basis of the preliminary design, the storage level increased to 139 m during the cofferdam power generation period and increased to 156 m in 2006, a year ahead of schedule. 175 m trial impoundment was achieved in 2008, five years ahead of schedule. Therefore, the comprehensive benefits of the TGP were materialized sooner. Take the 139 m storage level as an example. In the preliminary design, the operating water levels of the TGR during the cofferdam power generation period was 135 m and the reservoir had no regulation capacity. Given the fact that the river at Yichang, downstream from the Gezhou Dam, was scoured and undercut, and an estimated value of the undercut could occur in the riverbed downstream from the Gezhou Dam during the cofferdam power generation period, the compensation discharge calculation of the reservoir indicates that 810 million to 2.17 billion m^3 of water needed to be impounded in the reservoir after the flood, and the corresponding maximum water level needed to be 138.7 m. After measures taken in response to the issue of TGP reservoir area inundation involved with 139 m storage level, and safety rechecking of the hydroproject, the TGR impounded water to the level of 139 m in November 2003. In the dry season, shipping compensation was implemented so that the average daily minimum discharge flow of the TGR increased from the 2,800 m^3/s under natural conditions to 3,590 m^3/s and the navigational conditions in the lower reach were improved. Besides this, the 1.8 billion m^3 of flood control capacity of the reservoir at the 135-139 m water level made emergency dispatching during flood season possible, and relieved flood control pressure in the middle and lower reaches of the Yangtze River. During the flood season of 2004, the maximum inflow discharge of the TGR was 60,500 m^3/s, maximum outflow discharge

二、试验性蓄水方式取得成功

初步设计安排：三峡水库进入初期运行期以后，何时蓄水至175米，可根据移民安置、库尾泥沙淤积实际观测成果及重庆港泥沙淤积等情况相机确定，暂定6年。2008年，三峡水库移民工程顺利通过验收；泥沙方面的研究也表明，只有将蓄水位抬高至172米以上，才能实际观测到工程蓄水对重庆河段泥沙冲淤的影响。2008年，基于对泥沙实际监测成果、后期来沙量预测和水库淤积发展的分析，三峡水库以试验性蓄水的方式开始175米蓄水。利用试验性蓄水的契机，在充分论证的基础上，成功开展了三峡水库汛限水位浮动、中小洪水调度、汛末提前蓄水、泥沙减淤调度、生态调度等多项调度试验，在提升和拓展三峡工程综合效益的同时，也为正常运行期安全、科学调度积累了大量运行资料。中国工程院在《三峡工程试验性蓄水阶段评估报告》中指出，三峡水库在试验性蓄水期全面实现了三峡工程设计的功能要求，并积累了宝贵的经验。

三、优化调度管理，提升综合效益

自2003年三峡水库蓄水运用以来，水库运行环境较初步设计发生了较大变化。新的运行条件下，三峡水库按照初步设计的调度方式将难以实现既定的综合效益。为此，三峡水库开展了优化调度研究，并通过建立综合沟通协调机制，将优化调度研究成果应用于调度实践，逐步形成了"技术先行、沟通协调、运行实践、总结完善"的三峡水库优化调度模式。水库实施汛限水位浮动、中小洪水调度和提前蓄水等优化调度措施，理论上与初步设计相比增加了水库泥沙淤积。在入库泥沙大幅减少的背景下，三峡水库进一步采取了解决泥沙淤积分布的库尾减淤调度和减少三峡库区泥沙淤积的沙峰调度措施，探索建立了新的"蓄清排浑"模式。这种新的"蓄清排浑"模式在协调解决三峡水库优化调度与泥沙淤积关系的同时，使三峡工程初步设计的防洪、发电、航运、水资源利用等综合效益得到了提升和拓展。

56,800 m³/s, peak shaving approximately 4,000 m³/s, and impounded flood volume 495 million m³.

II. Trial Impoundment Method Proved a Success

Arrangement in preliminary design: During the preliminary operation period of the TGR, the time when storage can reach a level of 175 m could be determined according to resettlement progress, measured sediment accumulation at the reservoir tail and sediment accumulation in the Port of Chongqing. It was provisionally scheduled for six years. In 2008, the resettlement efforts of the TGR successfully passed acceptance. The research into the sediment problem indicates that only when the storage level increased to above 172 m can it be possible to observe how water impoundment would affect sediment scour and deposition in the Chongqing reach. In 2008, based on sediment monitoring results, prediction of future incoming sediment volume and analysis of sediment accumulation in the reservoir, the TGR started trial impoundment to the 175 m storage level. Given the opportunity of trial impoundment and on the basis of adequate demonstration, the TGR successfully conducted many dispatching tests, including fluctuation in the flood control level, minor-to-moderate flood dispatching, early impoundment before the end of the flood season, sediment accumulation reduction, and ecological dispatching. This enhanced and expanded the comprehensive benefits of the TGP and also gathered a large amount of operational data which could be used to ensure safety and rational dispatching in normal operation periods. The *Report on Assessment of Trial Impoundment Period of the TGP* published by the Chinese Academy of Engineering noted that the TGR fulfilled all the designed functions of the TGP and that it is a precious experience.

III. Optimized Dispatching Management and Enhanced Comprehensive Benefits

Since 2003 when the TGR started impoundment, the reservoir's operating environment changed significantly from the preliminary design. Under the new operating conditions, the dispatching model of the preliminary design of the TGR would not generate the expected comprehensive benefits. Therefore, the TGR conducted research into optimized dispatching. An integrated communication and coordination mechanism was established to apply the results of research in practical dispatching, and the TGR optimized dispatching model of "technological advancement, communication and coordination, practical operation, learning and

世界超级工程——中国三峡工程建设开发的实践与经验

A Mega Project in the World—Practice and Experiences in the Construction and Development of the Three Gorges Project in China

四、综合沟通协调机制逐步完善

　　三峡水库是一个具有综合性功能的水库，调度管理涉及防洪、发电、航运、供水、抗旱、生态、泥沙等多个方面，各方隶属于不同的主管部门，调度需求复杂繁多。只有协调好各方面的关系，才能充分发挥三峡工程的综合效益。为此，逐步建立并形成了三峡水库运行综合沟通协调机制。2003年4月，成立了由三峡集团负责、有关现场运行管理单位参加的三峡—葛洲坝梯级调度协调领导小组，定期或不定期协商处理有关现场运行管理中的问题，三峡—葛洲坝水利枢纽梯级现场运行管理单位的相互关系得到了妥善协调。三峡水库蓄水运用以后，不同时期的梯级枢纽调度规程、年度汛期调度运用方案及蓄水计划（含消落）和实时调度方案，都是在广泛征求水利、电网、航运、国土资源部门等方面的意见后，综合沟通协调的结果。水库日常调度中，三峡集团及时将水库调度信息传递给相关方面，在各方面的支持和配合下，三峡水库调度较好地协调了防洪、发电、航运、供水、泥沙、生态等各方面的管理关系。

improvement" was gradually established. The reservoir implemented optimized dispatching measures such as flood control level fluctuation, minor-to-moderate flood dispatching and impoundment ahead of schedule, and theoretically the reservoir sediment accumulation increased as compared with the preliminary design. With the incoming sediment reduced substantially, the TGR further implemented the reservoir's tail sediment reduction intended to solve the problem of sediment accumulation distribution and sediment peak dispatching intended to reduce sediment accumulation in the TGP reservoir area, and attempted to establish the new "impound clear water and discharge turbid water" model. This model could help balance the relation between optimized dispatching and sediment accumulation of the TGR, and also enhance and expand the comprehensive benefits encompassed in the preliminary design of the TGP, such as flood control, power generation, shipping and water resources utilization.

IV. Gradual Improvement of Integrated Communication and Coordination Mechanism

The TGR possesses comprehensive functions. Its dispatching and management involve many factors, including flood control, power generation, shipping, water supply, drought control, ecology, and sediment. These functions were assigned to different departments, and thus the dispatching demands were numerous and complex. Only with well-coordinated relations between the concerned parties, could the comprehensive benefits of the TGP adequately materialize. To this end, the integrated communication and coordination mechanism for operation of the TGR was gradually established. In April 2003, the CTG-Gezhouba Cascade Dispatching and Coordination Leading Group was instated, which was under the charge of CTG and included the members from the relevant onsite operation and management entities. This group regularly or irregularly held discussions to solve problems encountered during onsite operation and management, so the relation between the entities responsible for onsite operation and management of the CTG-Gezhouba hydroproject cascade system was properly coordinated. After the TGR started impoundment, the cascade hydroproject dispatching procedures in different periods, the flood season dispatching and operation scheme of each year, impoundment plan (including drawdown) and real-time dispatching scheme were all formulated after communication and coordination between the concerned parties, including the water conservancy department, power grid, Ministry of Water Resources, shipping department and the Ministry of Land and Resources. In routine dispatching of the reservoir, the CTG delivered the reservoir dispatching information to the concerned parties. With support and assistance of the concerned parties, the dispatching of the TGR properly balances the

五、联合调度促进三峡效益进一步提升

随着溪洛渡工程、向家坝工程建成投运,三峡水库单库优化调度研究与实践的经验逐步推广至上中游水库群,并开展了以三峡工程为核心的联合调度研究与实践工作。在联合调度模式下,三峡工程的综合效益将得到进一步提升。以城陵矶调度为例,单库调度下,三峡水库对城陵矶防洪补偿水位为155米,溪洛渡水库、向家坝水库配合三峡水库联合调度后,补偿水位可进一步提高至158米。

2016年汛期,正是在以三峡为核心的水库群联合调度模式下,三峡水库成功实施了城陵矶补偿调度,对避免城陵矶分洪、减轻长江中下游的防洪压力发挥了关键性作用。2020年汛期,面对21世纪最严峻汛情,长江流域梯级水库以三峡水库为核心,实施联合调度,成功应对5次编号大洪水,发挥巨大防洪效益。其中三峡水库累计拦蓄洪量305亿立方米,破历史记录。最大降低下游江段水位4米,平均缩短下游河段主要控制站水位超过警戒水位13.4天,避免荆江分洪区运用和损失。

2020年以后,随着金沙江下游乌东德、白鹤滩工程的陆续建成,联合调度格局将进一步扩大,以三峡水库为核心的长江上中游水库群联合调度综合效益将更加凸显。

administrative relationships between the entities responsible for flood control, power generation, shipping, water supply, sediment and ecology.

V. Combined Dispatching to Further Enhance the Benefits of the TGP

With the Xiluodu Project and Xiangjiaba Project commissioned, the research results and practical experience from optimized dispatching of the TGR have been applied in the reservoir group in the upper and middle reaches, and the joint dispatching research and practical efforts centered around the TGP have been conducted. In the combined dispatching model, the comprehensive benefits of the TGP will be further enhanced. Take dispatching for Chenglingji as an example. In dispatching of a single reservoir, the flood control compensation water level of the TGR intended for Chenglingji is 155 m. If the Xiluodu Reservoir and Xiangjiaba Reservoir assist the TGR in combined dispatching, the compensation water level can be increased to 158 m.

During the flood season of 2016, it was exactly because of the combined dispatching model of the reservoir group centered on the TGP that the TGR managed to execute compensation dispatching for Chenglingji. This played a key role in preventing flood diversion at Chenglingji and relieving flood control pressure in the middle and lower reaches of the Yangtze River. In the flood season of 2020, the most threatening flood by far in the 21 s century occurred. The Yangtze River basin cascade reservoirs centered around the TGR executed combined dispatching, managed to control five numbered major floods, and achieved impressive results in flood control. The TGR achieved 30.5 billion m^3 of total flood storage volume, which was a new historic record. The water level in the lower reach was decreased by a maximum of four meters, the duration in which the water levels at the main control stations in the lower reach exceeded the warning water level was shortened by 13.4 days on average, and thus use of the Jingjiang flood diversion area and consequent losses were avoided.

Since 2020, the Wudongde and Baihetan projects in the lower reach of Jinsha River were successively completed, so the combined dispatching model will be further expanded, and the comprehensive benefits of the reservoir group combined dispatching model centered on the TGR in the upper and middle reaches of the Yangtze River will be materialized to greater extent.

本章小结：

筑梦追梦一百年，勘察论证半世纪，建设运营廿六载。伴随着改革开放和民族复兴的历史进程，三峡工程已由梦想变为现实，在防洪、发电、航运和水资源利用等方面的巨大综合效益已经显现：镇锁洪魔，护佑荆江两岸；水利发电，构筑能源动脉；改善航运，畅通"黄金水道"；供水补水，涵养流域生态……三峡工程所发挥的效益，惠及民生所系的方方面面。三峡工程的成功建设和运行，构建了开发治理长江的新格局，推动了我国从水电开发大国向水电开发强国转变，展示了中国水利水电工程建设科技创新的辉煌成就，创造了人类改造大江大河的伟大奇迹，在中国经济和社会发展中产生了重大而深远的影响。

参考文献：

[1] 本书编委会. 百问三峡 [M]. 北京：科学普及出版社，2012.

[2]《中国三峡建设年鉴》编纂委员会. 中国三峡建设年鉴（2009）[J]. 宜昌：中国三峡建设年鉴社，2009.

[3]《中国三峡建设年鉴》编纂委员会. 中国三峡建设年鉴（2010）[J]. 宜昌：中国三峡建设年鉴社，2010.

[4] 水利部长江水利委员会. 长江流域综合规划（2012—2030）[EB]. 2012.

[5]《中国三峡建设年鉴》编纂委员会. 中国三峡建设年鉴（2020）[J]. 宜昌：中国三峡建设年鉴社，2020.

[6]《中国三峡建设年鉴》编纂委员会. 中国三峡建设年鉴（2021）[J]. 宜昌：中国三峡建设年鉴社，2021.

第七章 | 三峡工程的运行效益
Chapter 7　Operational Benefits of the TGP

Chapter Summary:

The dream coming true after a century, survey and demonstration in half a century, construction and operation in 26 years. In the historical background of the reform and opening-up and the national rejuvenation, the TGP has become a reality. The massive comprehensive benefits in terms of flood control, power generation, shipping and water resources utilization have been materialized: eliminate the threat of flood, and protect both banks of Jingjiang River; through hydropower generation, ensure energy security; improve shipping, clear the "golden waterway"; water supply and compensation, conservation of the ecosystem in the basin... The benefits of the TGP cover all aspects of people's livelihood. Successful construction and operation of the TGP has pushed control of the Yangtze River to a new level, and changed the hydropower development capability of China from excellent to outstanding. This project is the outcome of the technological innovation in water conservancy and hydropower of China and a great miracle in the history of mankind transforming great rivers and lakes, which brings far-reaching influence on economic and social development of China.

References:

[1] Editorial committee of this book. *Questions about the Three Gorges* [M]. Beijing: Popular Science Press, 2012.

[2] Editorial Board of *China Three Gorges Construction Yearbook. China Three Gorges Construction Yearbook* (2009) [J]. Yichang: Publishing House of Construction Yearbook of China's Three Gorges, 2009.

[3] Editorial Board of *China Three Gorges Construction Yearbook. China Three Gorges Construction Yearbook* (2010) [J]. Yichang: China Three Gorges Construction Yearbook Press, 2010.

[4] Changjiang Water Resources Commission under the Ministry of Water Resources. *The Comprehensive Planning for Yangtze River Basin* (2012-2030)[EB].2012.

[5] Editorial Board of *China Three Gorges Construction Yearbook. China Three Gorges Construction Yearbook* (2020) [J]. Yichang: China Three Gorges Construction Yearbook Press, 2020.

[6] Editorial Board of *China Three Gorges Construction Yearbook. China Three Gorges Construction Yearbook* (2021) [J]. Yichang: Publishing House of Construction Yearbook of China's Three Gorges, 2021.

> 阅读提示：

从三峡工程开工之初，国家就确立了"环境与工程建设同步"的指导原则。在工程论证、设计、建设和运行期间，国家采取了一系列保护性政策和措施，各级政府和部门围绕三峡生态与环境保护开展了大量的工作，投入了巨大的人力、物力和财力，为保护三峡生态与环境进行了不懈努力。

三峡工程生态保护政策和措施的实施，取得了明显成效。自蓄水以来，三峡工程全面运行并发挥着防洪、节能减排、"黄金水道"、保护生物多样性的综合效益，承担起"管理三峡，保护长江"的战略新任务。

From the very beginning of the Three Gorges Project (TGP), the state set forth a principle that called for "environmental protection and project construction in parallel." During the project's appraisal, design, construction and operation, the state implemented a range of protective policies and measures. Likewise, the governmental and authorities at all levels focused their efforts on the ecological and environmental protection of the TGP. They committed massive human, material, and financial resources to the herculean endeavor ecological and environmental protection in the TGP.

The TGP's ecological conservation policies and measures have proven remarkably effective. Since impoundment, the TGP has been operating at full capacity. Furthermore, it has generated massive benefits in terms of flood control, energy conservation, and emission reduction as expected, and has fulfilled its function as a golden waterway and a biodiversity conservation area.

Chapter 8 >>>>

第八章
三峡工程的生态保护
Ecological Conservation in the Three Gorges Project

第一节 三峡工程生态保护的论证和规划

三峡工程作为关系到千秋万代的大事，从工程论证阶段就同步开展对生态与环境影响的论证和规划，是国内生态与环境研究参与人员最多、研究最系统深入的工程。

一、三峡工程生态保护的论证

20世纪50年代，水利部长江流域规划办公室开展了长江流域规划和三峡工程设计论证工作，同时有针对性地开展了水文泥沙、水环境、水生生物、局地气候、陆生生物、移民安置、地质地震、人群健康、社会经济等数十个方面、近百个环境因子的基础研究，提出了初步成果并编入了长江流域规划要点报告。

从1979年开始，以长江水资源保护科学研究所和中国科学院为主的四十多个大专院校和科研机构通力合作，全面、系统地开展了多个三峡工程正常蓄水位方案的环境影响评价及对策研究工作，并建立了三峡工程生态与环境信息系统。此后，作为三峡工程生态与环境保护工作科学性的重要保障，随着工程的推进和认识的深入，三峡工程生态与环境研究工作越来越受到重视，越来越多的国内外生态与环境研究专家投身到与三峡相关的科研中来。

1985年，国家计划委员会和国家科学技术委员会成立了生态与环境论证专家组，对正常蓄水位150～180米方案的环境影响进行了评价。1986年，在国务院三峡工程论证领导小组的组织领导下，以国家科委、国家计委聘请的专家组为基础，增聘了部分专家，成立了长江三峡工程生态与环境专家组。专家组由生态、环境、水利等方面的55名专家组成。专家组于1988年1月

Chapter 8　Ecological Conservation in the Three Gorges Project

Section 1　Appraisal and Planning of Ecological Conservation for the Three Gorges Project

The TGP is a landmark project that will influence countless future generations. As such, its appraisal was conducted in parallel with the appraisal of its ecological and environmental impacts and the planning in this regard. This project had the largest number of personnel committed to ecological and environmental investigations, as well as the most systematic research among all the projects in China.

I. Appraisal of Ecological Conservation for the Three Gorges Project

In the 1950s, the Yangtze Valley Planning Office under the Ministry of Water Resources conducted design appraisal of the TGP, conducted basic research into dozens of aspects related to the project including hydrology, sediment, water environment, aquatic organisms, local climate, terrestrial organisms, resettlement, geology, seismic activity, population health, and social economy. Researchers also investigated nearly a hundred environmental factors and obtained the preliminary results that were included into the Yangtze Valley Planning Priorities Report.

Since 1979, over 40 universities and research institutions that were spearheaded by the Changjiang Water Resources Protection Institute and the Chinese Academy of Sciences collaborated closely in conducting comprehensive and systematic environmental impact assessments of multiple normal water level alternatives and related solutions. This collaborative endeavor eventually engendered the TGP Ecological and Environmental Information System. As progress was made and researchers' understanding of the project deepened, the ecological and environmental research of the TGP received more attention. Additionally, an increasing number of domestic and foreign ecological/environmental experts were committed to TGP-related research, which was a crucial factor in ensuring the soundness of the ecological and environmental protection efforts implemented for the TGP.

In 1985, the State Planning Commission and the State Scientific and Technological Commission instated the Ecological and Environmental Appraisal Expert Panel, which con-

完成了《长江三峡工程生态环境影响及对策论证报告》。1991年3月，专家组提出了生态与环境专题的预审意见，同年7月国务院三峡工程审查委员会审定了可行性研究阶段的评价成果。

1991年12月，根据国务院三峡工程审查委员会的要求，受三峡总公司筹建处委托，中国科学院环境评价部和长江水资源保护科学研究所联合编写了《长江三峡水利枢纽环境影响报告书》，经水利部认真审查、评议并提出修改补充意见后，报送国家环境保护总局审批。1992年2月，国家环境保护总局正式批准了这份报告书。经审查批准的环境影响报告书成为三峡工程生态环境保护的指导性文件，为开展三峡工程的生态环境保护工作提供了依据。此后，与三峡工程有关的环境影响研究并没有停止，而是随着工程的进展和认识的深入不断推进，并成为一项长期性的工作。

《长江三峡水利枢纽环境影响报告书》包括24类74个环境因子，对三峡工程的环境与资源的影响进行了系统、全面的分析和评价。根据对三峡工程生态与环境的影响的评价，该报告书提出了一系列对策建议。

ducted environmental impact assessment of the 150-180 m normal water level. In 1986, the Three Gorges Project Appraisal Leading Group of the State Council established the Three Gorges Project Ecological and Environmental Expert Panel. This group was comprised of the experts recruited by the State Scientific and Technological Commission and State Planning Commission, but also included some other environmental specialists. The expert panel consisted of 55 experts, who specialized in ecology, environmental protection, and water conservancy. In January 1988, the expert panel completed its *Report on the Ecological and Environmental Impact and Solutions for the Three Gorges Project*. In March 1991, the expert panel presented its preliminary review comments on the ecological and environmental subject. In July that year, the Three Gorges Project Review Committee of the State Council audited the appraisal results of the feasibility study stage.

In December 1991, as requested by the Three Gorges Project Review Committee of the State Council and entrusted by the Preparatory Office of the China Yangtze Three Gorges Project Development Corporation (CYTGPDC), the Environmental Assessment Division of Chinese Academy of Sciences and the Changjiang Water Resources Protection Institute jointly prepared the *Environmental Impact Statement of Three Gorges Hydroproject*. After a meticulous audit, deliberations, and revisions, this document was submitted to the State Environmental Protection Administration for approval. In February 1992, the State Environmental Protection Administration officially approved this document. The audited and approved Environmental Impact Statement became the guiding document for ecological and environmental protection with regard to the TGP and the basis for the ecological and environmental protection efforts in the project. Since then, the environmental impact research of the TGP continued unabated with the progress and deeper understanding of the project and eventually became a long-term endeavor.

The *Environmental Impact Statement of Three Gorges Hydroproject* covered 74 environmental factors of 24 categories, and encompassed systematic, comprehensive analysis and assessment of the environment and resource-related impacts of the TGP. On the basis of the ecological and environmental impact assessment of the TGP, this document offered a range of solutions and suggestions.

First, with consideration of the TGP's overall development, plans for managing and utilizing the land in the TGP reservoir area should be carefully formulated. This would mean rural and urban development, resettlement, resources development, water quality protection and environmental improvement must be a part of the overall plan. A comprehensive development plan that balanced economic, social and environmental benefits was also formulated. Next, a rational overall plan for environmental pollution control to be implemented in the TGP reservoir area was put in place. The document also suggested that the government in-

第一，结合三峡工程的总体开发，认真做好三峡库区国土整治和利用规划，将城乡建设、移民安置、资源开发、水质保护、环境整治等纳入总体规划中，制定出经济、社会和环境效益相统一的综合开发方案。第二，搞好三峡库区环境污染防治整体规划。第三，加强长江中上游林业建设，做好水土保持工作。第四，加强珍稀、濒危物种与资源保护。第五，加强文物保护和考古发掘工作。第六，优化水库调度，尽可能满足生态和环境保护与建设的要求。第七，三峡工程建成后，在发电收益中提取一定比例，建立三峡环境基金，用于生态和环境保护与建设。第八，继续开展三峡工程生态与环境科学研究与监测，建立健全三峡工程生态与环境监测网络。第九，建立健全三峡工程环境管理系统，制定和完善三峡工程环境保护法规。第十，加强环境保护的宣传、教育，提高环境保护意识。

《长江三峡水利枢纽环境影响报告书》总的评价结论认为：三峡工程对生态与环境的影响有利有弊，必须予以高度重视，只要对不利影响从政策、工程措施、监督管理以及科研和投资等方面采取得力措施并切实执行，使其减小到最低限度，生态与环境问题不致影响三峡工程的可行性。

二、三峡工程生态保护的规划

在三峡工程论证、设计、建设与运行过程中，以人为本、防治污染、保护生态、循环经济、自然和谐的规划设计理念一直贯穿始终，并随着三峡工程的开发和社会的关注，不断更新设计理念，调整、拓展设计内容。各项规划设计共同形成了长江三峡水利枢纽环境保护措施体系。

1. 枢纽工程生态保护的规划

1993 年，国务院三峡建委批准了《长江三峡水利枢纽初步设计报告（枢纽工程）》。该报告第十一篇"环境保护"内容包括水质、物种资源及栖息地、环境地质、泥沙和河道冲淤、施工区环境保护和生态与环境监测系统等方面的环境保护设计方案。

1995 年，国家环境保护局和国务院三峡建委办公室联合批准了《长江三峡工程施工区环境保护实施规划》。该规划对施工区污染治理与预防措施、水

tensify forestry development in the upper and middle reaches of the Yangtze River and maintain proper soil and water conservation. It also called for the heightening of protections for rare and endangered species and resources. The preservation and archaeological excavation of cultural relics was to be intensified, and the government was urged to optimize reservoir dispatching to maximally meet the requirements for both ecological and environmental protection and project construction. After the completion of the TGP, a part of the revenue generated by power generation was to be allocated to create a TGP environment fund that would be used for ecological and environmental protection. Ecological and environmental research and monitoring of the TGP was to continue, with the establishment and improvement of the TGP ecological and environmental monitoring network. This document also called for the establishment and improvement of the TGP Environmental Management System, as well as the enactment and strengthening of environmental protection laws and regulations regarding the TGP. Finally, there was a recommendation to intensify publicity and education on environmental protection and enhance the awareness of environmental protection.

The *Environmental Impact Statement of the Three Gorges Hydroproject* drew the following conclusions: The TGP brings both positive and negative ecological and environmental impacts, which require intensive attention. As long as contributive policies, engineering measures, supervision, managements, scientific research and investments are in place and faithfully implemented to minimize the negative impacts, the ecological and environmental issues will not compromise the feasibility of the TGP.

II. Planning the Ecological Conservation of the Three Gorges Project

The appraisal, design, construction and operation of TGP have always been guided by a planning and design philosophy that puts people foremost and focuses on pollution control, ecological protection, the circular economy and fostering natural harmony. As the TGP's development progressed and garnered increasing attention from society, the design philosophy for the project has been constantly updated, with the design concepts being adjusted and expanded. The plans and designs eventually materialized into the environmental protection measures of the TGP hydroproject.

1. Plan for the hydroproject's ecological conservation

In 1993, the TGPCC approved the *Report on Preliminary Design of Three Gorges Hydroproject (Hydroproject)*. Part 11 of the report, entitled "Environmental Protection," encompasses the environmental protection design schemes with regard to water quality, species resources and their habitats, environmental geology, sediment, river siltation, environmental

土保持措施、环境绿化措施、人群健康保护措施、环境监测等进行了系统的规划设计。

2008年,湖北省宜昌市人民政府批准了《三峡水利枢纽工程管理区保护与利用规划》,计划实施枢纽管理区水保绿化、污染防治、水生生物保护、交通等规划项目5大类共50项,重要的环境保护项目包括陈家冲渣场、长江珍稀鱼类保育中心等16个。

2. 移民安置生态保护的规划

1994年至1997年,湖北省的宜昌、秭归、兴山、巴东四县和重庆市的涪陵、万州等16区县分别编制了《长江三峡工程淹没处理及移民安置规划报告》。1998年,长江委长江勘测规划设计研究院汇总编制了《长江三峡工程水库淹没处理及移民安置规划报告》。该报告第十篇"移民安置区环境保护规划"中,包括土地资源开发利用环境保护、城集镇环境保护、工矿迁建企业环境保护、生态建设、人群健康保护、生态环境监测与管理措施等6个方面的内容。

为减轻三峡库区环境保护的压力,1999年国家对移民政策进行了重大调整:一是调整移民方针,由原先的靠后就地安置调整为本地安置与异地安置结合,并鼓励和引导更多的农村移民外迁安置;二是对污染严重、经济效益差、治理无望的小企业不再复建,而是加大关闭和破产力度。据此,国务院三峡建委办公室组织编制了《长江三峡工程库区湖北省移民安置区环境保护实施计划》和《长江三峡工程库区重庆市移民安置区环境保护实施计划》,对初步设计报告和移民安置规划报告提出的环境保护措施方案进行了调整,并对部分资金使用方向进行了调整。

protection in construction areas, and ecological and environmental monitoring systems.

In 1995, the State Environmental Protection Administration and the Office of the TGPCC jointly approved the *Implementation Plan for Environmental Protection in the Construction Area of the Three Gorges Project*. This document encompasses systematic plans and designs for pollution prevention anchoring control, soil and water conservation measures, environmental greening measures, population health protection, and environmental monitoring in the project construction area.

In 2008, the People's Government of Yichang, Hubei approved the *Plan for Protection and Utilization of the Three Gorges Hydroproject Management Area*. This document includes 50 plans across 5 categories regarding various aspects of the hydroproject, such as water protection, greening, pollution prevention and control, aquatic organism protection and traffic. The document covers 16 environmental protection related projects including the Chenjiachong Disposal Area and Changjiang Rare Fish Conservation Center.

2. Plan for ecological conservation during resettlement

From 1994 to 1997, 16 districts and counties, including Yichang, Zigui, Xingshan and Badong of Hubei and Fuling and Wanzhou of Chongqing, respectively formulated their own *Report on Planning for Treatment of Inundation and Resettlement in the Three Gorges Project*. In 1998, the Changjiang Institute of Survey, Planning, Design and Research, which is under the Changjiang Water Resources Committee, formulated its *Report on Planning for Treatment of Reservoir Inundation and Resettlement in the Three Gorges Project*. Part 10 entitled "Plan for Environmental Protection in Resettlement Area" in the report covers 6 aspects including environmental protection for land resources development and utilization, environmental protection in urban areas, environmental protection in relocation of industrial and mining enterprises, ecological development, population health protection, and monitoring as well as management for ecosystem and environment.

In order to relieve the environmental pressure of the TGP reservoir area, the state adjusted the resettlement policy in 1999. First, the resettlement principle was adjusted. The principle of relocating residents within the local region was changed to relocating them to another region. Additionally, more rural residents were encouraged and guided to relocate to another region. Second, small enterprises characterized by severe pollution, poor economic benefit and impossibility of improvement would not be rebuilt and instead were to be closed and dissolved. Under these new principles, the Office of the TGPCC formulated the *Implementation Plan for Environmental Protection in the Resettlement Area in Hubei in the Three Gorges Project Reservoir Area* and *Implementation Plan for Environmental Protection in the Resettlement Area in Chongqing in the Three Gorges Project Reservoir Area*. It also adjusted

3. 其他生态保护设计和规划

为长期、系统地观察三峡工程对生态与环境的影响,并为三峡库区及长江流域的生态与环境建设提供科学依据,1996年,国务院三峡建委办公室组织编制了《长江三峡工程生态与环境监测系统实施规划》。依据该规划,中国政府组建了由环保、水利、农业、林业、气象、卫生、国土、地震、交通、中国科学院、三峡总公司、湖北省、重庆市的有关部门和单位组成的跨地区、跨部门、多学科、多层次的三峡工程生态与环境监测系统。该监测系统根据工程进入不同阶段后生态与环境保护重点不同而调整,2005年进行了首次调整,2009年(工程由建设期全面转入运行期)进行了第二次调整,新增加了工程运行初期公众关注的消落区、水库经济鱼类、江湖生态环境监测等专题。

在三峡工程建设过程中,中央、地方政府,以及三峡总公司等,相继组织编制并实施了《三峡库区及其上游水污染防治规划(2001—2010年)》《三峡库区地质灾害防治总体规划》《三峡水库周边绿化带建设工程规划》《三峡库区及其上游水污染防治规划(修订本)》、"退耕还林"等专项规划、《长江三峡工程环境保护补偿项目实施计划》等。

2007年,国务院三峡建委办公室组织编制了《三峡工程生态环境建设与保护试点示范专项计划方案》(以下简称"7+1专项方案")并开展了试点示范工作。该方案主要包括消落区环境治理、支流水华应急处置与长效防治、农村截污处理(面源污染防治)、城镇截污处理(点源污染防治)、支流饮用水源安全保障建设、库岸带生态屏障建设、生物物种多样性保护等7个方面的生态建设任务,并对生态环境监测系统进行效能评估。

the environmental protection measures specified in the preliminary design report and resettlement planning report and changed the purpose of some funds.

3. Other ecological conservation designs and plans

The Office of the TGPCC formulated the *Implementation Plan for Ecological and Environmental Monitoring System of the Three Gorges Project* in 1996 to observe the ecological and environmental impacts of the TGP on a long-term and systematic basis and to establish the scientific basis for the ecological and environmental development in the TGP reservoir area and Yangtze River basin. In accordance with this plan, the Chinese government instituted the trans-regional, inter-departmental, multidisciplinary and multi-level TGP ecological and environmental monitoring system comprised of the relevant divisions and departments of the Ministry of Ecology and Environment, Ministry of Water Resources, Ministry of Agriculture, State Forestry Administration, Meteorological Administration, Ministry of Health, Ministry of Land and Resources, Earthquake Administration, Ministry of Transport, Chinese Academy of Sciences, CYTGPDC, Hubei, and Chongqing. This monitoring system was adjusted according to the priorities of ecological and environmental protection during different phases of the project. This implementation plan was adjusted for the first time in 2005. It was modified again in 2009 (at which point the project was in full operation after construction). The second adjustment included new issues concerning the public during the preliminary stages of operation, which were added into the monitoring system. It covered the drawdown area, commercial fish species in the reservoir, and river-lake ecological and environmental monitoring.

During construction of the Three Gorges Project, the central government, local governments, and the CYTGPDC formulated and implemented the *Water Pollution Prevention Planning for the Three Gorges Reservoir Area and Its Upper Reaches (2001-2010)*, *Overall Plan for Prevention and Control of Geological Hazards in TGP Reservoir Area*, *Plan for Construction of Green Belts in the Surrounding Areas of Three Gorges Reservoir*, *Water Pollution Prevention Planning for the Three Gorges Reservoir Area and Its Upper Reaches* (revised). They also initiated special plans for different issues, such as "returning farmland to forests," and initiated the *Implementation Plan for Environmental Protection Compensation in the Three Gorges Project*.

In 2007, the Office of the TGPCC formulated the *Special Plan for Pilot Project of Ecological Environment Development in the Three Gorges Project* (hereinafter referred to as "the 7+1 plan"), and implemented the pilot project. This plan mainly addressed 7 ecological endeavors including environmental improvement in the drawdown area, emergency response to water bloom in tributaries and long-term control, pollutant interception in rural areas (non-

第二节 三峡工程的生态保护机制、措施和实施

三峡工程的建设过程，也是生态环境保护意识逐步提高、生态环境保护理念逐步树立和践行的过程。在此过程中，三峡总公司根据"建设三峡，开发长江"历史使命，逐步形成了"建好一座电站，带动一方经济，改善一片环境，造福一批移民"的"四个一"水电开发理念和全过程环境管理理念，是新形势下坚持走可持续水电开发之路的企业行为典范。

一、三峡工程生态保护的管理体系和机制

在三峡工程建设之初的 1995 年，国务院三峡建委成立了全面协调枢纽工程区、水库淹没区、移民安置区及其他受三峡水利枢纽工程影响区域的生态与环境保护工作的三峡工程生态与环境保护协调小组。

协调小组内各成员单位在支援三峡工程生态与环境保护工作中的分工和侧重点为：国家环境保护局侧重于以三峡库区为主的污染治理和控制；水利部负责三峡库区的水土保持、水文泥沙水质观测和水资源保护工作；农业部负责移民安置中农业生态、乡镇企业污染问题和水生生物保护；林业部负责三峡库区的防护林体系建设和物种保护；卫生部负责坝区和库区的疾病监测、移民安置区的防病，以及健全三峡库区卫生防疫机构；交通部负责三峡库区航道管理与航运环境保护工作；中国科学院要依靠自身的综合科技优势，围绕三峡工程引起的主要环境问题，开展应用研究，提供实用技术；重庆市、湖北两省市负责本地区水库淹没处理和移民安置规划、计划和实施中的生态与环境保护各项措施的落实；国务院三峡建委办公室负责组织协调列入枢纽

point source pollution control), pollutant interception in urban areas (point source pollution control), protection of drinking water sources at tributaries, creation of an ecological barrier in reservoir bank area, and biodiversity protection. Additionally, it featured a performance evaluation of the ecological and environmental monitoring system.

Section 2 Ecological Conservation Mechanism and Measures for the Three Gorges Project and Their Implementation

The development process of the TGP also involves gradually raising awareness for ecological and environmental protection as well as gradually fostering and practicing a philosophy of ecological and environmental protection. In this process, the CYTGPDC, bearing in mind its monumental task of "building the TGP and developing the Yangtze River," gradually established a development philosophy for hydropower that centered on "building a power station, driving the economy of a region, improving the environment, and benefiting the resettled population." It also cultivated a philosophy of whole-process management. These two mindsets of development and management have become the model of corporate behavior for sustainable hydropower development in this new era.

I. Management System and Mechanism for Ecological Conservation of the Three Gorges Project

In 1995, the year construction of the TGP commenced, the TGPCC under the State Council instated the Three Gorges Project Ecological and Environmental Protection Coordinating Group, which is responsible for coordinating ecological and environmental protection efforts in the hydroproject area, the zone for reservoir inundation, the resettlement area and other areas impacted by the TGP.

The following is a list of the responsibilities and job priorities of the members of the Coordinating Group. The State Environmental Protection Administration is focused on pollution prevention and control in the TGP reservoir area. The Ministry of Water Resources handles

工程概算的生态与环境监测系统以及各类生态保护专项的实施;国务院三峡建委移民开发局负责协调三峡库区与移民安置有关的环境保护规划、计划;三峡总公司负责协调坝区的环境保护工作。

2001年9月初,三峡库区水污染和地质灾害防治工作领导小组成立。国家环保总局、国土资源部分别为三峡库区水污染和地质灾害防治领导小组组长单位。副组长和主要成员分别来自国家计委、财政部、建设部、国务院三峡建委办公室和移民开发局、水利部以及重庆市、湖北省、云南省、贵州省、四川省和长江委等部门。

为适应施工期环境保护工作需要,三峡集团在第三期工程建设过程中引入了工程环境监理,进一步完善了枢纽工程环境管理体系。管理体系中各部门和单位的分工明确、责任清晰,相继制定了指导工程生态与环境保护工作的制度、计划和技术规范,工程各阶段的环境保护协调管理工作成效显著。

soil and water conservation, sediment and water quality observation, and water resources protection. The Ministry of Agriculture is responsible for agricultural ecology in resettlement projects, pollution control over township enterprises, and aquatic organism protection. The State Forestry Administration covers shelter forest planting and species protection in the TGP reservoir area. The Ministry of Health is responsible for monitoring of disease in the dam and reservoir areas, disease prevention in the resettlement area, and improvement of the population's health, and epidemic prevention organizations in the TGP reservoir area. The Ministry of Transportation is responsible for waterway management and shipping environment protection in the TGP reservoir area. The Chinese Academy of Sciences conducts applied research and provides practical technologies to solve the main environmental issues in the TGP by leveraging its superior technology. Chongqing and Hubei are responsible for the treatment of reservoir inundation, resettlement planning, and the supervision of planning and implementation of the ecological and environmental protection measures in their respective jurisdictions. The TGPCC under the State Council is responsible for coordinating the implementation of the ecological and environmental monitoring system and the ecological conservation programs included in the hydroproject estimate. The Resettlement Development Bureau of the TGPCC under the State Council is responsible for coordinating the environmental protection plans and programs related to resettlement in the TGPCC reservoir area. The CYTGPDC is charged with coordinating the environmental protection efforts in the dam area.

The TGP Reservoir Area Pollution and Geological Hazard Control Leading Group was instated in early September 2001. The State Environmental Protection Administration and Ministry of Land and Resources serve as the leaders of the TGP Reservoir Area Pollution and Geological Hazard Control Leading Group. The deputy leaders and main members are from the State Planning Commission, Ministry of Finance, Ministry of Construction, the Office of the TGPCC and Resettlement Development Bureau under the State Council, Ministry of Water Resources, Chongqing, Hubei, Yunnan, Guizhou, Sichuan, and Changjiang Water Resources Committee.

In order to support environmental protection efforts during construction, the CTG introduced engineering environmental supervision in Phase III of the project, which further improved the hydroproject's environmental management system. The departments and divisions in the management system have clearly defined functions and responsibilities. Certain regulations, plans, and technical specifications have been formulated to guide the ecological and environmental protection efforts of the project, and the environmental protection coordination and management in each phase of the project have achieved remarkable results.

1. 枢纽施工区环境管理

三峡工程枢纽施工区环境管理实行的是三峡总公司统一管理与各施工单位分级管理的体制。根据相关环境保护法规、环境影响报告书批复意见要求，从1993年开始，三峡总公司逐步建立并完善了环境管理的组织体系，成立了"环境及文物保护委员会""环境保护处""坝区环境保护中心"等机构，负责三峡工程建设中的环境保护。

为确保工程施工与环境保护相协调进行，三峡总公司依据国家有关环境保护法规和经国家环境保护局与国务院三峡建委联合批准的《三峡工程施工区环境保护实施规划》，并结合工程建设的实际情况，积极开展环境保护工作，采取和实施了一系列环境控制措施和管理办法，使施工区的环境得到较好的控制与保护。

2005年，为适应国内大型水利水电项目施工期环境保护的新形势，从整体上加强对施工区的环境保护管理，三峡总公司决定在三期工程中施行环境监理制。由长江水资源保护科学研究所承担该项环境监理工作，在坝区设长江水资源保护科学研究所三峡工程环境监理部，按照三峡总公司的要求对施工区进行系统和规范的环境管理。

2. 移民安置区环境管理

三峡工程移民工作采用"中央统一领导，分省负责，县为基础"的管理体系。国务院三峡建委办公室移民安置规划司作为中央机构总体负责三峡库区移民搬迁安置工作。湖北省移民局和重庆市移民局作为省级移民工作的责任单位，三峡库区各县移民局作为移民工作的具体实施单位。

1. Environmental management in the hydroproject construction area

The environmental management of the hydroproject construction area in the TGP was under the centralized authority of the CYTGPDC, and each construction contractor was assigned a certain level of management authority. From 1993, in accordance with the applicable environmental protection laws and the official approval of the Environmental Impact Statement, the CYTGPDC gradually established and improved the organizational system for environmental management. It also instated organizations, such as the "Environment and Cultural Relics Protection Committee," "Environmental Protection Division," and "Dam Area Environmental Protection Center," to conduct environmental protection during the TGP construction.

In order to ensure the harmony between project construction and environmental protection, in accordance with the applicable environmental protection laws and the *Implementation Plan for Environmental Protection in the Construction Area of the Three Gorges Project* jointly approved by the State Environmental Protection Administration and the TGPCC, and with consideration of the actual conditions of the project, the CYTGPDC vigorously made efforts in environmental protection and implemented a range of environmental control measures and management methods, so that the environment in the construction area was properly controlled and protected.

In 2005, the CYTGPDC decided to implement the environmental supervision system in Phase III of the project to meet the new requirements for environmental protection in construction of the large water conservancy and hydropower projects in the country and to strengthen the overall environmental protection and management in the construction area. The Changjiang Water Resources Protection Institute performed environmental supervision. In the dam area, the TGP Environmental Supervision Department of the Changjiang Water Resources Protection Institute was instated to perform systematic and standardized environmental management of the construction area as per the requirements of the CYTGPDC.

2. Environmental management in the resettlement area

The resettlement endeavor in the TGP was put under the management system constituted of "central government leadership, provincial responsibilities, and county-level implementation". The Resettlement Planning Department subordinate to the Office of the TGPCC was the central authority in charge of relocation and settlement of the migrants in the reservoir area. The Hubei Immigration Bureau and Chongqing Immigration Bureau were the provincial authorities responsible for resettlement, while the immigration offices of the counties in the TGP reservoir area were the entities that conducted the resettlement efforts.

在三峡移民工作中，先后颁布了《长江三峡工程建设移民条例》《长江三峡工程建设移民监理规定（试行）》《长江三峡工程建设水库移民综合监理管理暂行办法》《关于进一步加强三峡工程水库移民综合监理工作的通知》等法规、制度，这些法规、制度均包含了移民安置工作中环境保护的要求，初步建立和形成了三峡移民环保法规体系，使移民安置区环保工作能够依法有序地顺利开展。

三峡工程移民过程中引进了移民综合监理，保障了移民工程实施进度、质量、投资的规范实施。

二、枢纽工程的生态保护措施和实施

1. 施工污染治理效果显著

施工区设 4 处砂石骨料生产系统和 6 处混凝土拌和系统，生产废水经配套建设的生产废水处理系统进行处理，加强环保设施的运行管理以确保处理效果，对沉渣、中水尽量回收利用于施工生产活动。施工区排水采用雨污分流制，试验性蓄水期建设了左岸乐天溪生活污水处理厂。干流监测结果显示，外排废水对施工江段水质影响较小。

采用湿法作业、密闭运输等主要手段控制施工大气污染，施工粉尘基本能得到有效控制，基本满足大气环境标准要求。通过施工布局优化，辅以优选设备、采用先进施工技术等措施，降低施工噪声强度，同时合理安排施工作业时间等以减轻对敏感对象的影响。施工现场噪声控制总体有效。建设施工区生活垃圾卫生填埋场工程，对渗滤液实施专项处理，处理出水水质基本满足标准要求。

For the purpose of resettlement in the TGP, certain laws and regulations were enacted, such as the *Regulations on Residents-Resettlement for the Yangtze River Three Gorges Project Construction, Regulations on Resettlement Supervision for the Yangtze River Three Gorges Project (Trial), Interim Measures for the Management of Comprehensive Supervision of Resettlement for the Yangtze River Three Gorges Project*, and *Notice on Further Strengthening Comprehensive Supervision of Resettlement for the Yangtze River Three Gorges Project*. These laws and regulations all encompassed the requirements for environmental protection during resettlement and preliminarily constituted the TGP resettlement and environmental protection law system that ensured the successful implementation of the environmental protection efforts in the resettlement area.

For the purpose of resettlement in the TGP, the comprehensive resettlement supervision system was introduced, guaranteeing fast progress and high-quality, legitimate investments for the resettlement efforts.

II. Ecological Conservation and Its Implementation in the Hydroproject

1. Remarkable effectiveness of pollution control during construction

The construction area was furnished with 4 sand and stone aggregate production systems and 6 concrete mixing systems. The industrial waste water was treated by the waste water treatment system dedicated to the project, and the operation and management of the environmental protection facilities were intensified to ensure effective wastewater treatment. The sediment and reclaimed water were recycled and used for the construction work to the maximum extent. The construction area was furnished with a rainwater-sewage separation system. During the trial storage period, the Letianxi Domestic Wastewater Treatment Plant was built on the left bank. According to the results of trunk stream monitoring, the discharged wastewater had only a minor impact on the water quality in the construction area river segment.

The approaches such as wet operation and enclosed transport were used to control atmospheric pollution during construction. The construction dust was essentially under effective control and the applicable atmospheric environment standards were essentially met. Measures such as an optimized construction layout, carefully selected equipment, and advanced construction technologies reduced noise intensity during construction, and the construction schedule was logically arranged to mitigate the impact on sensitive objects. Noise control at the construction site was generally effective. A sanitary landfill was placed at the construction site to receive domestic waste, the leachate was treated in a dedicated treatment

2. 工程区新增水土流失得到有效控制

采取工程措施与植物措施治理新增水土流失、修复工程区生态功能。施工结束后，以封山育林方式保护施工区内的原生植被，适度恢复乡土灌草优势种，提高与周边环境的同质性。施工迹地植被覆盖度超过95%，重构绿地生态系统，逐步恢复工程占地区生态功能。

3. 水文泥沙情势影响缓解措施总体得到落实

水库遵循"蓄清排浑"调度原则，根据实际情况，对运行调度指标进行了调整。开展了汛期沙峰排沙调度试验和消落期库尾泥沙减淤调度试验，研究落实减缓淤积的优化调度方案，初步分析表明，汛期沙峰调度对减轻水库泥沙淤积有一定效果。但由于影响水库冲淤变化的因素很复杂，效果有待进一步分析。

通过炸礁、疏浚和筑坝等治理措施，变动回水区航道治理措施得到落实。变动回水区航运条件得到明显改善，但局部库段在枯季库水位消落时出现淤积碍航情况，通过疏浚等措施保证了通航条件。通过枯期补偿调度和护底整治工程落实了确保葛洲坝枢纽通航的措施。水利部门实施了长江中下游河势控制工程，对缓解河势变化起到了积极作用，交通部门实施了坝下游碍航浅滩整治工程，基本落实了环境影响报告提出的治理措施。

system, and the quality of the treated water largely met the applicable standards.

2. Effective control of additional water and soil loss in the project area

Both engineering and botanical measures were taken to control the additional water and soil loss and restore the ecological functions in the project area. After construction was completed, hills were set apart for tree growing as an effective method to protect the native vegetation in the construction area, and the dominant species of shrub and herbaceous plants were appropriately restored to enhance their homogeneity with the ambient environment. The vegetation coverage of the construction area exceeded 95%, the green space ecosystem was reconstituted, and the ecological functions of the project occupied land areas were gradually restored.

3. Implementation of mitigation measures against the impact of hydrological sediment

Under the dispatching principle of "impound clear water and discharge turbid water," the dispatching criteria for the reservoir were adjusted. The flood season sediment peak desilting dispatching test and drawdown period reservoir tail sediment reduction dispatching test were conducted to investigate and implement the optimized dispatching scheme intended to mitigate sediment accumulation. According to the preliminary analysis, the sediment peak dispatching in flood season was effective in mitigating sediment accumulation in the reservoir. However, the factors affecting the scouring and silting of the reservoir were quite complex, so further analysis was needed.

Through the methods such as rock blasting, dredging, and damming, measures to improve the waterway in the varying backwater zone were implemented. The shipping conditions in the varying backwater zone were noticeably improved. However, the sediment accumulation in partial reservoir segments that occurred during reservoir water level drawdown in the dry season impeded shipping, so dredging had to be conducted to clear the waterway. Compensation dispatching in the dry season and river bottom protection ensured that the Gezhouba hydroproject was clear to support shipping. The water conservancy department implemented the river regime control in the middle and lower reaches of the Yangtze River and positively changed the river regime. The transport department removed the obstructive shoals and essentially implemented the improvement measures specified in the environmental impact report.

4. 水环境保护措施体系较完善，水质状况基本稳定

三峡库区城镇生活污水和生活垃圾处理纳入移民安置工程统一组织实施。在对受淹没影响工业企业开展搬迁规划过程中，淘汰不符合区域环境保护要求的落后企业 1 345 家，占搬迁企业总数的 82.4%，以优化三峡库区工业布局、减少工业排污。

水库蓄水前分四期实施库底清理工作，环评要求清理的库区固废全部规范清理并处置，累计清理和处置生活垃圾 367.28 万吨、工业固体废物 333.04 万吨、危险废物 4.93 万吨、安全送贮废放射源 22 枚。危险工业固废处置设施建设规范，运行管理状况良好。三峡库区农村开展了"一池三改"、养殖场污染治理和农村截污处理工程（"7+1"专项计划）等一系列面源污染防治工程，项目运行状况总体良好。

主管部门通过制定船舶污染防治管理的规章制度，规范和强化船舶污染治理。船舶污染防治主要通过对油污水、洗舱废水、船舶生活污水和垃圾实施处理，三峡库区未发现航运引起的油污染事故。三峡总公司组织实施清漂工作，对打捞上岸的漂浮物进行分类和转运，采用焚烧获取热值、综合利用及送至垃圾填埋场进行无害化处理等方式加以处置，未造成水体的二次污染，并维护了三峡库区的水域景观。

在三峡工程建设过程中，按照国家水资源保护及饮用水安全的总体部署，结合三峡工程环境保护的要求，三峡库区两省市政府开展了辖区范围内水源地保护区划定和规范化建设工作，在库区干流及一级支流回水区划定了城市饮用水水源地 81 个和集镇饮用水水源地 65 个。回水区内的城市水源绝大部分以Ⅱ类和Ⅲ类水质为主，水质总体保持稳定。回水区内的集镇水源水质劣于城市水源水质；各级政府将保障饮用水水源地安全作为落实最严格水资源管理制度的重要内容，着力加强重要饮用水水源地安全保障达标建设工作，坝下游有 14 个沿江水源地列入国家重要饮用水源地名录。

4. Sophisticated system of environmental protection measures and largely stable water quality

Disposal of the urban domestic wastewater and domestic waste in the TGP reservoir area was included into the general plan for the resettlement endeavor. During relocation of the industrial enterprises affected by inundation, 1,345 outmoded enterprises failing to meet the regional environmental protection requirements were dissolved. These accounted for 82.4% of the relocated enterprises, and their relocation helped to optimize the industrial structure in the TGP reservoir area and minimize industrial pollution discharge.

Prior to impoundment in the reservoir, reservoir bottom cleaning was conducted in four phases. The solid wastes that must be removed as specified in the EIA report were all removed and disposed of. In total, 3.6728 million tons of domestic waste, 3.3304 million tons of industrial solid waste, and 49,300 tons of hazardous waste were removed and disposed of, and 22 spent radioactive sources were safely transported and stored. The hazardous industrial solid waste disposal facilities were built according to standard and were operated and managed properly. The villages in the TGP reservoir area implemented non-point source pollution control programs such as "modification of biogas digester, corral, toilet, and kitchen," farm pollution control, and rural sewage interception ("7+1" plan), and the programs generally operated well.

The competent authorities enacted the regulations on the prevention and control of ship pollution to standardize and strengthen ship pollution control. Ship pollution prevention and control was mostly implemented through disposal of oil polluted water, tank cleaning wastewater and the domestic wastewater and garbage from ships. No oil pollution accidents resulting from shipping were detected in the TGP reservoir area. The CYTGPDC conducted floating waste collection, classified and transported the collected floating objects, and disposed of such materials through harmless methods such as combustion that generates heat, comprehensive utilization, or conveyance to a landfill. No secondary pollution occurred in the water bodies and the water scenery in the TGP reservoir area was preserved.

During the construction of the TGP, in accordance with the national strategy for water resources protection and drinking water security, and with consideration of the environmental protection requirements of TGP, the governments of the province and municipality involved in the TGP reservoir area conducted demarcation and standard development of the water source protection zones in their jurisdictions. They demarcated 81 city drinking water sources and 65 town drinking water sources in the backwater areas along the trunk stream and primary tributaries in the reservoir area. The vast majority of the city water sources in the backwater areas had Class II or Class III water quality, and their water quality remained stable

针对原预测不足的三峡库区支流"水华"现象，国务院三峡建委办公室及三峡总公司在组织开展水华发生机制、富营养化成因、控制技术对策等方面科学研究的同时，实施了处置技术试点项目，研究成果对三峡库区富营养化和水华状况的评价、预测及防控具有积极的意义。2009年以来，三峡总公司还建设运行了三峡水库水华监测网络并对库区12条支流开展持续监测，掌握水库运行水华发生特点和演变规律，并为防控支流水华的生态调度提供支撑。

5. 从流域角度统筹水生生态保护，措施体系完整、保护效果逐步显现

施工期，调整砂石料源为人工砂石料，有效减免虎牙滩长江河床天然砂砾料场开采施工对该江段中华鲟产卵活动的影响。围堰形成初期，对基坑内的鱼类开展人工救护、驱赶和放归保护。

先后建立国家级和省级自然保护区11个以保护水生生物重要栖息地，形成了覆盖滩涂、河流、河口、湖泊等多种水域生态系统类型的保护区网络。自然保护区环境现状总体良好，栖息地生态功能得以维护。农业部及沿江各省市相关部门通过建立水产种质资源保护区、珍稀特有鱼类救护中心、濒危物种专项救护、建立濒危水生野生动物经营利用等举措，建立起了较为完整的水生生物多样性和濒危物种保护体系。

据不完全统计，2002年至2012年期间，沿江各地共投入增殖放流资金约2.6亿元，放流鱼苗约4.7亿尾，其中珍稀保护物种主要为中华鲟、胭脂鱼、达氏鲟、松江鲈等，经济鱼类主要为"四大家鱼"。跟踪监测结果显示，人工增殖放流能有效补充自然种群资源，一定程度上恢复和补充了长江渔业资源。

overall. The water quality of the town water sources in the backwater areas was inferior to that of the city water sources. The governments at all levels deemed the security of drinking water sources as a priority in implementing the most stringent water resource management system and made great efforts to maintain the security of the important drinking water sources according to the standards. Fourteen water sources along the river downstream of the dam were included into the national list of important drinking water sources.

To solve the problem of algal blooms along the tributaries in the TGP reservoir area that had previously been inadequately estimated, the Office of the TGPCC and the CYTGPDC organized and conducted scientific research on the occurrence mechanism of algal blooms, the causes of eutrophication, and the technologies and solutions needed to solve this problem. It also implemented a pilot project for treatment technology. The research results played a positive role in evaluation, prediction, prevention, and control of the eutrophication and algal blooms in the TGP reservoir area. Since 2009, the CYTGPDC has also developed and operated the Three Gorges Reservoir Algal Bloom Monitoring Network, conducted continuous monitoring over the 12 tributaries in the reservoir area, learned the characteristics and evolution pattern of algal blooms during reservoir operation, and provided support for the ecological dispatches intended to prevent and control algal blooms in the tributaries.

5. Aquatic ecological conservation, protection system completion, and protection effects from the perspective of the entire basin

During the construction period, artificial sand was used as an aggregate to effectively alleviate the impact on the spawning of Chinese sturgeon imposed by mining activity in the natural gravel quarry along the riverbed of the Yangtze River at Huyatan. During the early stage of cofferdam construction, the fish species in the foundation pits were rescued, expelled, and released or protected through human intervention.

Eleven national or provincial nature reserves were established to protect the critical habitats of aquatic organisms, and a protection network was created, covering many aquatic ecosystems including shoals, rivers, estuaries, and lakes. The nature reserves are currently in benign environmental conditions and the ecological functions of the habitats are preserved. The Ministry of Agriculture and the relevant authorities of the provinces and cities along the river have created a complete aquatic organism biodiversity and endangered species protection system through approaches such as aquatic germplasm resources protection zones and rare and special fish species rescue centers, as well as through the rescue of endangered species and management and utilization of endangered aquatic wildlife.

From 2002 to 2012, according to incomplete statistics, approximately RMB 260 million was invested in the breeding and releasing of fish along the river. Approximately 470 million

为了减轻捕捞强度、保护物种资源和水域生态环境，沿江各省市相关部门实施了长江流域春季禁渔制度、严厉打击非法捕捞行为、加强政策法规建设、推进渔政管理能力建设等渔政管理措施。

6. 通过实施植被保护、物种保护等措施，保护了三峡库区生物多样性

通过湖北宜昌大老岭国家森林公园（天宝山）植物多样性保护工程、兴山龙门河常绿阔叶林自然保护工程和丰都世坪植物多样性保护工程的实施，保护三峡库区地带性植被——常绿阔叶林。在巫山小三峡风景区内划定了动物保护区，并实施严格的封闭式保护，保护了野生动物生境。各级政府对三峡库区的生态保护高度重视，先后新建了自然保护区和森林公园等。

对疏花水柏枝、荷叶铁线蕨、丰都车前、宜昌黄杨、鄂西鼠李 5 种珍稀特有植物，采取了野外就地保存、迁地保护和设施保存等多途径进行保护。对三峡库区古大树采取了排查、建档、围栏保护和迁地保护等措施，就地保护和移栽的古大树保护状况较好。

此外，国务院三峡建委办公室及三峡总公司还组织开展了消落区治理研究和试点示范工作。

7. 工程涉及区域人群健康保护状况良好

施工区和移民安置区建立了较为完善的人群健康保护措施体系，三峡工程建设和水库蓄水以来，人群健康状况总体良好，未发现传染病暴发或流行迹象，法定传染病发病率处于正常水平，血吸虫病控制在历史最低水平，无重大突发公共卫生事件报告。

newly hatched fish were released, including rare and protected species such as the Chinese sturgeon, Chinese high-fin banded shark, Dabry's sturgeon, and Trachidermus fasciatus, as well as commercial fish species dominated by the "four Chinese carps". According to the tracking and monitoring results, artificial breeding and releasing can effectively replenish the natural population resources and, to a certain extent, restore and supplement the fishery resources of the Yangtze River.

In order to reduce fishing intensity and protect the species resources and aquatic ecological environment, the relevant authorities of the provinces and cities along the Yangtze River enacted closed fishing along the Yangtze River basin, rigorously cracked down on illegal fishing, improved policies and laws, and worked to enhance fishery administration competency.

6. Preservation of biodiversity in the TGP reservoir area through vegetation and species protection

The Plant Diversity Protection Project of Dalaoling National Forest Park (Tianbao Mountain) in Yichang, Hubei, the Xingshan Longmen River Evergreen Broad-leaf Forest Protection Project, and the Fengdu Shiping Plant Diversity Protection Project were implemented to protect the evergreen broad-leaf forest, a vegetation zone in the TGP reservoir area. A wildlife sanctuary was demarcated in the Wushan Small Three Gorges Scenic Area, and enclosed protection was implemented to protect the wildlife habitat. All levels of government focused on ecological conservation in the TGP reservoir area and created new nature reserves and forest parks.

The 5 rare plant species, including Myricaria laxiflora, Adiantum reniforme, Fengdu plantain herb, Buxus ichangensis, and Exi buckthorn were protected in situ, relocated, or preserved in protective facilities. The large ancient trees in the TGP reservoir area were identified and archived and then enclosed or transplanted. The protected and transplanted large ancient trees are now well protected.

In addition, the Office of the TGPCC and the CYTGPDC conducted research into the control of the drawdown area and implemented a pilot project for this purpose.

7. Health protection for the population in the project affected areas

A sophisticated population health protection system was instituted in the construction and resettlement areas. Since the impoundment of the TGP and impoundment in the reservoir, the people in those areas have been physically healthy, no infectious disease or epidemic has been detected, the incidence rate of notifiable diseases has been at a normal level, the incidence rate of schistosomiasis has been kept at the lowest level in history, and no major

8. 及时组织开展景观及文物保护，保护效果达到预期目标

三峡库区自然景观保护管理机制健全，制度较完善。截至 2013 年年底，三峡库区已建立国家级风景名胜区 3 个、省级风景名胜区 16 个、国家森林公园 7 个、省级森林公园 21 个。通过配套工程建设，在保护原生景观同时，提升了自然景观的景观游憩和自然资源保护能力。

对受影响的 1 087 处文物分别实施原地保护、搬迁保护、留取资料三类保护措施。长期开展地质环境保护，专项防治三峡库区地质灾害。从 2001 年开始，国家先后投入 113 亿元进行了二期、三期三峡库区地质灾害规模性集中治理，开展了地质灾害的工程治理、搬迁避让、监测预警、应急抢险、科学研究等工作，有效保护了三峡库区地质环境，减少了地质灾害的发生。治理工程提高了灾害体整体的稳定性，达到防治效果。

9. 中游平原湖区和河口区环境保护要求和措施基本得到落实

相关部门和单位组织开展了湖泊湿地保护与修复、物种保护、管理执法、生态调查与研究等工作。地方各级政府在洞庭湖区、鄱阳湖区、四湖地区等中游平原湖区开展了土地整理、退田还湖、退耕还渔、农田水利建设等综合利用和治理工作。地方政府实施了蓄淡避咸水库修建、围垦促淤工程建设和土壤盐渍化防治等项目；中国科学院等单位组织开展了河口地区环境及生态变化情况研究工作。

public health emergency has been reported.

8. Anticipated protection goals achieved through the timely protection of landscapes and cultural relics

The natural landscape protection mechanism in the TGP reservoir area is well functional and supported by sophisticated regulations. By the end of 2013, the reservoir area had established 3 national scenic areas, 16 provincial scenic areas, 7 national forest parks, and 21 provincial forest parks. Through supporting projects, the native landscape was protected, and the capacity of these areas to receive tourists and protect natural resources has also been enhanced.

The 1,087 Cultural Heritage sites affected by the project were protected through three methods, including in-situ preservation, relocation, and database creation. Geological environmental protection has been implemented on a long-term basis to specifically prevent and control geological hazards in the TGP reservoir area. Since 2001, the state invested RMB 11.3 billion in the centralized large-scale control of the geological hazards in the TGP reservoir area during Phase II and Phase III of the project and made various efforts regarding geological hazard control, such as engineering control, relocation, evasion, monitoring, warning, and emergency rescue. These measures have effectively protected the geological environment of the TGP reservoir area and reduced the occurrence of geological hazards there. The hazard control measures enhanced the overall stability of the hazardous objects and resulted in the prevention and control of geological hazards.

9. Implementation of environmental protection requirements and measures for the plain lake areas and central estuary environment

The relevant departments and entities made efforts such as the protection and restoration of lake wetlands, species protection, administrative law enforcement, and ecological surveys and research. The local governments organized comprehensive utilization and control related endeavors, such as land consolidation, returning farmland to lakes, returning farmland to fisheries, and agricultural water conservancy development in the plain lake areas in the middle reaches of the Yangtze River, including the Dongting Lake area, Poyang Lake area, and four-lake area. The local governments also implemented projects such as the construction of reservoirs to "impound fresh water and discharge salty water," the construction of cultivation and silting projects, and the prevention/control of soil salinization. Entities including the Chinese Academy of Sciences organized and conducted research into the environmental and ecological changes of the estuary areas.

三、移民安置区的生态保护措施和实施

在移民安置过程中,根据《国务院办公厅关于三峡工程库区移民工作若干问题的通知》要求,淘汰落后工业产能,改进生产工艺,提升工业污染防治能力和水平;在农村移民安置区,对25度以上坡耕地实施退耕还林,开展城集镇污染防治。通过实施各项移民安置环保措施,结合各级政府组织实施的专项环保规划,大幅改善了三峡库区及移民安置区的环境质量状况。

1. 加强农村污染防治和生态环境保护

结合"长防""长治""退耕还林"等工程实施了坡改梯、水保林、经果林、封山育林、小型农田水利工程等农村土地资源开发利用环境保护措施,遥感数据对比分析结果显示三峡库区森林覆盖率显著提高、水土流失趋势得到遏制,移民安置区生态环境明显改善,一定程度上体现了土地资源开发利用环境保护措施的有效性。

2. 实施农村污染防治

结合"7+1专项方案""生态家园富民工程"等项目实施了农村"一池三改"工程;针对农村畜禽养殖污染实施了沼气池示范治理工程。农村"一池三改"工程的实施既处理了农村生活污水,又给农户带来清洁能源,减少了薪柴使用量,保护了三峡库区森林资源,但由于沼气池原料不足,部分农村沼气池处于闲置状态。

III. Ecological Conservation Measures and Implementation for Resettlement Areas

During the resettlement process, in accordance with the requirements of the *Notice of the General Office of the State Council on Several Issues Regarding Resident Resettlement in the Reservoir Area of Three Gorges Project*, the outdated industrial capacity was eliminated, the production processes was improved, and the industrial pollution prevention and control competency was enhanced. In the rural resettlement area, the sloped farmland areas with a >25° gradient were all reverted to forest land and urban pollution control was implemented. The resettlement related environmental protection measures and the special environmental protection programs implemented by the governments at various levels have substantially improved the environmental quality of the TGP reservoir area and resettlement areas.

1. Intensified pollution control and ecological and environmental protection in rural areas

In conjunction with the endeavors such as "long-term prevention," "long-term control," and "returning farmland to forests," other measures were taken for the purpose of rural land resource utilization and environmental protection, such as terracing sloping land, soil and water conservation forests, economic fruit woods, setting apart hills for tree growing, and small-scale farmland water conservancy projects. According to the comparison and analysis of remote sensing data, the forest coverage of the TGP reservoir area significantly increased, the trend of water and soil loss was curbed, and the ecological environment in the resettlement areas was markedly improved. To a certain extent, this proves that the land resource utilization and environmental protection measures are effective.

2. Rural pollution control

In conjunction with the "7+1 plan" and the "enriching people in an eco-friendly homeland" projects, the "modification of biogas digester, corral, toilet, and kitchen" project was also implemented. To solve the problem of pollution caused by livestock breeding, the standard biogas digester management project was implemented. The "modification of biogas digester, corral, toilet, and kitchen" project disposed of the rural domestic wastewater, brought clean energy to farmer households, reduced the use of wood fuel, and helped protect the forest resources in the TGP reservoir area. Due to the short supply of biogas digester fuel, however, some of the rural biogas digesters are unused.

图 8.1　城集镇迁建的同时兴建的生活污水处理厂（摄影：袁国平）

3. 迁建城集镇配套建设污染防治设施

城集镇迁建的同时，实施了集中供水水源地的建设和保护；结合国家专项，建成城集镇生活污水处理厂60座，总处理规模69.41万吨每天；建成生活垃圾填埋场36座，总处理规模3 871.8吨每天；建成危险废物处置场5座，处置能力58万吨每年；污染防治措施满足移民安置环境保护规划的要求，处理效果基本满足排放要求。移民安置区污水排放量逐年增大，但污染物排放量有所消减，安置区污染防治措施总体有效。

4. 工矿企业迁建实施产业结构调整，提高污染防治能力

根据产业政策调整，工矿企业迁建由"技改迁建"变为"结构调整"，对污染严重、产品无市场和资不抵债的企业实施了破产关闭和一次性补偿销号。迁建企业结合技术升级改造，贯彻"以新带老"原则，对原有污染源进行治理，企业产生的危险废物均委托有危险废物处理资质的单位进行处理，典型抽样企业调查表明企业污染防治设施运行可靠，基本满足达标排放要求。

Figure 8.1　A Domestic Wastewater Treatment Plant Built During the City/Town Relocation Process (Photographed by Yuan Guoping)

3. Pollution control facilities built during the city/town relocation process

In conjunction with the city/town relocation process, the development and protection of the centralized water supply sources was implemented. In accordance with the national plan, 60 urban domestic wastewater treatment plants have been built, with a total treatment capacity of 694,100 t/d, 36 domestic waste landfills have been built, with a total disposal capacity of 3,871.8 t/d, and 5 hazardous waste disposal sites have been built, with a disposal capacity of 580,000 t/a. The pollution control measures meet the requirements of the resettlement related environmental protection plan and their results meet the emission standard. The wastewater discharge in the resettlement area has increased year by year. However, the pollutant discharge has decreased, and the pollution control measures in the resettlement areas are effective overall.

4. Enhanced pollution control capacity through industrial restructuring by relocating industrial and mining enterprises

Given the changes in industrial policies, the policy for industrial and mining enterprises changed from "technological upgrading and relocation" to "restructuring," in which the enterprises that were responsible for severe pollution, whose products had no market and that were in the state of insolvency, were bankrupted, closed, and deregistered after one-time compensation. The relocated enterprises also performed technological upgrades. Under the principle of "pollutant reduction through technological upgrading," the existing pollution sources were eliminated, and the hazardous waste generated by the enterprises were treated by entities qualified in the treatment of hazardous wastes. According to random inspections of typical enterprises, the enterprise pollution control facilities have operated reliably and essentially meet the emission standards.

5. Implementation of environmental protection measures for special restoration projects

The Environmental Impact Statement and the environmental protection plan for the resettlement area specified no requirements regarding special restoration projects. According to the historical background, with a large demand for funding and short design and construction period, a small number of special restoration projects were included in the environmental im-

5. 专项复建工程环境保护措施得到落实

环境影响报告书和移民安置区环境保护规划均未对专项复建工程提出相关要求。在当时资金需求量大、设计建设周期短等特定的历史时期下，少数专项复建项目执行了环境影响评价制度，但复建项目在专项设计中对施工期、运营期环境保护提出了相关要求，施工期和运行初期基本落实了废水、废气、噪声和弃渣等污染防治措施，并开展了水土保持和植被恢复工作，资料收集和现场核查结果表明，污染防治和生态保护效果良好。

第三节 三峡工程的生态保护成效

任何水利水电工程的建设都会对生态与环境有所扰动，三峡工程也不例外。但同时，三峡工程对改善生态与环境所做出的巨大贡献，也应得到更加客观的认知和评价：通过筑坝治水减少或杜绝洪涝和干旱造成的生态灾难和人民生命财产的损失，这就是最大的生态效益。此外，三峡工程对减轻洪水对长江中下游平原生态与环境的破坏、减缓洞庭湖的淤积萎缩等，也产生了重大而积极的影响。巨大的减排效益，对减缓全球气候变暖有很大作用。

一、防洪是最大的生态保护

防洪是兴建三峡工程的出发点。上游干流及中游支流洪水来量大，中游缺乏有足够容积的调洪、滞洪场所，加上河道宣泄能力不足，是造成长江洪水成灾的主要原因。长江历史变迁中，云梦泽和洞庭湖曾先后发挥天然水库滞蓄洪水的作用，但随着近代以来洞庭湖泥沙持续淤积，水面面积和容积日渐萎缩，滞蓄洪水的能力被削弱，加剧了洪灾防治难度。三峡水库建成后，能有效调节并控制宜昌以上洪水来量，减少下泄流量，大大提升长江洪水调蓄能力，使长江中下游免遭洪水灾害。

pact assessment system. However, the special design of the restoration projects specified the requirements regarding environmental protection during construction and operation periods. During the construction period and early days of the operation period, the pollution control measures for waste water, waste gas, noise, and waste slag were implemented and soil and water conservation and vegetation restoration were conducted. According to the collected data and site surveys, the pollution control and ecological conservation efforts have been quite effective.

Section 3 Effectiveness of Ecological Conservation in the Three Gorges Project

Any water conservancy or hydropower project will cause an ecological or environmental disturbance, and the TGP is no exception. However, the TGP has made tremendous contributions to ecological and environmental improvement, and thus deserves objective understanding and assessment. Damming and water control have reduced or even completely eradicated the ecological disasters caused by flooding and droughts and averted the loss of life and property of people in the area. This is its greatest ecological benefit. In addition, the TGP has ameliorated the ecological and environmental disruption of the plains in the middle and lower reaches of the Yangtze River induced by flooding and mitigated silt accumulation and shrinking in Dongting Lake, which is also a major positive impact. The great benefits of emissions reduction are also helpful in mitigating global warming.

I. Prioritizing Ecological Conservation through Flood Control

Flood control was the original purpose of the TGP. The trunk stream in the upper reaches and the tributaries in the middle reaches generated large floods. However, the middle reaches lacked sufficient flood regulation and flood detention space, and the river lacked flood discharge capacity. This is the primary cause of flood hazards along the Yangtze River. Throughout the history of the Yangtze River, the Yunmeng Lake and Dongting Lake functioned as natural reservoirs capable of floodwater detention and floodwater storage. In modern times, the sediment accumulation in Dongting Lake resulted in the deterioration of

防洪是三峡工程最大的生态效益。水利工程通过防洪减灾，可以有效避免洪灾造成的生态环境恶化，维持生态平衡，包括避免洪灾引起水质和卫生条件恶化；避免土地被冲毁、淤压，导致沙化荒废；避免地下水位上升，引起土地沼泽化或次生盐碱化；减免林、草被淹死亡，生态环境恶化等。三峡工程的修建，使广袤富庶的华中平原不再受洪水威胁，生态秩序走向良性循环，平原湖区生态达到新的相对稳定，有效避免因洪水泛滥而引发一系列社会、环境问题。通过对上游来水的控制，可以减少汛期分流入洞庭湖的洪水和泥沙，有效减轻洪水对洞庭湖区的威胁，延缓洞庭湖泥沙淤积速度，延长洞庭湖的寿命；由于水库调节作用，枯水期下泄流量增加，有助于提高坝下游河道污水稀释化，改善水质，减轻污染。

三峡工程的建成，标志着以三峡工程为骨干的长江中下游防洪体系基本形成，处于长江上游来水进入中下游平原河道的"咽喉"，可控制长江防洪最险的荆江河段95%的洪水来量，所起到的削峰、错峰的作用，对长江上游特大洪水的调节作用，是其他任何工程都不能替代的。截至2021年，三峡水库累计拦洪63次，总蓄洪量1 900多亿立方米，干流堤防未发生一起重大险情，保证了长江中下游的安全稳定，降低了防汛成本。据中国工程院关于三峡工程试验性蓄水阶段评估的估算，三峡工程多年平均防洪效益为88亿元，工程防洪减灾效益显著。

二、绿色电力的生态效益

三峡水电站（含电源电站）2021年全年累计生产清洁电能1 036亿千瓦时，打破此前南美洲伊泰普水电站于2016年创造并保持的1 030.98亿千瓦时的单座水电站年发电量世界纪录，有效缓解了华中、华东地区及广东省的用电紧张局面，为电网的安全稳定运行发挥了重要作用；按每千瓦时电量可产生13.8元GDP推算，1 036亿千瓦时电量可支撑我国约1.43万亿元GDP，为我国节能减排做出了重要贡献。

1. 增发效益

通过加强分析预报来水，向国调、华中网调提供及时准确的梯级电站出

the water surface area, and the lake volume continued to decrease. Consequently, the floodwater detention and floodwater storage capacity was diminished, making flood control more difficult. The TGP, once completed, can effectively regulate and control flooding upstream Yichang, reduce discharge flow, enhance the floodwater regulation and storage capacity of the Yangtze River, and protect the middle and lower reaches of the Yangtze River from flood disasters.

Flood control is the greatest ecological benefit of the TGP. The water conservancy project can prevent or eliminate flood disasters, thus effectively averting ecological environment deterioration caused by flood disasters. It can also maintain ecological balance and preclude the deterioration of water quality and sanitary conditions caused by flood disasters. In addition, it can protect the land from scouring and silting, which would lead to desertification; prevent rise of the groundwater level, which would lead to paludification or secondary salinization of land; and reduce or eliminate the inundation of forest and grass lands, which would lead to ecological environment deterioration. The TGP shelters the vast and productive plains in Central China from the threat of floods, pushes the ecological order into a virtuous cycle, provides relative stability for the ecology of the plain lake areas, and effectively precludes a range of social and environmental issues caused by flooding. By controlling the upstream water volume, the TGP can reduce flooding and the amount of sediment flowing into Dongting Lake during flood season, substantially mitigate the threat of flooding in the Dongting Lake area, slow the sediment accumulation in Dongting Lake, and prolong the life of the lake. Thanks to regulation through the reservoir, the discharge flow during the dry season has increased, which helps dilute the wastewater in the rivers downstream of the dam, improves water quality and abates pollution.

Completion of the TGP signifies that the flood control system is essentially formed in the middle and lower reaches of the Yangtze River, with the TGP being the foundation of the system. The water from the upper reaches of the Yangtze River flows into the "throat" upstream of the rivers and plains along the middle and lower reaches of the river. The project can control 95% of the flood volume of the Jingjiang River, which is the most dangerous part of the Yangtze River flood control system. The peak shaving, peak shifting, and catastrophic flood regulation functions of the TGP in the upper reaches of the Yangtze River can never be replaced by another project. By 2021, the TGP had accomplished flood retention 63 times, and its total flood storage volume was over 190 billion m^3. The protective embankments along the trunk stream experienced no major flood risk. This guarantees the safety and stability of the middle and lower reaches of the Yangtze River and reduces the cost of flood control. According to the TGP trial storage period assessment conducted by the Chinese Academy of Engineering, the average annual flood control benefit of the TGP was RMB 8.8

力预报、合理控制水库水位、及时清除拦污栅前漂浮物、重复利用库容，及时调整电站出力使电站机组弃水期处于出力最大状态，枯水期处于效率最高状态等措施，三峡水电站发电量显著增加，相比初步设计的调度方式，2003年至2019年累计增发电量658亿千瓦时。

2. 调峰效益

三峡水电站具有快速启停机组、迅速自动调整负荷的良好调节性能，为电力系统的安全稳定运行提供了可靠保障。2003年至2019年，三峡水电站结合自身能力积极参与电网系统调峰运行，平均最大调峰容量为412万千瓦，有效缓解了电力市场供需矛盾，改善了调峰容量紧张局面，促进了电网安全稳定运行。

此外，三峡水电站地处华中腹地，电力系统覆盖了长江经济带，在全国互联电网格局中处于中心位置，对全国电网互联互通起到关键性作用，成为"西电东送"的中通道，实现了华中与华东、南方电网直流联网，与华北电网交流联网，形成了水火互济运行的新格局。

billion, and the project generated significant benefits in flood control and disaster reduction.

II. Ecological Benefits of Green Electrical Power

In 2021, the Three Gorges Hydropower Station (TGHS) (including the power source station) produced a total of 103.6 billion kWh of clean electric energy, which exceeded the 2016 world record of annual energy output of single hydropower station, 103.098 kWh, set by the Itaipu Hydroelectric Power Station in South America. The TGHS effectively relieves the electrical power shortage in Central China, East China, and Guangdong Province and plays a key role in the safe and stable operation of the power grid. Based on the standard of RMB 13.8 GDP per kWh, the 103.6 billion kWh energy output is equivalent to approximately RMB 1.43 trillion GDP for China and is a major contributor to energy conservation and emission reduction in China.

1. Output increase benefits

By strengthening the analysis and forecast of incoming floods, the national dispatch center and Central China grid dispatch center are able to provide a timely and accurate output forecast of the Cascade Hydropower Station, the water level of the reservoir can be controlled accordingly, floating objects that have accumulated can be removed, the storage capacity can be reused, and the power station output can be adjusted in a timely manner to ensure the units are at maximum output during a period with surplus water and at maximum efficiency during the dry season. The TGHS energy output has increased markedly. Compared with the dispatching model from the preliminary design, the energy output accumulatively increased by 65.8 billion kWh from 2003 to 2019.

2. Peak regulation benefits

The TGHS has quick startup-shutdown units and is capable of automatically adjusting the load quickly, providing reliable support for the safe and stable operation of the power system. From 2003 to 2019, the TGHS, based on a realistic understanding of its capacity, participated in peak regulation of the power grid system. Its maximum peak regulation capacity was 4.12 million kW, which substantially ameliorated the supply-demand contradiction in the electricity market, relieved a shortage of peak regulation capacity, and contributed to the safe and stable operation of the power grid.

In addition, the TGHS is located in the interior of Central China, its power system covers the Yangtze River Economic Belt, and it is at the center of the national interconnected power grid. Due to these factors, it plays a crucial role in the creation of the national

3. 节能减排效益

作为世界上总装机容量最大的水电站，三峡水电站总装机容量达2 250万千瓦，设计年均发电量882亿千瓦时，是我国最大的清洁能源生产基地。三峡水电站每年发出的清洁电能，相当于少燃烧3 000万吨标准煤或2 500万吨原油，减少约1亿吨二氧化碳排放以及大量的粉尘、二氧化硫、氮氧化物等污染物排放，带来巨大的节能减排生态效益。据中国工程院关于三峡工程建设第三方独立评估，按照碳排放交易价格估算，2003年至2016年折算二氧化碳减排效益达491亿元。

2021年，三峡水电站生产的清洁电力，相当于节约标准煤3 175.8万吨，减排二氧化碳8 685.8万吨、二氧化硫1.94万吨、氮氧化物2.02万吨，节能减排效益显著。三峡水电站为国家逐步实现碳达峰、碳中和目标做出积极贡献。

三、"黄金水道"的重要作用

2010年10月，三峡水库蓄水至175米运行后，渠化重庆以下川江航道里程600多公里，结合实施三峡库区碍航礁石炸除工程，消除了坝址至重庆间的139处滩险、46处单行控制河段和25处重载货轮需牵引段，三峡库区航道年通过能力由1 000万吨提高到5 000万吨，实现了全年全线昼夜通航，万吨级船队可由上海直达重庆，单位航运成本降低三分之一以上。

interconnected power grid, functions as a channel in the "west-to-east power transmission" project, facilitates the DC grid connection between the Central China Grid, East China Power System, and China Southern Power Grid, enables AC connection with the North China Power Grid, and helps create an energy system comprised of both hydropower and thermal power.

3. Energy conservation and emission reduction benefits

The TGHS, the hydropower station with the largest installed capacity in the world, has a total installed capacity of 22.5 million kW and a designed annual average energy output of 88.2 billion kWh, making it the largest clean energy production base in China. The clean energy produced by the TGHS every year is equivalent to reducing the consumption of standard coal by 30 million tons or crude oil by 25 million tons. It also means that carbon dioxide emission is reduced by approximately 100 million tons, and the emission of dust, sulfur dioxide, and nitrogen oxide is reduced by a large margin. This signifies massive ecological benefits in energy conservation and emission reduction. According to a third-party independent assessment of TGP conducted by the Chinese Academy of Engineering, the carbon dioxide emission reduction benefit from 2003 to 2016 was RMB 49.1 billion, if estimated by the carbon emission trading price.

In 2021, the clean electric energy produced by the TGHS was equivalent to reducing standard coal consumption by approximately 31.758 million tons, which means the emission of carbon dioxide was reduced by 86.858 million tons, sulfur dioxide by 19,400 tons and nitrogen oxide by 20,200 tons, which is a clearly significant benefit for energy conservation and emission reduction. The TGHS has made great contributions toward achieving the country's carbon peak and carbon neutrality objectives.

III. Important Function of the "Golden Waterway"

In October 2010, after the storage level of the Three Gorges region increased to 175 m, over 600 km of the Chuanjiang River downstream of Chongqing was canalized. The obstructive rocks in the TGP reservoir area were removed by blasting. From the dam site to Chongqing, 139 obstructive shoals, 46 one-way controlled segments, and 25 segments in which heavy-duty freighters needed traction were eliminated. The annual transit capacity of the waterway in the TGP reservoir area increased from 10 million tons to 50 million tons, and day-and-night navigation became a reality. A 10,000-ton fleet could sail all the way from Shanghai to Chongqing. The unit shipping cost was reduced by more than one third.

After this improvement in the shipping conditions, the average energy consumption per

通航条件改善后,船舶单位平均能耗显著降低。据测算,与蓄水前相比,长江水运能耗降低约37%。2003年至2010年,重庆市运输船舶共节约燃油近239万吨,相当于减排二氧化碳717万吨、二氧化硫9.56万吨、氮氧化物12.7万吨。按照目前年货运量1亿吨计算,三峡库区每年航运能耗降低约8.5万吨标准煤,相当于减排二氧化碳21.8万吨;与公路运输相比,每年可节省约200万吨标准煤,减排二氧化碳513.4万吨。

截至2021年12月,三峡船闸累计过闸货运量超过16亿吨,长江航道成为名副其实的"黄金水道",为长江经济带建设和发展提供了重要支撑。

四、对中华鲟和三峡珍稀植物的保护

1982年水利部批准成立中华鲟研究所,2008年该研究所整体从葛洲坝集团划归中国长江三峡集团公司管理,这是我国首个因大型水利工程兴建而设立的珍稀鱼类科研机构。此外,中国科学院水生生物研究所、长江水产研究所等也开展了大量科研工作。1983年,我国人工繁殖中华鲟成功,并于1984年首次将幼鲟放流入长江;2009年,人工繁殖"子二代"中华鲟获得成功。

图 8.2　长江流域"活化石"——中华鲟(摄影:王绪波)

1984年至2021年,三峡集团中华鲟研究所连续37年开展了64次中华

vessel decreased considerably. As calculated, the energy consumption for water transport in the Yangtze River decreased by approximately 37% after impoundment. From 2003 to 2010, Chongqing transport vessel fuel consumption was reduced by nearly 2.39 million tons, which is equivalent to reducing emission of carbon dioxide was reduced by 7.17 million tons, sulfur dioxide by 95,600 tons and nitrogen oxide by 127,000 tons. According to the current annual freight volume of 100 million tons, shipping in the TGP reservoir area has reduced standard coal consumption by approximately 85,000 tons, which is equivalent to a reduced carbon dioxide emission of 218,000 tons. Compared with road transportation, the project can reduce standard coal consumption by approximately 2 million tons and carbon dioxide emission by 5.134 million tons.

By December 2021, the accumulative freight volume passing through the TGP ship lock was 1.6 billion tons. The Yangtze River is genuinely worthy of its reputation as the "golden waterway" and has become an important support for the development of the Yangtze River Economic Belt.

IV. Protection of Chinese Sturgeon and Rare Plants in the Three Gorges

In 1982, the Ministry of Water Resources authorized the founding of the Chinese Sturgeon Research Institute. In 2008, the administrative authority of this institute was transferred from the China Gezhouba Group Corporation to the China Three Gorges Corporation. It is China's first rare fish species research institute to be instated specifically for a large water conservancy project. In addition, the Institute of Hydrobiology of the Chinese Academy of Sciences and the Yangtze River Fisheries Research Institute have also conducted a large amount of research regarding this topic. In 1983, the artificial fecundation of Chinese sturgeon was accomplished in China. In 1984, the juvenile sturgeons were released into the Yangtze River. In 2009, the artificial fecundation of the second filial generation of Chinese sturgeon was accomplished.

⊃ Figure 8.2   A "Living Fossil" in the Yangtze River Basin—Chinese Sturgeon (Photographed by Wang Xubo)

From 1984 to 2021, the Chinese Sturgeon Research Institute subordinate to the CTG released Chinese sturgeon 64 times over 37 consecutive years, with over 5.04 million Chinese sturgeon being released, including over 40,000 Chinese sturgeon of the second filial generation. This action greatly assisted in replenishing the Chinese sturgeon population resources and enabling the sustainable multiplication of Chinese sturgeon. In August 2020, the Three

鲟放流活动，中华鲟放流数量已达504余万尾，其中放流"子二代"中华鲟4万余尾，为补充中华鲟种群资源、实现中华鲟可持续繁衍生息发挥了重要作用。2020年8月，三峡集团长江三峡珍稀特有鱼类保育中心在三峡坝区建成并投入运行，该中心的主要保育对象为中华鲟等长江珍稀特有鱼类，中心的建成投运提升了中华鲟的保育能力。

1992年，部分科研单位对三峡库区的珍稀植物进行跟踪观测，掌握了这些植物的分布情况和原生环境。随后，科研人员根据每种植物的生长特性，对它们进行植物野外生长环境模拟，开展批量繁殖。2017年，三峡集团在三峡苗圃研究中心的基础上成立长江珍稀植物研究所，每年下拨500万元人民币支持科研和保护工作，加大对可能受到工程影响的三峡库区陆生、水生植物的保护力度。

长江三峡珍稀特有鱼类保育中心和长江珍稀植物研究所经过多年努力，攻克了中华鲟人工繁殖技术难题，确保了中华鲟物种续存；实现了长江中上游指标性物种（圆口铜鱼、长鳍吻鮈）规模化驯养繁殖；开展了鱼类增殖放流，有效补充了鱼类种群资源，一定程度上恢复鱼类资源量。截至2020年年底，已迁地保护珙桐、红豆杉、荷叶铁线蕨、疏花水柏枝、伯乐树、桫椤等特有珍稀资源性植物136科458属1181种2.5万余株，建成全国种类最多、面积最大的三峡特有珍稀植物保育基地；已取得12项国家发明专利和1项实用新型专利，通过组培、扦插、播种等方式繁育特有珍稀资源性植物苗木18万余株；完成厚皮香和红豆树等疑似新种的物候期观测工作；完成三峡濒危特有珍稀资源性植物标本采集与制作422种近900份，完成三峡濒危珍稀特有资源性植物种子野外采集20种近80余个居群，保存了一定数量的种子和遗传资源，维护了三峡库区生物多样性。

图8.3　荷叶铁线蕨在武汉植物园迁地保护后生长情况

Gorges Rare and Endemic Fishes Conservation Center subordinate to the CTG was completed and commissioned in the TGP dam area. This center mainly preserves rare and endemic fish species in the Yangtze River, such as Chinese sturgeon. Completion and operation of this center has enhanced the capacity for Chinese sturgeon conservation.

In 1992, research institutes tracked and observed rare plant species in the TGP reservoir area and learned their physical locations and native environment. Later on, the researchers simulated a wild growth environment for each plant based on their characteristics and conducted mass propagation. In 2017, the CTG founded the Yangtze River Rare Plants Research Institute by restructuring the Three Gorges Nursery Research Center. Later on, the CTG allocated RMB 5 million to support research and protection of the plants every year, and intensified protection of the terrestrial and aquatic plants that are potentially affected by the TGP.

Through years of hard work, the Three Gorges Rare and Endemic Fishes Conservation Center and the Yangtze River Rare Plants Research Institute have solved the technical challenges in the artificial fecundation of Chinese sturgeon, thus ensuring the subsistence of the species. They have achieved large-scale breeding of the index species (Coreius guichenoti and Rhinogobio ventralis) in the middle and upper reaches of the Yangtze River and have bred and released the fish, which has substantially replenished the fish population resources and, to a certain extent, restored fish resources. As at the end of 2020, over 25,000 endemic and rare resource plants of 1,181 species in 458 genera in 136 families, including Davidia involucrata, Adiantum reniforme, Myricaria laxiflora, Bretschneidera sinensis nada, and Alsophila spinulosa, had been transplanted and protected. In addition, the Three Gorges Endemic and Rare Plants Conservation Base, which has the largest number of plant species and covers the largest area in the country, had been established, with 12 national invention patents and 1 utility model patent granted and over 180,000 seedlings of endemic and rare resource plants cultivated through tissue culture, cutting and seeding. The phenological period observation of suspected new species, such as Ternstroemia gymnanthera and Ormosia hosiei, had also been completed, and nearly 900 specimens of 422 endangered endemic and rare resource plant species in the Three Gorges have been collected and prepared. Additionally, the field acquisition of seeds for nearly 80 populations of 20 endangered endemic and rare resource plant species in the Three Gorges has been completed, a number of seeds and genetic resources have been preserved, and the biodiversity in the TGP reservoir area has been maintained.

⊙ Figure 8.3　Transplanted Adiantum Reniforme Growing in the Wuhan Botanical Garden

本章小结：

三峡工程的论证、设计、建设和运行，是人类认识自然规律、利用自然规律、改造客观世界以求达到人与水和谐共处的过程。三峡工程顺应并推动了中国特色生态文明的历史发展，培育形成了水电全生命周期的生态环保理念，建立并丰富完善了生态保护与修复技术体系。持续稳定运行并发挥出巨大的生态环保效益，树立了世界水电可持续发展的成功典范。三峡工程是生态效益巨大的生态工程，并将在推进长江绿色发展、建设美丽中国中持续发挥积极作用。

参考文献：

［1］本书编委会. 百问三峡［M］. 北京：科学普及出版社，2012.

［2］中国科学院环境评价部，长江水资源保护科学研究所. 长江三峡水利枢纽环境影响报告书（简写本）［M］. 北京：科学普及出版社，1996.

［3］《中国三峡建设年鉴》编纂委员会. 中国三峡建设年鉴（2020）［J］. 宜昌：中国三峡建设年鉴社，2020.

［4］《中国三峡建设年鉴》编纂委员会. 中国三峡建设年鉴（2021）［J］. 宜昌：中国三峡建设年鉴社，2021.

Chapter Summary:

The appraisal, design, construction, and operation of the Three Gorges Project were a part of humankind's endeavor to achieve harmony between people and water by understanding and utilizing the natural law and changing the objective world. For the TGP, which is in line with and beneficial for the development of an ecological civilization with Chinese characteristics, the ecological and environmental protection philosophy covering the full life cycle of the hydropower project has been established and the ecological conservation and restoration technology system has been created and improved. The project has operated continuously and stably, generated massive ecological and environmental protection benefits, and set a successful example for the sustainable development of hydropower projects. The TGP is an eco-friendly project that provides massive ecological benefits and will continue to play a positive role in promoting the green development of the Yangtze River and building China into a country with a good environment.

References:

[1] Editorial committee of this book. *Questions about the Three Gorges [M]*. Beijing: Popular Science Press, 2012.

[2] Environmental Assessment Division of the Chinese Academy of Sciences, Changjiang Water Resources Protection Institute. *Environmental Impact Statement on the Yangtze River Hydroproject* (simplified edition) [M]. Beijing: Popular Science Press, 1996.

[3] Editorial Board of the *China Three Gorges Construction Yearbook. China Three Gorges Construction Yearbook* (2020) [J]. Yichang: China Three Gorges Construction Yearbook Press, 2020.

[4] Editorial Board of the *China Three Gorges Construction Yearbook. China Three Gorges Construction Yearbook* (2020) [J]. Yichang: China Three Gorges Construction Yearbook Press, 2021.

> 阅读提示：

三峡地区自古就是早期人类和巴楚先民繁衍生息的场所，近现代更成为中原连接西南的咽喉要冲。大量人类文化与文明活动的文物遗存，构成了延续至今的历史文化宝库。

三峡水库蓄水后，难以估量的丰富遗存或将永没水下。为确保这些中华民族的文化密码不致失落，数千位文物考古专家历时十余年，实施了波澜壮阔的抢救保护工程，取得了具有重大历史意义的丰硕成果。

三峡水库建成蓄水后，一些自然人文景观已被淹没或迁建，或许少了些湍流激荡、历史沧桑，但高峡平湖横空出世，又为"百里画廊"平添了别样风采。

The Three Gorges region is where early humans including the Bachu ancestors lived and thrived in ancient times. In modern times, it is the strategic passage connecting the Central Plains and Southwest China. A large quantity of cultural relics left by previous civilizations still remain there and constitute a treasure trove of historical and cultural heritage.

After the impoundment of the Three Gorges Reservoir, an immeasurable amount of cultural heritage may become permanently submerged. To ensure the survival of the cultural treasures of the Chinese nation, thousands of cultural archaeologists spent decades on a magnificent endeavor to preserve the cultural relics and obtained substantial achievements of great historical value.

After the completion and impoundment of the Three Gorges Reservoir, some of the natural and cultural landscapes have been submerged or relocated. The historical landscapes accompanying the raging water in the river may be lost, but the "great lake in the gorge" has become another unique scene in the "long art gallery."

Chapter 9 >>>>

第九章
三峡工程的文物和自然人文景观保护

Cultural Relics and Natural and Cultural Landscapes of the Three Gorges Project

第一节 三峡工程的文物保护规划

中国是世界四大文明古国之一，地面和地下都遗存有大量文物，这些历史文物是中华民族的宝贵遗产，是中华民族灿烂文化的见证。尤其是三峡库区，早在200万年以前，已居住着中国最古老的"巫山人"，积淀了丰厚的巴楚文化，蕴藏着大量的地下、地面文物。

一、三峡地区文物的早期发掘

三峡地区包括东起湖北宜昌、西至重庆的长江及其支流流经的地域。长江南北两岸分别有众多支流汇入。由于南部大娄山紧逼江侧而地势陡峭，北侧则是大巴山余脉，地势相对平缓，因此其北部支流较大，如嘉陵江、小江、草堂河、朱衣河、梅溪河、大宁河、龙船河、香溪等。而南岸除乌江外，其余均为一些较小的溪流。其东端为巫山，北靠大巴山山麓，南依云贵高原北缘，是中国东部和西部面向海洋和面向亚洲腹地的两大地理单元的重要结合部之一，本身形成一个相对独立的地理单元。这个地区主要是丘陵山地，极少是平原，形成独特的山谷地貌，对这个地区古代人类的生存生活产生了深刻的影响。同时这个地区又是沟通四川盆地和江汉平原的咽喉地带，同样也对这个地区古代人类的生存生活产生了重要的影响。

三峡地区文物考古工作开始较早。19世纪下半叶，一些外国传教士、探险家如布朗（J. C. Brown）、贝伯（E. C. Baber）等在重庆地区发现一些石器。1925年至1926年，中亚探险队的美国学者纳尔逊（N. C. Nelson）在三峡地区调查石器地点37处（其中12处地点采集到陶片）。20世纪30年代，美国传教士埃德加（J. H. Edgar）也在三峡地区采集到一些石器。这些仅限于地面

Section 1 Preservation Plan for Cultural Relics in the Three Gorges Project

China, one of the Four Great Ancient Civilizations, is home to a large quantity of surface and underground cultural relics. Such historical relics are part of the valuable heritage of the Chinese nation and a witness of the splendid Chinese culture. The TGP reservoir area, in particular, was the homeland of the ancient Wushan people 2 million years ago. It retains ample elements of the Bachu culture and still possesses a large quantity of underground and surface cultural relics.

I. Early Excavation of the Cultural Relics in the Three Gorges Region

The Three Gorges region is located along the Yangtze River and its tributaries that starts in Yichang, Hubei in the east and extends to Chongqing in the west. Many tributaries converge into this region at both the southern and northern banks of the Yangtze River. Dalou Mountain towers along the southern bank and features steep terrain. On the northern bank is the lower part of Daba Mountain, thus the terrain is relatively gentle. The northern tributaries include large rivers, e.g., the Jialing River, Xiaojiang River, Caotang River, Zhuyi River, Meixi River, Daning River, Longchuan River, and Xiangxi River. On the southern bank are brooks, with the Wujiang River being the only exception. At the eastern end of this region is Wushan Mountain, with the foot of Daba Mountain to the north and Yunnan-Guizhou Plateau to the south. This region is one of the key junctions between the geographical unit in East China that faces toward the ocean and the geographical unit in West China that faces toward the heartland of Asia. The region itself is a relatively separate geographical unit. This region is dominated by hills and mountains and has few plains. It features the unique valley landform that had a profound influence on the subsistence and livelihood of the ancient people in this region. This region is also a strategic passage between the Sichuan Basin and Jianghan Plain, which also greatly influenced the subsistence and livelihood of the ancient people in the area.

采集工作。20世纪30年代至40年代抗日战争时期，一些到此避难的中国学者也做过一些调查、勘测、发掘、搜集等工作。

20世纪50年代中期，因为有了兴建三峡水库的动议，文物考古工作也随之启动。50年代末，中国科学院考古研究所（1977年后隶属中国社会科学院）、长江流域规划办公室考古队及湖北、四川两省文物机构等单位，对重点地段进行调查与发掘，其较重要的发掘有巫山大溪遗址、忠县井沟遗址和西陵峡沿江遗址等。

20世纪70年代，为配合葛洲坝工程的建设，重点对西陵峡地区古代遗址进行考古发掘。这期间还对沿江的洪水、枯水题记进行了调查。20世纪80年代后，随着三峡工程坝址的选定，在坝区范围内对中堡岛、朝天嘴、杨家湾、柳林溪等遗址进行了重点发掘。这些年来，对西陵峡地区古文化遗存发掘得较多，而对隶属重庆市区域内古文化遗存发掘得较少。

总之，在三峡工程最后决策之前，三峡地区的考古发掘工作规模不大，对涉及淹没的文物古迹的了解不够全面和深入。

二、三峡工程文物保护的规划

1.《三峡文物保护规划》的出台

1993年，国家文物局组织全国文物工作者，深入三峡库区开展了调查和试掘，并着手有关基础资料的收集、整理。随后，湖北、四川两省文物部门又在宜昌、万县设立了三峡工程文物保护工作站，具体进行调查、勘测、发掘等文物保护工作。这些前期准备工作为以后科学规划三峡文物保护奠定了坚实基础。

1994年3月，按照国务院三峡建委办公室的要求，国家文物局指定原中国历史博物馆（现中国国家博物馆）和原中国文物研究所（现中国文化遗产研究院）成立"三峡工程库区文物保护规划组"，负责规划的制定工作。规划组先后组织全国30家文物保护研究机构和大专院校的300余位专业人员，对三峡淹没区和迁建区展开大规模调查、勘测、发掘工作，这是中华人民共和国成立以来为配合基本建设工程进行的最大规模的文物调查。

Chapter 9 Cultural Relics and Natural and Cultural Landscapes of the Three Gorges Project

Archaeological surveys of the cultural relics in the Three Gorges region started very early. In the second half of the 19th century, certain foreign missionaries and explorers, such as J.C. Brown and E.C. Baber, discovered a number of stone artifacts in Chongqing. From 1925 to 1926, an American scholar who was part of the Central Asia Expedition named N.C. Nelson investigated 37 stone artifact sites in the Three Gorges region (including 12 sites from which pottery shards were collected). In the 1930s, the American missionary J.H. Edgar also found certain stone artifacts in the Three Gorges region. These were only the surface surveys. During the War of Resistance against Japanese Aggression in the 1930s and 1940s, certain Chinese scholars who took refuge in that region also engaged in investigation, survey, excavation, and collection.

In the mid-1950s, an archaeological survey of the cultural relics commenced after the motion for construction of the Three Gorges Reservoir was proposed. In the late 1950s, the Institute of Archaeology of the Chinese Academy of Sciences (became a subordinate entity of the Chinese Academy of Social Sciences in 1977), the Archaeological Team of the Yangtze Valley Planning Office, and the cultural relics preservation entities of Hubei and Sichuan conducted surveys and excavations at key sites. Important discoveries include the Daxi Site in Wushan Mountain, Jinggou Site in Zhongxian County, and the riverside site in Xiling Gorge.

In the 1970s, in order to support the construction of the Gezhou Dam, archaeological excavation focused on the ancient sites in the Xiling Gorge. At that time, the flood and drought problems along the river were also investigated. Since the siting of TGP dam was finalized in the 1980s, excavation focused on the ancient sites in Zhongbao Island, Chaotianzui, Yangjiawan, and Liulinxi, which were within the dam area. In recent years, archaeologists have unearthed a large number of ancient cultural sites in Xiling Gorge, but fewer in the areas under Chongqing jurisdiction.

In short, archaeological excavation in the Three Gorges region did not happen on a large scale before the final decision on the TGP was reached, and there was no comprehensive or profound understanding of the cultural relics and historic sites to be submerged.

II. Preservation Plan for Cultural Relics in the Three Gorges Project

1. Release of the *Plan for Preservation of Cultural Relics in the Three Gorges*

In 1993, the National Cultural Heritage Administration organized the country's cultural heritage experts to conduct an in-depth investigation and trial excavation in the TGP reservoir

规划组坚持"保护为主、抢救第一"的方针,"重点保护、重点发掘""不改变文物原状(地面文物)""最大限度地抢救,力争把损失减少到最小"等原则,同时集思广益,先后召开了8次专家座谈会,听取意见和建议。

1996年6月,规划组制定完成了《长江三峡工程淹没及迁建区文物古迹保护规划》(以下简称《三峡文物保护规划》),经评估审核和修改完善,并经国务院三峡建委批准后,2000年在三峡库区正式实施。这是迄今为止我国规模最大、涉及范围最广、参与人数最多的文物保护规划。它开拓了一条符合三峡库区文物分布状况的保护模式,并在全国文物保护工作中开创了"先规划,后实施"的文物保护新路子。

2.《三峡文物保护规划》的内容

《三峡文物保护规划》共计32册280万字,包括总报告、分省(市)报告、分县报告、《三峡工程库区地面文物保护规划经费概算细则》和《〈长江三峡工程淹没及迁建区文物古迹保护规划〉有关内容的修订与补充》等。主要内容有:

第一,探明了三峡淹没和迁建区的文物"家底"。三峡水库淹没区和移民迁建区共规划文物保护项目1 087项(湖北库区335项、重庆库区752项),其中需要保护的地面文物364处,需要保护的地下文物723处,比规划前的文物多了近十倍;规划勘探面积3 163万平方米,发掘面积187万平方米。

第二,对淹没和迁建区文物进行了科学的价值评估和保护措施分类。根据各文物和文物点的价值、保护单位级别、社会影响和保存状况等,依据地下、地面文物的特点,制定了不同等级的保护措施。地下文物的保护,根据文物的渊源、年代、类别、状况、规模、地理位置、价值等做了详尽的阐述和科学分类,形成了明晰的文物"清单"。地面文物则根据文物价值、类别、质地、形式、位置和保存状况等,以原地保护、搬迁保护、留取资料等不同保护方式,分别对每一处文物或文物点制定了保护方案。

第三,制定了与工程进度相符的文物保护规划。按照三峡工程建设进度的要求,根据文物所在高程、发掘面积、耗费时间等具体实物指标,制定了与蓄水计划相符的保护规划,包括具体的保护方案和完成时间,各年度的投资计划、人员投入计划等。

area and begin collecting and processing basic data. Later on, the cultural relics preservation entities of Hubei and Sichuan instated TGP-related cultural relics preservation workstations in Yichang and Wanxian for the purpose of cultural relic investigation, survey, and excavation. These early preparations laid a firm foundation for the planned preservation of cultural relics in the Three Gorges.

In March 1994, as requested by the Office of the TGPCC, the National Cultural Heritage Administration instructed the former Museum of Chinese History (now the National Museum of China) and the former National Institute of Cultural Heritage (now the Chinese Academy of Cultural Heritage) to instate the "Three Gorges Project Reservoir Area Cultural Heritage Planning Group" to formulate the plans. The planning group organized over 300 professionals from 30 cultural relics preservation and research entities, universities, colleges across the country to conduct a large-scale investigation, survey, and excavation in the TGP inundation areas and resettlement areas. This was the most extensive cultural heritage investigation undertaken in support of a capital construction project since the founding of the People's Republic of China.

The Planning Group adhered to the policy of "protection is priority, rescue first" and principles such as "focus on protection, focus on excavation of priority sites," "keep cultural relics in pristine state (surface cultural relics)," and "maximize rescue, minimize loss." They also held 8 expert symposiums to listen to opinions and suggestions from experts.

In June 1996, the Planning Group completed the *Plan for Preservation of Cultural Relics and Historic Sites in the Inundation Areas and Resettlement Areas in the Three Gorges Project* (hereinafter referred to as "the *Plan for Preservation of Cultural Relics in the Three Gorges*" or "the Plan"). The Plan was evaluated, audited, revised, and improved. Then it was approved by the TGPCC under the State Council. In 2000, the Plan was officially implemented in the TGP reservoir area. This cultural heritage preservation plan is the largest in scale, scope, and number of participants in China. It initiates a preservation model that is beneficial for the cultural heritage in the TGP reservoir area and opens up the new "planning before implementation" approach for cultural heritage preservation that can be applied to cultural heritage preservation efforts across the country.

2. The contents of the *Plan for Preservation of Cultural Relics in the Three Gorges*

The *Plan for Preservation of Cultural Relics in the Three Gorges* consists of 2.8 million Chinese characters in 32 volumes and encompasses the general report, province (city) specific reports, county specific reports, *Specifications for Budget Estimate in the Plan for Surface Cultural Relics Preservation in the Three Gorges Project Reservoir Area*, and the *Revision*

第四,对重点保护项目制定了专题保护规划。白鹤梁水文题刻、张桓侯庙、石宝寨地处淹没区,在受淹文物中,属最重要的地面文物。它们历史悠久、规模宏大、保存完好,具有特殊的历史、文化、艺术价值,已列为国家级文物保护单位。对这三处文物的保护,规划组极为谨慎,特委托专业性较强的高等院校分别进行了专题性的重点规划。

第五,编制了民族民俗文物保护规划。当时,人们对于民族民俗文物的认识还没有形成文物的概念,更没有形成保护意识。规划组立足于民族民俗文物与其他有形文物具有同等保护意义的发展思路,制定了《三峡民族民俗文物保护规划》,这是我国第一部针对民族民俗文物制定的专题保护规划,具有前瞻性和预见性。如今,民族民俗文物已被冠名为非物质文化遗产,取得了与物质文化遗产相同的保护地位。

第六,制定了博物馆建设的发展规划。规划组从促进三峡文化事业发展的角度,将文物再利用的问题进行了预期规划,制定了以重庆中国三峡博物馆为核心,以发展三峡库区博物馆建设为覆盖的博物馆建设规划。这项规划对三峡文化事业的发展具有巨大的促进和启示作用。

and Supplement for the Plan for Preservation of Cultural Relics and Historic Sites in the Inundation Areas and Resettlement Areas in the Three Gorges Project. The main contents include:

First, the actual cultural relics existing in the inundation areas and resettlement areas in the TGP were identified. The Plan covered 1,087 cultural heritage preservation sites in the reservoir inundation zone and resettlement areas in the TGP reservoir area (335 in the Hubei reservoir area and 752 in the Chongqing reservoir area). These included 364 surface sites and 723 underground sites to be protected. This number was ten times the number of sites identified before the plan. The planned survey area was 31.63 million m^2, and the excavation area was 1.87 million m^2.

Second, a scientific value assessment of the cultural relics in the inundation and resettlement areas was conducted and protection measures were classified. Based on the value, protection classification, social influence, and preservation condition of each cultural relic and heritage site, as well as the characteristics of the underground and surface cultural relics, the protection measures of various levels were formulated. Regarding protection of the underground cultural relics, the history, age, category, condition, scale, geographic location, and value of the cultural relics were comprehensively described and classified, and an exhaustive list of cultural relics was created. As for the surface cultural relics, a specific preservation plan for each cultural relic and heritage site was formulated based on the value, category, quality, form, location, and preservation condition of the cultural relics, and various protection approaches were used, such as in-situ protection, relocation, and database creation.

Third, a cultural relics preservation plan was formulated to synchronize with the project schedule. Based on the schedule of the TGP and the material indexes such as elevation, excavation area, and time cost of the cultural relics, a preservation plan in line with the impoundment plan was formulated. The plan encompassed the actual preservation scheme, completion time, the investment plan, and workforce allocation plan.

Fourth, special preservation plans were formulated for the priority sites. The White Crane Ridge Inscriptions, Zhang Fei Temple, and Shibao Village were located in inundation areas and were the most important surface cultural relics among the to-be-inundated sites. These sites had long histories and large size, were intact, and possessed special historical, cultural, and artistic value. Due to these characteristics, they were classified as official national protected sites. The Planning Group was extremely cautious with these 3 sites, so they commissioned professional colleges to formulate special key plans.

Fifth, a plan for the preservation of ethnic and folk cultural heritage was formulated. At the time of planning, most people did not deem ethnic and folk heritage as "cultural heritage," let alone consider the need to protect it. Recognizing that the ethnic and folk cultural heritage had equal value as the tangible cultural heritage, the Planning Group formulated the

第二节 三峡工程文物保护的成效

1996年至2007年年底,全国20多个省、自治区、直辖市70多个单位的文物专家、科研和考古工作者,以及10多所高等院校的教授和专家学者,共计1 000多人云集三峡库区,投入举世瞩目的文物抢救保护工程。最终,经过他们的艰苦努力,完成了既定目标,既妥善保护了文物,又保障了三峡工程的建设,使三峡文物的损失降到了最低限度。

一、圆满完成了规划目标

1 137处文物得到了有效保护,比规划任务的1 087处多完成了50处,这些超额完成保护的文物点包括:坝区22处,湖北库区6处,重庆库区22处。

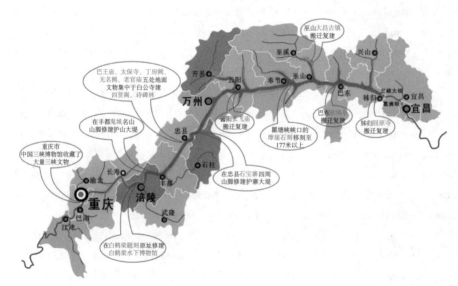

图9.1 三峡库区部分重点文物保护示意图

Plan for Preservation of Ethnic and Folk Culture Heritage. This was China's first special preservation plan formulated specifically for ethnic and folk cultural heritage and was very forward-looking in nature. Now, ethnic and folk cultural heritage has been named "intangible cultural heritage" and is given the same level protection as tangible cultural heritage.

Sixth, a plan for the development of museums was formulated. From the perspective of promoting the cultural undertakings in the Three Gorges region, the Planning Group made a long-term plan to reuse the cultural relics and formulated a museum development plan that is centered on the China Three Gorges Museum in Chongqing and serves the development of the museums in the TGP reservoir area. This plan has been immensely beneficial and enlightening in the development of the cultural undertakings in the Three Gorges region.

Section 2 | Effectiveness of Cultural Heritage Protection in the Three Gorges Project

From 1996 to the end of 2007, over 1,000 personnel, including cultural heritage experts, scientists, and archaeologists from over 70 entities in 20 provinces, autonomous regions, and municipalities, as well as professors and experts from over 10 institutions of higher learning, gathered in the TGP reservoir area and committed themselves to the cultural heritage rescue undertakings that attracted worldwide attention. Thanks to their hard work, they eventually accomplished their objectives. The cultural heritage was properly protected, the TGP was accomplished on schedule, and the loss of cultural heritage in the Three Gorges region was minimized.

I. Satisfactory Realization of Planned Objectives

A total of 1,137 cultural heritage sites were properly protected, 50 sites more than the planned 1,087 sites. The extra protected sites included 22 sites in the dam area, 6 sites in the Hubei reservoir area, and 22 sites in Chongqing reservoir area.

◌ Figure 9.1 Schematic Diagram for Some of the Key Cultural Heritage Sites in the TGP Reservoir Area

在地下文物保护中，772处文物得到了不同方式的保护，出土文物24.7782万件（套）。其中，湖北库区出土文物11.4694万件（套）（含坝区），重庆库区出土文物13.3088万件（套）。在这些出土文物中，一般文物18.5391万件（套）（含标本），较珍贵文物62 391件（套），较珍贵文物占出土文物总量的比例为25.18%，其中，湖北库区22 370件（套），重庆库区40 021件（套）。完成发掘面积177.8512万平方米，其中，湖北库区47.546万平方米，重庆库区130.3052万平方米；完成勘探面积1 219.8408万平方米，其中，湖北库区198.6万平方米，重庆库区1 021.2408万平方米。

在地面文物保护中，365处文物得到了妥善保护。其中，原地保护项目63处，搬迁保护项目132处，留取资料169处，仿古新建1处。这是我国在同一区域、同一时间段内出土文物数量最多，地面文物保护项目最丰富的文物保护工程。

二、初步构建了完整的三峡文化历史序列

在文物保护的具体措施上，根据各文物点的不同特点，有重点地对高家镇、烟墩堡等60余处旧石器遗址进行了发掘，建立了距今十万年以来旧石器文化的年代框架，填补了三峡地区缺少旧石器文化遗存的空白；发现了哨棚嘴、魏家梁子等文化遗存，将瞿塘峡以西地区新石器文化的年代推至距今7 000年以前，填补了渝东地区新石器时代文化的空白；通过对巴人遗址、墓地的发掘，了解了从功能布局到建筑、冶金、盐业、窑业等多方面的历史信息，打开了研究巴人历史文化的神秘之门；通过对大量墓地、遗址的发掘，极大地丰富了自夏商周以来三峡库区的考古发现，排列出了完整的三峡文化序列，为研究三峡地区文化发展、文明进程、环境变迁、社会状况的演变积累了大量实物资料。

Among the underground sites, 772 sites were protected in various ways, and 247,782 pieces (sets) of cultural relics were unearthed. These included 114,694 pieces (sets) of cultural relics unearthed in the Hubei reservoir area (including the dam area) and 133,088 pieces (sets) unearthed in the Chongqing reservoir area. These unearthed cultural relics included 185,391 pieces (sets) of ordinary cultural relics (including specimens) and 62,391 pieces (sets) of precious cultural relics, which account for 25.18% of the total number of unearthed cultural relics and include 22,370 pieces (sets) from the Hubei reservoir area and 40,021 pieces (sets) from the Chongqing reservoir area. The completed excavation area was 1,778,512 m^2, including 475,460 m^2 in the Hubei reservoir area and 1,303,052 m^2 in the Chongqing reservoir area. The surveyed area was 12,198,408 m^2, including 1.986 million m^2 in the Hubei reservoir area and 10,212,408 m^2 in the Chongqing reservoir area.

Among the cultural heritage sites, 365 underground sites were properly protected. These included 63 in-situ protected sites, 132 relocated sites, 169 sites for which a database was created, and 1 site rebuilt based on the ancient form. This cultural heritage preservation project had the largest number of cultural relics unearthed in the same region and same time period in China and had the largest number of protected surface sites.

II. Preliminary Creation of the Three Gorges Cultural and Historical Sequence

Various protection techniques were used based on the characteristics of the cultural heritage sites: over 60 Paleolithic sites in the key locations in Gaojia Town and Yandunbao have been unearthed, and the chronological framework of the Paleolithic culture over the past one hundred thousand years has been established, filling in the gaps in the Paleolithic culture in the Three Gorges region. The cultural heritage sites such as Shaopengzui and Weijialiangzi and the history of the Neolithic culture in the region to the west of Qutang Gorge was dated back to 7,000 years ago, filling in the gaps in the Neolithic culture in eastern Chongqing. Through excavation of the historic sites and tombs of the ancient Ba people, they obtained historic information on many aspects, including the functional layout, architecture, metallurgy, salt industry, and ceramic industry and unveiled the mysteries of the history and culture of the Ba people. Through the excavation of many tombs and historic sites, they added a large number of archaeological discoveries regarding the Xia, Shang, and Zhou dynasties in the TGP reservoir area, established a complete cultural sequence of the Three Gorges region, and collected a large quantity of materials that were useful in the research into the cultural evolution, civilization progress, environmental change, and evolution of social conditions in the Three Gorges region.

三、出土了大量珍贵而有价值的文物

图 9.2 柳林溪遗址石雕人像

在出土的 24 余万件文物中,珍贵文物有 1.3 万件。其中柳林溪遗址距今 7 000～6 000 年的石雕人像,株归朝天嘴遗址距今 7 800～6 900 年的小口罐,中堡岛遗址的石环、玉璜、宋代瓷器,小幺姑沱遗址的六朝时期石羊等都具有极高的史料和研究价值。巴东旧县坪遗址的宋代官府区、居民区、宗庙区和墓葬区在国内第一次全面揭示了宋代县城面貌,被列入 2002 年全国十大考古新发现的名录。

众多出土文物中,有些除具有一定的研究价值外,还具有极高的观赏和艺术价值。例如:奉节出土的新石器时期的磨制钻孔石铲,反映了距今 5 000 年前后峡江地区新石器时期娴熟的石器加工技艺水平。在忠县乌杨镇出土的汉魏时期的乌杨石阙,是目前我国唯一一件通过考古发掘出土的石阙。商代的三羊尊,战国时期的虎钮錞于、蟠螭纹提梁壶等都是在三峡库区出土的古代青铜器中的精品,其铸造工艺已达到了较高的水平。在云阳旧县坪遗址出土的汉代景云石碑是十分珍贵的文物,在存世不多的汉代石碑中,有碑文的非常少见;而汉代景云石碑有清晰隶书碑文达 13 行 367 字,是已发现文字最多、最清晰的汉碑精品之一。

图 9.3 云阳旧县坪遗址出土的汉代景云碑

四、保护了大批历史文化遗产和遗迹

三峡地区是人类活动较早的地区,一些珍贵文化遗存反映了人类在三峡地区活动的方方面面。如东汉时期的石阙和唐、宋、元、明、清时期的摩崖造像、碑碣、诗文题刻等,均是珍贵的文化遗存,具有很高的研究和观赏价值;白鹤梁枯水水文题刻和宋代以来数十处洪水水文题刻,是世界上最丰富

III. Unearthing Large Numbers of Precious Cultural Relics

Over 240,000 cultural relics were unearthed, including 13,000 precious cultural relics. The 6,000–7,000-year-old stone statue at the Liulinxi site, the 6,900–7,800-year-old small-opening pot at the Chaotianzui site in Zigui County, the stone rings, semi-annular jade pendant, and Song dynasty porcelain at the Zhongbao Island site, and the stone sheep from the Six Dynasties at the Xiaomegutuo site are all highly valuable relics for historical research. The Jiuxianping site in Badong County, which encompasses the Song Dynasty government sector, residential area, ancestral temple area, and cemetery area, is the first site in China that comprehensively reflects the features of the county towns of the Song Dynasty and was thus included on the list of the top ten new archaeological discoveries in China in 2002.

⊂ Figure 9.2　Stone Statue at the Liulinxi Site

Some of the numerous unearthed cultural relics possess a certain research value, as well as very high ornamental and artistic value. For example, the ground perforated stone shovel from the Neolithic Age unearthed in Fengjie indicates that stone artifacts were skillfully fabricated in the Xiajiang region about 5,000 years ago. The Wuyang Stone Statue from the Han and Wei dynasties unearthed in Wuyang Town, Zhongxian County is the only stone statue discovered through archaeological excavation in China. The Three-sheep Goblet from the Shang dynasty and the Tiger Percussion Instrument and the Lifting Handle Teapot with Pan and Chi patterns from the Warring States Period are quality bronze ware unearthed in the TGP reservoir area, and they were fabricated by highly advanced casting technology. The Jingyun stone tablet from the Han dynasty unearthed at the Jiuxianping site in Yunyang is a precious artifact because very few surviving stone tablets from the Han dynasty still carry inscriptions. The inscription on the Jingyun stone tablet from the Han dynasty contains 367 Chinese characters of the clerical script style in 13 lines, which makes it one of the Han tablets with the largest number of clearly legible Chinese characters.

⊂ Figure 9.3　Jingyun Stone Tablet from the Han Dynasty Unearthed at the Jiuxianping Site in Yunyang

IV. Preservation of Large Quantities of Historical and Cultural Heritage and Relics

Human activity appeared in the Three Gorges region very early. Some of the precious cultural relics found there indicate many aspects of human activity in the region. For ex-

的古代水文题刻，除具有"水下碑林"和"世界第一水文观测站"之称外，还是研究长江水文变化不可多得的历史资料。目前，对于长江百年、千年一遇枯水最低水位线的划定，仍以该题刻的水位标志为依据；大宁河栈道等数处古代栈道、纤道，是世界上规模最大的古代航运遗迹，是人们研究和了解古代峡江地区航运及交通运输、社会状况不可多得的实物资料；云阳的白兔井等一大批盐井，记录了当地千百年来的盐业发展史；皇华城等城防设施是人们了解当时社会状况及城防建设的实物佐证。

五、重点保护和集中复建了一些文物古迹

白鹤梁水文题刻、张桓侯庙、石宝寨是三峡地区最著名的文物古迹，也是三峡地区最早被国务院确定为全国重点文物保护单位的文物，它们地处淹没区，是重点保护对象。

对于白鹤梁水文题刻的保护采取了兴建"水下博物馆"的方案，该方案以科学的方法将水的压力释放，形成无水压型"水下博物馆"。这是一套在技术和保护理念等方面都领先于世界的保护方案，是目前世界上唯一一座水下博物馆，观众可在数十米的水下，安全和近距离地观赏水中的石刻艺术。

张桓侯庙是典型的寺庙型古建筑，古朴的楼、亭、阁、殿、廊等主体建筑被原样搬迁，搬迁后的建筑体例和格局几乎没有改变，自然环境面貌也基本选择在与原环境面貌相似的地域。

图9.4　整体搬迁后的张桓侯庙

ample, the stone statue from the Eastern Han dynasty and the bas-reliefs on precipices, stone tablets, and poetry inscriptions from the Song, Yuan, Ming and Qing dynasties are all precious cultural relics that possess high research and ornamental value. The dry season inscription at the White Crane Ridge and dozens of flood season inscriptions made since the Song dynasty are the most extensive ancient hydrological inscriptions in the world. They are reputed as an "underground forest of steles" and the "world's first hydrological observation station." They also encompass hard-to-get historical data that indicate the hydrological changes of the Yangtze River. At present, the once-in-a-century and once-in-a-millennium minimum dry season water levels of the Yangtze River are still designated based on the water level markers in these inscriptions. Several plank roads and track roads, including the Daning River Plank Road, are the largest ancient shipping relics in the world and are rare physical materials that help us study and understand the ancient shipping, traffic, and social conditions in the Xiajiang region. A large number of salt wells including the Baitu Well in Yunyang witnessed the local salt industry for thousands of years. The city defense relics at Huanghua City are physical proof of the social conditions and city defense techniques in ancient times.

V. Priority Given to Preservation and Restoration of Some Key Cultural Relics

The White Crane Ridge Inscriptions, Zhang Fei Temple, and Shibao Village are the most famous cultural relics in the Three Gorges region and are the earliest officially protected key sites in that region as designated by the State Council. They were deemed key protection objects because they are located in the inundation area.

The White Crane Ridge Inscriptions have been protected by the solution of building an "underground museum." In this solution, water pressure is released through scientific means to create an "underground museum" without water pressure. This is a very advanced protection approach both technically and conceptually. In this underground museum, which is the only one in the world, visitors can safely view the stone inscription art up close dozens of meters underwater.

The Zhang Fei Temple is a typical ancient temple-style building. The main ancient structures such as the tower, pavilion, loft, hall, and corridor have been relocated, but their architectural style and layout are virtually unchanged. They have been placed in an area with an environment similar to their original environment.

⊂ Figure 9.4 Zhang Fei Temple Relocated in Its Entirety

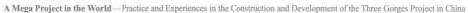

围堤护坡是经众多专家多次论证后确立的石宝寨保护方案，围堤是为了将水拦住，护坡则是为了预防水对山体基岩的冲刷淘蚀。这是一套科学的保护方案，保证了石宝寨的安全。

在秭归、巴东、兴山、巫山、奉节、云阳、万州、丰都、忠县、涪陵等区县，兴建了文物保护复建区，一些富有地方特色的古民居、古建筑、古寺庙等已统一迁入。目前，这些复建区已基本建成，并对社会开放，成为各区县文化凝聚和文化旅游的集结地，正在发挥传承和弘扬国家传统文化的作用。

六、建设了博物馆和文物搬迁复建区

三峡文物保护工程中出土了大量的珍贵文物，为保护、利用这批文物，必然会促进博物馆的建设。早在20世纪90年代编制的规划报告中，即对博物馆建设编制了专项规划。

2005年6月，重庆中国三峡博物馆落成并对外开放。2009年，白鹤梁水下博物馆落成，该博物馆是原址保存的记录千年以来的枯水石刻的博物馆。宜昌博物馆新馆主体工程也已落成。近些年来，陆续建成开放的博物馆还有兴山民俗博物馆、巫山博物馆、夔州博物馆、重庆三峡移民纪念馆、开州博物馆、忠州博物馆、云阳博物馆等，这些博物馆集保存、研究、展示于一体，免费对公众开放，成为当地爱国主义教育基地。

在三峡工程淹没区，一批地面文物将被淹没，包括宗教建筑、石质文物、古民居、古桥梁、古城墙等132项，需要搬迁保护。根据属地管理和"集中复建，统一管理"原则，各区县的搬迁文物均在本区县内选择适当地点，统一规划复建，形成了秭归凤凰山古建筑群、巴东狮子包古建筑群、兴山古夫民居、巫山江东嘴文物复建区、夔州古城文化博览园、云阳青龙古建筑群、丰都小官山古建筑群、忠县文化生态保护区等文物搬迁复建区，成为三峡地区新的文化景点。

A dike and revetment protection plan was developed for Shibao Village after multiple appraisals from experts. The dike keeps the water out while the revetment is meant to protect the mountain bedrock from being scoured and eroded by water. This is a scientific protection approach that guarantees the safety of Shibao Village.

In the districts and counties including Zigui, Badong, Xingshan, Wushan, Fengjie, Yunyang, Wanzhou, Fengdu, Zhongxian, and Fuling, cultural heritage protection and restoration areas were built to contain certain ancient residences, buildings and temples rich in local characteristics. At this time, these restoration areas have been basically completed and opened to the public. They are cultural centers and tourist attractions in these districts and counties and also fulfill the function of preserving and promoting the traditional culture of the country.

VI. Construction of Museums and Cultural Heritage Relocation and Restoration Areas

While conducting cultural heritage protection during the TGP, a large quantity of precious cultural relics was unearthed. The need to protect and utilize such cultural relics inevitably led to the development of museums. The plans and reports prepared in the 1990s contained special plans for museum development.

In June 2005, the China Three Gorges Museum in Chongqing was completed and opened to the public. In 2009, the White Crane Ridge Underwater Museum was completed. This museum was constructed to preserve the dry season inscriptions at their original site. The main works of the new Yichang Museum site have also been completed. Public museums completed in recent years also include the Xingshan Folk Museum, Wushan Museum, Kuizhou Museum, Chongqing Three Gorges Immigration Memorial, Kaizhou Museum, Zhongzhou Museum, and Yunyang Museum. These museums integrate the functions of preservation, research, and exhibition and are open to the public free of charge. They have also become local patriotism education bases.

In the TGP inundation area, a number of surface sites would be submerged, including 132 sites such as religious buildings, stone artifacts, ancient residences, ancient bridges, and ancient city walls that need to be relocated. Under the principles of apanage management and "concentrated restoration and centralized management," the cultural relics of the districts and counties were relocated to proper locations within their jurisdictions under integrated planning for restoration. The completed cultural heritage relocation and restoration areas include the Zigui Fenghuangshan Ancient Architectural Complex, Badong Shizibao Ancient Architectural Complex, Xingshan Gufu Folk House, Wushan Jiangdongzui Cultural Heritage

七、增加了一些国家级文物保护单位

在对三峡文物进行保护的初期,三峡库区仅有白鹤梁水文题刻为国家级文物保护单位,通过对三峡库区文物的保护,以秭归凤凰山文物复建区、云阳张桓侯庙、忠县石宝寨、丁房阙、无铭阙、奉节白帝城等为代表的文物,其价值得到了充分开发和认识,保护级别也得到了提升,被国务院批准为全国重点文物保护单位。这些文物保护单位级别的提升,加大了三峡文化事业发展的优势。

第三节 三峡工程的文化保护规划及成效

三峡地区地处鄂、渝、陕、川、滇、黔、湘六省(市)交界处,是巴文化的重要活动区域,由于该区域是中国古代西南地区与华中地区的重要交通通道和移民通道,各种文化在此交融,形成了独具特色的区域历史文化,积淀了丰富多彩的非物质文化资源。三峡地区非物质文化遗产不仅数量多、类型广,而且历史久、影响大。口传历史、表演艺术、风俗习惯以及工艺竞技等所蕴含的价值取向、思维方式、审美情趣和文化性格,对整个中华文明的形成与发展产生过广泛影响和重大贡献。

Restoration Area, Kuizhou Ancient City Cultural Museum, Yunyang Qinglong Ancient Architectural Complex, Fengdu Xiaoguanshan Ancient Architectural Complex, and the Zhongxian Cultural and Ecological Conservation Area. They are now new cultural attractions in the Three Gorges region.

VII. Addition of Officially Protected National Sites

In the early stages of cultural heritage preservation during the TGP, only the White Crane Ridge Inscriptions in the TGP reservoir area was designated as an officially protected national site. Through the cultural heritage preservation efforts in the TGP reservoir area, the values of the cultural heritage sites represented by the Zigui Fenghuangshan Cultural Heritage Restoration Area, Yunyang Zhang Fei Temple, Zhongxian Shibao Village, Dingfang Palace, Wuming Palace, and Fengjie Baidi City were adequately developed and understood, their protection class was raised, and they were designated by the State Council as officially protected national sites. The raised protection class of these officially protected sites has reinforced the advantage of the TGP cultural undertakings.

Section 3 | Planning for and Effectiveness of Cultural Preservation in the Three Gorges Project

The Three Gorges region is at the junction between Hubei, Chongqing, Shaanxi, Sichuan, Yunnan, Guizhou, and Hunan and is an important region that bore the human activities of the Ba culture. This region used to be an important thoroughfare and immigration passage between southwestern China and central China in ancient times, so various cultures met and integrated in this place, making it a region with a unique historical culture. As a result, ample and diverse non-physical cultural resources have accumulated in the area. The intangible cultural heritage in the Three Gorges region is defined by its large quantity and numerous types, as well as by its long history and great influence. The value orientation, mode of thinking, aesthetic taste, and cultural character embodied in the oral history, performance art, local customs, and skill contests have had an extensive influence on and made major contributions to the creation and development of the Chinese civilization.

一、三峡工程文化保护的规划和成效

按照国家对非物质文化遗产的分类标准,三峡地区国家级非物质文化遗产主要分为两类。

第一类是指各种以非物质形态存在的与群众生活密切相关、各族人民世代相承的传统文化表现形式,包括口头传统(含作为文化载体的语言)、传统表演艺术、民俗活动和礼仪与节庆、有关自然界和宇宙的民间传统知识和实践、传统手工艺技能等以及与之相关的器具、实物、手工制品等。三峡地区符合这类标准的主要有:下堡坪民间故事(宜昌市夷陵区);青林寺谜语(宜都市);兴山民歌(湖北省兴山县);宜昌丝竹(宜昌市夷陵区);枝江民间吹打乐(湖北省枝江市);土家族撒叶尔嗬(湖北省长阳土家族自治县);走马镇民间故事(重庆市九龙坡区);石柱土家啰儿调(重庆市石柱土家族自治县);川江号子(重庆市);南溪号子(重庆市黔江区);木洞山歌(重庆市巴南区);吹打(接龙吹打,重庆市巴南区;金桥吹打,重庆市万盛区);梁平癞子锣鼓(重庆市梁平县);龙舞(重庆市铜梁县);川剧(重庆市);灯戏(梁山灯戏,重庆市梁平县);梁平木版年画(重庆市梁平县);秀山花灯(重庆市秀山土家族苗族自治县)等18项。

第二类是指按照民间传统习惯定期举行的传统文化活动或集中展现传统文化表现形式的场所,如庙会、歌圩、传统节日庆典等,兼具空间性和时间性。三峡地区符合这类标准的主要有:土家族撒叶尔嗬(湖北省长阳土家族自治县)、端午节(屈原故里端午习俗,湖北省秭归县)、龙舞(重庆市铜梁县)、灯戏(梁山灯戏,重庆市梁平县)、秀山花灯(重庆市秀山土家族苗族自治县)等5项。

三峡水库蓄水后,承载着"无形文化遗产"和"活标本"的居民体将要迁移,将要脱离近水环境和整体的凝聚氛围,依附在他们身上的传统文化习俗和由此滋生的相关文物及血脉关系也将被打乱或消亡。在听取多方意见后,规划组委托中央民族大学对淹没区民族民俗文物的保护进行了专题规划,形成了《民族民俗文物保护规划报告》。报告以征集民俗文物、记录迁移前的民间习俗和传统生活习惯及生产状况为主,对古今关联的族群关系,则采用DNA测试的方法,寻找与古人关联的后裔群体。

Chapter 9 Cultural Relics and Natural and Cultural Landscapes of the Three Gorges Project

I. Planning for and Effectiveness of Cultural Preservation in the Three Gorges Project

According to the national standard, the intangible cultural heritage in the Three Gorges region is classified into two types.

The first type encompasses the traditional cultural expressions that exist in intangible form, such as those that are closely related to the livelihood of the people and are passed on from generation to generation by the people of all ethnic groups, including oral tradition (including the languages that are deemed as cultural carriers), traditional performance art, folk activities, etiquette, festivals, traditional folk knowledge and practices related to nature and the cosmos, traditional handicraft skills and tools, material objects, and handiwork related to these elements. The cultural heritage in the Three Gorges region meeting this standard includes: the folktales of Xiabaoping (Yiling District, Yichang); riddles of the Qinglin Temple (Yidu); Xingshan folk songs (Xingshan County, Hubei); traditional stringed and woodwind instruments of Yichang (Yiling District, Yichang); woodwind and percussion music of Zhijiang (Zhijiang, Hubei); the Funeral Dance of the Tujia people (Changyang Tujia Autonomous County, Hubei); folktales of Zouma Town (Jiulongpo District, Chongqing); Tujia Luoer songs of Shizhu (Shizhu Tujia Autonomous County, Chongqing); working songs of Chuanjiang (Chongqing); Nanxi songs (Qianjiang District, Chongqing); Mudong folk songs (Banan District, Chongqing); wind and percussion instruments (Jielong instruments, Banan District, Chongqing); Jinqiao instruments (Wansheng District, Chongqing); Liangping Laizi gongs and drums (Liangping County, Chongqing); Dragon Dance (Tongliang County, Chongqing); Sichuan opera (Chongqing); lantern opera (Liangshan lantern opera, Liangping County, Chongqing); Liangping New Year wood block print (Liangping County, Chongqing); and the Xiushan festive lantern (Xiushan Tujia and Miao Autonomous County, Chongqing). There are a total of 18 cultural heritage items.

The second type encompasses the traditional cultural activities that are regularly organized by traditional customs or the places where traditional cultural expressions are exhibited, e.g. temple fairs, singing events, and traditional festival celebrations. These are both spatial and temporal. The cultural heritage in the Three Gorges region meeting this standard include 5 cultural heritage items, i.e. the Funeral Dance of the Tujia people (Changyang Tujia Autonomous County, Hubei), Dragon Boat Festival (Dragon Boat Festival custom at Qu Yuan's hometown, Zigui County, Hubei), Dragon Dance (Tongliang County, Chongqing), lantern opera (Liangshan lantern opera, Liangping County, Chongqing), and the Xiushan festive lantern (Xiushan Tujia and Miao Autonomous County, Chongqing).

Due to impoundment in the Three Gorges region, the residential bodies possessing the

三峡库区在非物质文化遗产的保护上着力颇多，也取得了很好的成效，其中湖北省宜昌市的做法就非常值得借鉴和推广。2003年湖北省宜昌市在三峡地区内率先建立了非物质文化遗产资源数据库，充分利用现代化手段对非物质文化遗产进行真实、系统和全面的记录，分类建立档案和数据系统。通过普查，全面了解和掌握本地域各民族非物质文化遗产资源的种类、数量、分布状况、生存环境、保护现状。截至2009年三峡工程建设完成，湖北省宜昌市已完成12个大项、70个子项、62个民族民间艺术种类、128名民族民间文化代表、2 470名民族民间艺人的普查建档工作。现已成形的视频数据总量达2.5T，图片数据量达500G，文字数据达2G，各县（市、区）采集数据近1.8T，高标准、高起点地建立"三峡·宜昌民族民间文化资源特色数据库"。

二、高峡平湖的三峡风光依然迷人

1. 三峡之美

古老而神奇的长江三峡是大自然赋予人类的鬼斧神工之作，是天然而成的绝美山水画廊。自2003年6月三峡水库开始蓄水后，人们难免质疑：三峡的美景是否将不复存在？

"intangible cultural heritage" and "living specimens" will be relocated and separated from the near-water environment and the collective atmosphere. The traditional cultural customs carried by these objects and the cultural heritage and inheritance relations will be disrupted or die out. After collecting opinions from many parties, the Planning Group commissioned Minzu University of China to prepare special plans for the preservation of the ethnic and folk cultural heritage in the inundation area and formulate the *Report on Planning for Ethnic and Folk Cultural Heritage Preservation*. The report is focused on collecting folk cultural heritage and recording the pre-relocation folk customs, traditional lifestyle, and economic conditions. A DNA testing method is used to identify the descendants related to the ancient people to form a connection between the modern and ancient ethnic groups.

Intensive efforts were made to preserve the intangible cultural heritage in the TGP reservoir area, and these efforts brought excellent results. The practices used in Yichang, Hubei are worth learning and applying. In 2003, Yichang, Hubei created the first intangible cultural heritage resources database for the Three Gorges region, used modern technology to record the intangible cultural heritage in an authentic, systematic, and comprehensive manner, and created organized archives and data systems. Through general surveys, they comprehensively learned and recorded the types, quantity, distribution, environment and preservation conditions of the intangible cultural heritage resources of the local ethnic groups. By 2009 when the TGP was completed, Yichang, Hubei had completed the general surveying and archiving of 12 preservation programs, 70 preservation sub-programs, 62 ethnic and folk art types, 128 ethnic and folk cultural representatives, and 2,470 ethnic and folk artists. The database now contains 2.5 T of video data, 500G of image data and 2G of textual data. The counties (cities and districts) have collected nearly 1.8 T of data, creating a high-standard, high-quality "Three Gorges Yichang Ethnic and Folk Cultural Resources Specialized Database."

II. The Scenery of the "Great Lake in the Gorge" in the TGP Area is still Charming

1. The beauty of the Three Gorges

The ancient and amazing Three Gorges region is a marvelous creation of nature that features a splendid natural landscape. Since June 2003 when the Three Gorges region started impoundment, people had to ask: will the beautiful landscape of the Three Gorges disappear?

On October 26, 2010, the TGP successfully achieved the objective of a 175 m trial storage level, answering that question explicitly: the original scenery is still attractive and even enhanced by new views. The Three Gorges, even in the state of impoundment, still features a

世界超级工程——中国三峡工程建设开发的实践与经验

A Mega Project in the World—Practice and Experiences in the Construction and Development of the Three Gorges Project in China

2010年10月26日,三峡工程成功实现试验性蓄水175米目标。这个疑团终于有了明晰的答案——原有风光依然迷人,新增景观更添姿色。蓄水后的三峡,仍旧是一幅旖旎壮丽的山水画卷。

三峡水库蓄水至175米后,雄奇险峻的瞿塘峡和幽深秀丽的巫峡水位升高50～60米,水面上升和变宽可能使峡谷感减弱,但峡谷两岸高达1 000～1 500米的山峰并没有明显变矮。人们仍然需要引颈仰视才能一睹"神女"的风采。乘船进入两岸峡谷中,仍然给人以"峰与天关接,舟从地窟行"的感觉。虽然瞿塘峡入口处的"孟良梯""古栈道"等景观已被淹没,但"粉壁墙"上的摩崖题刻"夔门天下雄""瞿塘天下险"等已被原样移刻在177米高程以上,游客仍可尽情观赏。

图9.5 进入两峡峡谷中的游轮(摄影:袁国平)

以逶迤秀美著称的西陵峡分为东、中、西三段,西陵峡的东段和中段的下半段在三峡大坝下游,不受蓄水影响,保持了原汁原味的峡谷风貌。三峡蓄水后,西陵峡西段水位升高近90米,其间的牛肝马肺峡、兵书宝剑峡下部已被淹没,峡谷感有所减弱,但是原来只能远远仰望的牛肝马肺峡,现在和

charming and magnificent natural landscape.

When the storage level of the Three Gorges region reached 175 m, the magnificent and steep Qutang Gorge and the deep and fascinating Wuxia Gorge saw their water levels rise by 50–160 m. The increased water level and surface width may make the gorges look less like gorges, but the 1,000–1,500 m high peaks did not become noticeably shorter. People still have to look up to admire the Goddess Peak. When on a ship sailing through the gorges, you still see the marvelous view of "peaks reaching the skies and boats roaming in caves." The "Mengliang Staircase" and "Ancient Plank Road" at the inlet of Qutang Gorge have been inundated, but the inscriptions "Great Kuimen Gate" and "Precipitous Qutang Gorge" on the precipices have been moved to above the 177 m line, so tourists can still relish these views.

Figure 9.5　Cruise Ship in the Gorges (Photographed by Yuan Guoping)

The Xiling Gorge, renowned for its meandering and beautiful river, is divided into the eastern, middle, and western segments. The lower half of the eastern and middle segments is downstream of the Three Gorges Dam and is therefore unaffected by the impoundment. The gorge landscapes are perfectly preserved. After impoundment of the TGP, the water level in the western segment of Xiling Gorge rose by nearly 90 m. The lower parts of Niugan Mafei Gorge and Bingshu Baojian Gorge have been inundated and look less like gorges, but the Niugan Mafei Gorge that used to be admired from a distance is now closer to tourists. In addition, the "Niugan Mafei" and "Bingshu" (hanging coffins of the Ba people) have been removed and preserved in the Fenghuangshan Park in Zigui County, making them easier for tourists to view.

2. Beautiful great lake in the gorge

In the "long art gallery" of the Three Gorges, the scenery of "great lake in the gorge" created by the Three Gorges Dam is yet another miracle that adds to the marvelous view of the Three Gorges. The backwater from the Three Gorges region flows to Chongqing. The average water depth is over 50 m, and the average width is 1,576 m. The water flow becomes slower, and the river surface becomes wider. Rapids and turbulent water changing to the "great lake in the gorge" is the most obvious change of the Three Gorges landscape. The TGP reservoir area is surrounded by a high concentration of tourist attractions. The natural landscapes are dominated by brooks and deep valleys and the cultural landscapes are mostly represented by the Tujia folk customs and boat tracker culture. Since impoundment, it is very easy to reach these locations by water. Additionally, thanks to the project area core landscapes developed around the dam and the perfectly preserved lower part of Xiling Gorge, the tourism resources of the new Three Gorges have been made more diverse and more attractive.

○ 图 9.6　三峡大坝的平湖景象

游人们更亲近了。同时，"牛肝马肺"和"兵书（即巴人悬棺）"已被取下，存放于秭归县凤凰山公园，更利于游人参观。

2. 高峡平湖之美

在百里三峡画廊中，三峡大坝造就的高峡平湖横空出世，使三峡雄姿另添别样风采。三峡水库回水至重庆，平均水深 50 多米，平均宽度 1 576 米，水流由急变缓，江面由窄变宽，由激流险滩到高峡平湖，是三峡风景最直观的变化。三峡库区周边旅游景点十分集中，自然风光以溪流、幽谷见长，人文景观以土家民俗、纤夫文化最为浓郁。蓄水之后从水路到达这些景点十分便利，加上新开发的以大坝为中心的工程核心景区，以及保持原貌的西陵峡下半段，新三峡的旅游资源更加丰富多彩、引人入胜。

三峡水库蓄水后，原来三峡深处靠纤夫拉纤才能驶入船只的"世外桃源"，因长江支流水域加宽，成为令人惊喜的新景观。巫峡境内的神女溪、鳊鱼溪、禹王河、大溪河沿岸，以险峰碧水著称，蓄水前这些小峡谷滩浅石多，船只难以进入。蓄水后，游客乘游艇即可长驱直入，一览其秀美风光。在重庆巫山县，从前尺余宽的小溪，如今可划着竹筏漂流，使人流连山水之间而独领神韵。

"大河有水小河满"，三峡水位上升，将更多游客带进了小三峡上游大宁河的更深处。大宁河流域有大昌古镇、剪刀峡、荆竹峡、月牙峡、庙峡等风景区，沿途悬崖上、石洞中共安放着 200 多具悬棺。游客不但可以乘船游览，还可寻找古人凿壁悬棺的遗迹，猜度古人运送棺木的技法，感受先人的智慧和神奇的殡葬文化。

3. 大坝之美

目前，三峡坝区已成为三峡旅游中神韵独具、最可观赏的景区之一。登上坛子岭一览，三峡大坝如水上长城横卧江面，气势如虹；往来船舶从三峡船闸中穿梭如织，蔚为壮观。漫步在截流纪念园，游客可以观赏到当年截流施工的遗迹和截流使用过的大型机械，进而想象当年建设者兴修水利、改造

After impoundment in the Three Gorges region, the remote scenery deep inside the Three Gorges that could only be reached by tow boat are now scenic marvels accessible to all people thanks to the widened water surface in the tribu-

Figure 9.6 Great Lake at the Three Gorges Dam

taries of the Yangtze River. The Shennv Creek, Bianyu Creek, Yuwang River, and Daxi River inside Wuxia Gorge are renowned for their steep peaks and green waters. Before impoundment, these small gorges had many obstructive shoals and rocks that denied accessibility by boat. After impoundment, tourists can reach these locations by yacht and admire the amazing scenery along the way. In Wushan County, Chongqing, the creeks that used to be a few inches wide can now be enjoyed by bamboo raft so that you can relish the unique natural scenery.

When big rivers have water, the small ones are filled. The rise in the water level in the Three Gorges allows more tourists to reach further into the Daning River upstream of the Small Three Gorges. The Daning River basin is home to the scenic spots such as Dachang Ancient Town, Jiandao Gorge, Jingzhu Gorge, Yueya Gorge, and Miaoxia Gorge. Over 200 hanging coffins are kept at the cliffs and stone caverns along the river. In addition to sightseeing by boat, tourists can also look at the hanging coffins up close, speculate about the technique used by the ancient people to transport the coffins, and admire the wisdom of the ancient people and the unique funeral and interment culture.

3. Beauty of the dam

At this time, the TGP dam area is one of the most unique and appealing views in the Three Gorges. If you stand on Tanzi Ridge, you can admire the majestic watery scene of the Three Gorges Dam, which looks like a "waterborne Great Wall." The vessels sailing through the TGP ship lock are also quite a sight. In the Interception Memorial Garden, tourists can see the river closure site and the large machinery used to close the river. Then they can imagine the great endeavor of building the water conservancy project and reshaping the natural terrain. When there is a flood discharge during flood season, lucky tourists can see up close how the huge roaring waters gush from the holes and create massive splashes and mist in the river. This is truly a magnificent and breath-taking sight.

On cool, sunny days, traveling by cruise ship on the river, feeling the cool breeze, ad-

山河的宏大场面。倘若在汛期碰巧遇上大坝泄洪，游客可以近距离地观赏到江水如一条条巨龙咆哮着从大坝泄水孔喷涌而出，跃入江面，水花四溅，水雾弥漫，真切地感受那波澜壮阔、雷霆万钧、动人心魄的恢宏气势。

　　风和日丽之际，伴一缕清风登游船而巡江，饱览两岸胜景，感受江山壮美，一直是长江三峡游最令人神往的首选。如今，乘船游览三峡不再拘泥于固有的行驶路线，受制于开停船的时间，游客完全可以按照自己的兴趣和爱好，自由选择游览方式与线路，实现了由"照单上菜"到"自主选择"的个性化旅游方式的转变，游览三峡可选择的余地更加自如，空间更加广阔。

本章小结：

　　三峡文物保护是一项配合三峡水利枢纽工程建设的文物保护工程，开创了"先规划、后实施"的文物保护先河，成功制定了我国第一部系统的文物保护规划——《三峡文物保护规划》，为三峡文物保护工作的顺利进行奠定了基础。经过二十多年的保护历程，数百处文化遗址的发掘，大量珍贵文物的出土，为解决历史问题和学术研究提供了实物资料，长江流域也是中华民族发祥地的观点得到了进一步印证。雄奇壮丽的三峡风光，通过文物保护，使许多带有古代元素的人文景观更加丰富多彩。

参考文献：

［1］本书编委会. 百问三峡［M］. 北京：科学普及出版社，2012.

［2］郝国胜. 三峡文物保护研究［M］. 北京：科学出版社，2018.

［3］国家文物局，国务院三峡工程建设委员会办公室. 三峡文物保护［M］. 北京：科学出版社，2018.

［4］王孔敬. 三峡地区国家级非物质文化遗产保护研究［J］. 重庆三峡学院学报，2011，27(02).

miring the great scenery on both banks, and relishing the magnificent landscapes have always been the preferred experience for tourists in the Three Gorges. Now, sightseeing by boat in the Three Gorges is no longer limited to the established routes or constrained by the departure and mooring schedule. Tourists can freely choose the transport and route to their preferences, so the predetermined tour style has changed into a customized "choose your own" tour style. Tourists are now allowed more options and a larger tour space.

Chapter Summary:

The preservation of cultural relics in the Three Gorges is a cultural heritage preservation undertaking conducted in support of the TGP hydroproject. It has set a precedent for "planning before implementation" for cultural heritage preservation, and the *Plan for Preservation of Cultural Relics in the Three Gorges*, China's first systematic cultural heritage preservation plan was successfully enacted, laying a firm foundation for the preservation of cultural relics in the Three Gorges. Through more than two decades of preservation efforts, hundreds of cultural heritage sites were discovered and a large quantity of precious cultural relics were unearthed. This afforded the physical materials required for resolving historic mysteries and conducting academic research. The viewpoint that the Yangtze River basin is also a birthplace of the Chinese nation has been further substantiated. The marvelous and magnificent scenery in the Three Gorges is supplemented and amplified by the preservation of many cultural landscapes containing ancient elements.

References:

[1] Editorial committee of this book. *Questions about the Three Gorges* [M]. Beijing: Popular Science Press, 2012.

[2] Hao Guosheng. *Study on Preservation of Cultural Relics in the Three Gorges* [M]. Beijing: Science Press, 2018.

[3] National Cultural Heritage Administration, Office of Three Gorges Project Construction Committee under the State Council. *Preservation of Culture Heritage in the Three Gorges* [M]. Beijing: Science Press, 2018.

[4] Wang Kongjing. *Study on Preservation of National Intangible Cultural Heritage in Three Gorges* [J].Journal of Chongqing Three Gorges University, 2011, 27(02).

> 阅读提示：

三峡工程的国际合作贯穿于三峡工程整个生命周期。三峡工程在国际合作中以开放合作、互利共赢的原则，与国际咨询机构、设计单位、设备供货商，以及有影响力的国际行业组织、环保组织等建立了良好的互动合作关系。三峡工程通过国际合作产生的重大成果，奠定了其在全球水电行业中的重要地位，成为行业的标杆，推动了行业的技术进步和理念创新。以三峡工程为代表的中国水电建设能力和技术标准也得到了国际社会的普遍认可。

International cooperation of ran through the whole life cycle of the Three Gorges Project. Based on the principles of opening up, cooperation and mutual benefit, the Three Gorges Project established good cooperative relations and interactions with international advisory organizations, design units, equipment suppliers, influential international trade organizations, and environmental groups working in international cooperation. The major achievements made by the Three Gorges Project through international cooperation laid the foundation for its important position in the global hydropower industry. It has become a model project in the industry, and has promoted technological progress and conceptual innovation in the industry. China's capability and technical standards in the construction of hydropower stations represented by the Three Gorges Project are recognized the world over.

Chapter 10 >>>>

第十章
三峡工程的国际合作和国际影响

International Cooperation and the Global Influence of the Three Gorges Project

第一节 三峡工程中的国际合作

三峡工程的国际交流与合作始于规划、设计、论证时期，并贯穿于工程建设与运行管理的全过程。从孙中山先生提出设想，到中华人民共和国成立后数十年的论证，以及工程建设和运行管理期间，三峡工程的国际合作与交流从未间断。通过引进世界先进的设备、技术、管理，实现一流的工程建设、一流的施工质量、一流的运行管理等目标。通过形式多样的国际合作与交流，一方面汲取了国际大型水电工程建设经验，为解决三峡工程建设中出现的难题提供了有益的借鉴；另一方面也促进了中国水电科技和管理水平的提升。三峡工程以全方位开放的姿态置于全世界的关注之下，带动了全球水利水电事业的发展。

一、三峡工程设计论证期的国际合作

三峡工程从初步构想伊始，便包含进行国际合作的建议。1918年，孙中山先生撰写了《国际共同发展中国实业计划》，第一次提出开发长江水力资源，并希望通过国际合作实现建设三峡的目的。

1944年4月，时任国民政府经济顾问的美籍专家潘绥提出《利用美贷筹建中国水力发电厂与清偿贷款方法》(《潘绥报告》)，建议由美国贷款9亿美元并提供设备，在三峡修建一座装机容量为1 000万千瓦的水电厂和年产500万吨化肥的化肥厂，用向美国出口化肥的办法还债。1944年10月，主持过当时世界最大水利工程——美国田纳西水电站设计工作、时任美国内政部垦务局总工程师萨凡奇博士提出了《扬子江三峡计划初步报告》("萨凡奇计划")，该报告提出以发电为主，兼有防洪、航运、灌溉功能的综合利用方案，被视

Chapter 10 International Cooperation and the Global Influence of the Three Gorges Project

Section 1 | International Cooperation in the Three Gorges Project

International exchange and cooperation started right from the planning, designing and demonstration of the Three Gorges Project, and ran through the whole course of construction and operation management. This continued uninterrupted, from the initial conception put forward by Mr. Sun Yat-sen to the demonstration spanning several decades after the founding of the PRC, and its construction and operation management. The introduction of world-class equipment, technology and management helped China to realize such goals as first-rate engineering, construction quality and operation management. Diversified international cooperation and exchange made the project to draw on the construction experience of foreign large-scale hydropower projects to solve the issues encountered in the construction of such an ambitious project. In addition, they helped China develop its expertise in hydropower technology and management. The Three Gorges Projects drew the attention of the whole world by adopting a wide-ranging approach to opening up. This has driven the water conservancy and hydropower development across the world.

I. International Cooperation at the Design and Demonstration Stage

The preliminary conception of the Three Gorges Project contained the suggestion of international cooperation. In 1918, Mr. Sun Yat-sen wrote *The International Development of China*, which proposed the development of the Yangtze River's the waterpower resources the first time, with the aim of building a hydropower station at the Three Gorges through international cooperation.

In April 1944, the American expert G. R. Paschal, the then economic advisor for the Kuomintang (KMT) government, compiled the *Report on Building a Dam-type Power Plant in China with the U.S. Loan and the Way to Repay the Loan* (also known as the *Paschal Report*). The report advised that the U.S. give China a loan of 900 million USD and provide

世界超级工程——中国三峡工程建设开发的实践与经验
A Mega Project in the World—Practice and Experiences in the Construction and Development of the Three Gorges Project in China

为当时水利工程的一大创举。该报告被送达时任美国总统罗斯福（Franklin Delano Roosevelt），并向新闻界透露，在国际上引起巨大轰动。1946年9月，按设计合同，国民政府遴选技术人员赴美国内政部垦务局参加设计，中方先后有60余名技术人员分别参与水文组（及经济研究）、泥沙研究组、大坝组、电力组、机械组的设计工作。

中华人民共和国成立后，中国政府高度重视长江流域治理和三峡工程建设工作。1955年6月，以马林诺夫斯基为代组长的苏联专家组来到武汉，中苏合作全面开展长江流域规划工作。1955年到1960年期间，应中国政府邀请，先后有126位苏联专家来华协助、指导与规划长江水利建设和三峡工程等项目的技术工作。1984年到1994年期间，美国方面共派来华专家70人次，对三峡工程具体设计和实施中的40多个项目进行了短期咨询，此外还提供了各有关专业大量有参考价值的技术规范、设计手册、指南导则等技术资料。

为了在更广的范围内论证三峡工程的技术可行性和经济、财务合理性，中国政府希望委托资深的国际咨询公司独立编制一份符合国际惯例、能为国际金融机构接受的三峡工程可行性研究报告。中国政府的这一愿望得到了加拿大政府的支持，决定赠款聘请国际咨询公司编制这一报告。两国政府就此于1984年11月和1985年10月两次签署了谅解备忘录。1986年6月，中国

the equipment needed for China to build a hydropower plant with an installed capacity of 10 million kilowatts and a chemical fertilizer factory with an annual output of 5 million tons of fertilizer on the Three Gorges; it was proposed that China repay the debt by exporting fertilizers to the US. In October 1944, Dr. Savage, who had been in charge of the design of the Tennessee Hydropower Station in the U.S., the largest water conservancy project in the world at that time, was the then chief engineer of the Bureau of Reclamation under the U.S. Department of the Interior. He proposed the *Initial Report on the Yangtze River Gorges Plan* (also known as the *Savage Plan*), which put forward a comprehensive harnessing scheme that encompassed flood control, navigation and irrigation. This report was considered an innovation for water conservancy projects at that time. It was sent to Franklin D. Roosevelt, the then U.S. President, and announced to the press, causing a global sensation. In September 1946, the KMT government selected technicians and sent them to the Bureau of Reclamation under the U.S. Department of the Interior to take part in the design according to the design contract. More than 60 technicians from the Chinese side participated in the design of the hydrological group (including undertaking economic research), the silt research group, the dam group, the electric power group and the mechanical group.

After the founding of the People's Republic of China in 1949, the Chinese attached great importance to the management of the Yangtze River basin and the construction of the Three Gorges Project. In June 1955, the Soviet Union expert team, led by the acting group leader, P. I. Malinowski, came to Wuhan. China and the Soviet Union launched the comprehensive cooperative planning of the Yangtze River. From 1955 to 1960, 126 Soviet Union experts came to China to assist, guide and plan the technological work in the construction of water conservancy and projects like the Three Gorges Project on the Yangtze River at the invitation of the Chinese government. From 1984 to 1994, the U.S. sent 70 experts to China, carrying out short-term consultations for the concrete design of the Three Gorges Project and more than 40 other projects under construction. In addition, they provided large amounts of technical data, including technical specifications, design manuals, and instructions and guidelines in relevant specialties, which could be used for future reference.

To demonstrate the Three Gorges Project's technical, financial and economic feasibility in a wide way, the Chinese government hoped to commission a senior international consultation company to compile an independent feasibility study for the Three Gorges Project, in line with international practice and the requirements of international financial institutions. This aspiration gained the support of the Canadian government, which granted the funds to cover the international consultation company compiling such a report. The governments of the two countries signed a memorandum of understanding in November 1984 and October 1985 for this purpose. In June 1986, the Ministry of Water Resources and the Power Industry

水电部和加拿大国际开发署代表两国政府正式聘用加拿大扬子江联营公司负责按国际通用标准与中国平行编制三峡可行性研究报告。加拿大共派出各专业的专家89批541人次，除到现场察看外，还在长办工作达3 630人日。长办、三峡总公司还派出专家18批51人次在加拿大工作约2 970人日。为配合加方专家在中国国内的工作，中方派出450余名专家学者予以协助；为了加方专家能广泛听取各种意见，中方组织国内有关部委、省市、高等院校和县乡政府等200多个单位参加讨论、考察，展现了全方位深层次的合作。CYJV的研究深度和广度相对中国1983年和1985年的报告均有大的提升，对中国国内重编可行性研究报告发挥了促进作用。

二、三峡工程建设期的国际合作

三峡工程是当今世界上在建规模最大的水电工程，所涉及的技术领域之多、所面临的技术难题之多也是世界上首屈一指的。三峡工程建设期的国际合作主要是引进施工设备、机电设备、外国智力，等。

与国际一流施工设备供货商合作，解决施工难题。三峡大坝施工具有混凝土浇筑强度高，金属结构安装量大，工期紧，任务重等特点。三峡主体工程混凝土总量高达2 800多万立方米，特别是在二期施工阶段需要连续实施高强度浇筑，选择何种浇筑手段，从论证设计一直到国家"六五""七五"科技攻关，都将其列为重大技术课题。从1996年开始，三峡总公司面向全球分三次采购"三峡二期工程和厂房混凝土施工主要设备"，来自丹麦、美国、德国、法国及国内的7家设备厂商参与投标。经专家评标，三峡总公司从世界各地选购了当时世界上最先进的浇筑设备，包括美国罗泰克公司生产的4台塔带机、日本三菱公司与法国POTAIN公司联合生产的2台顶带机、德国克虏伯PWH公司生产的2台摆塔式缆机。三峡工程采用以塔（顶）带机为主，以摆塔式缆机和高架门机为辅的混凝土浇筑施工方案，将混凝土水平运输与垂直运输、仓面布料功能融为一体，实现了混凝土从拌和楼到仓面连续、均匀、高强度的工厂化"一条龙"作业，使三峡工程从1999年至2001年连续三年打破了巴西伊泰普水电站创造的混凝土浇筑世界纪录，为三峡二期工程三大目标的全面实现立下了汗马功劳。

of the PRC and the Canadian International Development Agency (CIDA) formally engaged the CIPM Yangtze Joint Venture (CYJV), a consortium led by Canadian International Project Managers Ltd. (CIPM) and sponsored by the CIDA, on behalf of the governments of the two countries to be responsible for compiling this report according to international and Chinese standards. Canada sent 541 experts specialized in different fields in different 89 batches. In addition to on-the-spot inspections, they also worked at the Yangtze River Office for 3,630 man-days. The Yangtze River Office and China Three Gorges Corporation sent 51 experts in 18 batches to Canada and worked 2,970 man-days there. The Chinese side sent over 450 experts to assist Canadian experts in doing their work in China; to help the Canadian experts be better informed, the Chinese side also organized the participation of more than 200 units, including the relevant domestic ministries and commissions, institutions of higher learning, government departments at provincial/municipal/county/town level in discussions and investigations. As a result, the CYJV study was much deeper and wider compared with the reports produced by China in 1983 and 1985, and it facilitated the re-compilation of the feasibility study report in China.

II. International Cooperation During the Construction of the Three Gorges Project

The Three Gorges Project is the largest hydropower project so far in the world. The number of technological fields it involved and complex technical problems it faced were unprecedented. International cooperation during the construction of the Three Gorges Project mainly took the form of the introduction of construction equipment, electromechanical equipment and foreign knowledge.

The difficulties met in the construction were solved through cooperation with the world-class construction equipment suppliers. The construction of the Three Gorges Dam was characterized by high strength concrete placement, large numbers of metal structures, a tight time schedule and a high number of tasks. The total quantity of concrete for the main part of the Three Gorges Project exceeded 28 million cubic meters. Continuous high strength concrete placement was necessary in Phase II. The choice of the concrete placement method was listed as a major technological difficulty from the demonstration and design stage until the programs of technological breakthroughs during the Seventh Five-year Plan and Sixth Five-year Plan periods. Since 1996, China Three Gorges Corporation made three separate international purchases of the main equipment for the concrete construction for Phase II and factory buildings of the Three Gorges Project. Seven equipment manufacturers from Denmark, the US, Germany, France and China tendered for the bid. Based on expert evaluation, China Three

世界超级工程——中国三峡工程建设开发的实践与经验
A Mega Project in the World —Practice and Experiences in the Construction and Development of the Three Gorges Project in China

机电设备国际招标，造全球最大装机机组。三峡水电站总装机32台，单机最大容量70万千瓦，是世界上总装机容量最大的水电站。在三峡工程建设初期，因国内机电设备制造商缺乏制造特大型机组的技术和经验，三峡工程电站永久性设备需要从国外引进。1997年，三峡工程左岸电站14台70万千瓦水轮发电机组面向国际公开招标，世界上所有制造过超大型发电机组的公司都在邀请之列。最后，14台机组分别由两个供货集团中标，中标总价为7.4亿美元。法国阿尔斯通和瑞士ABB公司中标8台机组，德国福伊特、加拿大GE公司和德国西门子组成的VGS联营体中标6台。1999年9月，三峡左岸电站高压电气设备国际采购公开开标。德国西门子公司承担三峡左岸电站15台主变压器供货任务，瑞士ABB公司中标39个间隔550千伏GIS供货合同。除此之外，左岸电站机组调速系统、励磁系统和计算机监控系统均采用了国际招标。这些世界一流设备陆续装备在三峡水电站，全面发挥效益。

引进国外智力，促管理与国际接轨。针对三峡工程建设面临的技术挑战

Gorges Corporation purchased the most advanced concrete placement equipment in the world at that time from different countries, including four tower conveyors made by Rotek in the US, two conveyors jointly made by Mitsubishi in Japan and Potain in France, and two luffing cable-cranes made by Krupp PWH in Germany. The Three Gorges Project adopted a concrete placement construction scheme which mainly relied on tower conveyors, and is supported by luffing cable-cranes and gantry cranes. They integrated the horizontal transportation, vertical transportation and concrete placement, and realized a complete sequence of operation from the concrete batching plant to the placement area, featuring continuous, even and high-intensity operation. The Three Gorges Project broke the concrete placement world record for three consecutive years from 1999 to 2001, breaking that set by Itaipu Dam in Brazil. The equipment made great contributions to the fulfillment of three major goals in Phase II of the Three Gorges Project.

The electromechanical equipment was purchased by way of international competitive bidding, and formed the largest installed generator units in the world. A total of 32 sets of generator units were installed in the Three Gorges Project with the largest capacity of single unit being 700,000 kW, making it the hydropower station with the biggest installed capacity in the world. At the early stage of the construction, because domestic electromechanical equipment manufacturers lacked the technology and experience to make this size generator units, the permanent hydropower station equipment needed to be imported from foreign countries. In 1997, international bids were invited for the 14 water turbogenerator units of 700,000 kW for the left bank hydropower station. All the companies which had made such enormous generator units were invited to bid; in the end, two suppliers won the bidding for the 14 units. The total price of the winning bids was 740 million USD. A collaborative bid between Alstom in France and ABB in Switzerland won the bidding for eight units, and the VGS consortium, consisting of Voith in Germany, GE in Canada and Siemens in Germany, won the bidding for the remaining six units. In September 1999, public bids were invited for the high-voltage electric equipment for the left bank hydropower station of the Three Gorges Project from across the globe. Siemens in Germany supplied 15 main transformers while ABB in Switzerland was awarded the contract for supplying 39 sets of 550 kilovolts gas insulated switchgear (GIS). In addition, the speed governing system, excitation system and computer monitoring system for the generator units of the left bank hydropower station were all purchased by way of international competitive bidding. The above-mentioned world-class equipment was installed at the hydropower station of the Three Gorges Project and generated comprehensive economic benefits.

Foreign expertise was introduced to align the management with international standards. To meet the technological challenges in the construction of the Three Gorges Project and deal

及质量、安全等方面的隐患，三峡总公司先后与美国、挪威、法国、日本、德国等国家的公司签订了二十多个咨询合同，聘请了数十位外国专家前来提供咨询。合同总金额约为550万美元，内容涉及国际招标文件审查、高边坡支护、大型混凝土施工设备安装与维护、混凝土施工质量控制、焊接质量控制以及安全管理等方面。通过参照国际上同类项目的成功经验和公司间的双向交流经验，三峡总公司建立并掌握了广泛的技术合作渠道，在三峡工程的实践中，引进了国际上先进、成熟的技术和管理经验，为攻克三峡工程建设的技术挑战，保证工程建设质量提供了有力的保障。

为了使三峡工程建设与国际惯例接轨，三峡总公司利用加拿大政府EDC贷款，与加拿大MONENCE AGR公司联合开发三峡工程管理信息系统，即前文所提TGPMS。MONENCE AGR公司有着丰富的项目管理经验，该公司的MPMS信息系统能覆盖项目管理的所有环节。三峡总公司在MPMS系统基础上结合中国项目管理的特点，进行了大量的客户化改造，形成具有自主知识产权的三峡工程信息管理系统（TGPMS）。TGPMS系统是对中国工程项目传统管理方式、管理理念的一次革命。

在三峡升船机试通航前，中国2次派员前往德国尼德芬诺、吕内堡等升船机学习、借鉴德方升船机运行维护管理的先进经验，4次邀请德方专家到中国进行考察并开展技术交流。通过交流和互访，升船机管理处对升船机的技术了解更为深入，运行维护管理队伍更为专业，为实现三峡升船机"建管结合、无缝衔接"的目标打下了基础。

with unexpected dangers in quality and security, China Three Gorges Corporation signed more than 20 consultation contracts with firms in the US, Norway, France, Japan and Germany, and engaged several dozen foreign consultants. The total cost of these contracts was 5.5 million USD. This covered the review of international bidding documents, high slope retaining, the installation and maintenance of large-scale concrete construction equipment, quality control for concrete construction, quality control for welding and safety management. Through referring to the successful experiences of similar projects in foreign countries and two-way experience exchange between firms, China Three Gorges Corporation set up and mastered extensive channels for technological cooperation. Through the experience of completing the Three Gorges Project, it introduced advanced and developed global technology and management experience, which provided a strong guarantee for overcoming the technological challenges of the Three Gorges Project and ensuring construction quality.

To align the construction of the Three Gorges Project with international practice, China Three Gorges Corporation utilized the EDC loans given by the Canadian government to jointly develop a management information system with Monenco Agra Inc. in Canada, namely TGPMS outlined in the preceding chapters. Monenco Agra Inc. had rich experience in project management, and its MPMS information system can cover all aspects of project management. China Three Gorges Corporation made a large number of customer-oriented modifications according to the characteristics of China's project management and on the basis of the MPMS system, forming the Three Gorges Project Management System (TGPMS) which was awarded independent intellectual property rights. The TGPMS system is a revolution in the methods and ideas of traditional project management in China.

Before the navigation test run of the dam's ship lifter started, China twice sent personnel to Niederfinow and Luneburg in Germany to learn from the German side's advanced experience in the maintenance and management of ship lifters, and invited the German experts to China for investigation and technological exchange for four times. Through exchange visits, the Ship Lifter Management Office gained a deeper understanding of ship lifter technology and created a more professional operation maintenance and management team, laying a foundation for realizing the goal of creating a seamless connection between construction and the management of the ship lifter.

三、三峡工程运行期的国际合作

三峡总公司先后与法国 EDF 公司、委内瑞拉 EDELCA 公司、加拿大 HYDRO QUBEC 公司、挪威 NordPool 公司、美国内政部垦务局、巴西 ITAIUPU 电站签订技术交流和人员培训协议，先后派出 80 多人赴上述公司进行电厂运行及管理、电力企业运营、财务管理和竞争性电力市场等方面的培训，有助于管理好三峡电厂。参加中加大学工业合作伙伴计划，培养了能源管理研究生 15 人，并请加拿大有关企业专家到公司举办风险管理、财务管理培训。与美国工业研究院联合举办项目管理培训班，与国家外专局合作进行 PMP 培训，与 GE 公司合作进行六西格玛培训。与日本前田公司、韩国水资源公社、马来西亚科学院签定了互换工程师研修协议，派遣了数名工程师前往研修。

2002 年 4 月，三峡电厂生产管理信息系统（EPMS）采购合同签订。三峡电厂从 IFS 公司引进了 EPMS 系统。该系统是发电企业生产管理的信息平台，借助 EPMS 系统，三峡电厂可以实现施工维护、物料、财务管理与成本分析、人力资源、安全等 10 个方面的自动化管理，真正达到无人值守（少人值班）的水平。

三峡升船机通航运行以来，不断攻克运行维护技术难题，通航效益稳步提升，也为德国升船机的建设运行提供了借鉴。随着交流的不断深入，中德双方每年开展往来互动，交流领域从升船机、船闸技术扩展到港口建设、气候变化、环境保护等方面。双方逐渐形成了平等交流、互相学习、共同进步的良好氛围。三峡升船机的工程师们还多次受邀到国际航运协会及英国、澳大利亚等国家和地区开展国际互动交流，分享航运管理、绿色发展、应急搜救等经验，提供世界级的三峡智慧、三峡方案。

III. International Cooperation During the Operation of the Three Gorges Project

China Three Gorges Corporation signed agreements for technological exchanges and personnel training with EDF in France, EDELCA in Venezuela, Hydro-Québec in Canada, NordPool in Norway, the Bureau of Reclamation under the U.S. Department of the Interior, and Itaipu hydropower station in Brazil. It also sent more than 80 people to the above-mentioned organizations for training in the operation and management of the power plant, operation of the electric enterprises, financial management and competitive electric market, which was beneficial to the management of the power plant of the Three Gorges Project. 15 students graduated from postgraduate study in energy management under the Canada-China University-Industry Partnership Program; the experts from relevant Canadian enterprises were also invited to the China Three Gorges Corporation to offer training sessions in risk management and financial management. The Three Gorges Project joined with the Industrial Research Institute in the US to offer project management training courses. It also conducted PMP training in cooperation with the State Administration of Foreign Experts Affairs and conducted Six Sigma training in cooperation with GE. It signed research and study agreements for exchange engineers with Maeda Corporation in Japan, Korea Water Resources Corporation, and the Academy of Sciences in Malaysia, and several engineers were sent to these places for research and study.

In April 2002, the power plant of the Three Gorges Project signed a procurement contract for the enterprise process modeling system (EPMS). It introduced the EPMS system from IFS. This system is an information platform for the production management of power generation enterprises. Relying on the EPMS system, the Three Gorges Project power plant was able to realize automatic management in ten aspects including construction maintenance, supplies, financial management and cost analysis, human resources and safety. Unattended operation (or needing fewer people on duty) has been achieved.

Since the ship lifter was put into operation in the Three Gorges Project, its economic benefits have steadily risen and the technological difficulties in operation maintenance have been overcome one by one. The construction and operation of German ship lifters has also drawn on the experiences of the Three Gorges Project. With the exchange constantly deepening, the China and Germany have continued their annual exchange visits, and have expanded their remit from only ship lifter and lock technology to now include the building of ports, climate change and environmental protection. A good atmosphere for exchange on an equal footing, mutual learning and common progress has been gradually formed. The engineers of the ship lifter of the Three Gorges Project have been invited to the World Association for Waterborne Transport Infrastructure (PIANC) and countries and regions like the UK and

1993年至2009年17年的建设期间，三峡总公司、国家电网公司和长江设计院先后访问了40多个国家，参加100多次国际行业会议、国际学术组织会议等，与世界知名公司开展了广泛交流。一批具有国际视野的有关专业的高端人才得到培养，有的已被选入国际学术组织的领导机构，增加了中国工程科技界在国际上的话语权。三峡总公司还广泛开展与国际行业协会、流域管理机构、国际环保组织的交流，引进水电可持续开发和环境保护的先进理念与实践经验。

第二节 三峡工程的国际影响

一、三峡工程受到了国际社会好评

三峡工程是迄今为止世界上规模最大的水利枢纽工程，2013年被国际咨询工程师联合会（FIDIC）评为"百年重大土木工程项目卓越成就奖"。作为一项造福人类的伟大工程，三峡工程在防洪、发电、航运和水资源综合利用等方面的巨大效益受到国际社会好评。

2010年7月，中国长江流域遭遇特大洪水袭击，国外媒体在报道此次灾害的同时，三峡工程在此次洪灾中的作用，也成为国外媒体关注的焦点。《中国三峡工程报》以"外媒关注三峡防洪作用"为题，对国外媒体的评价进行了综合报道。[①] 美国彭博社相关报道称，"世界上最大的中国三峡大坝遏制了12年来中国最大洪水的水势，缓解了中国中部地区的洪水灾情"；"加拿大在线"称，"三峡工程驯服了长江最近十多年来罕见的洪峰，堪称现代发展的奇迹，还有效产出大量急需的电能、发展航运、减少水害"；新加坡《联合早报》报道称，"三峡工程在今年防洪中的巨大作用，可能会减轻人们对它的指责"。

除了显著的防洪效益，三峡工程的发电、航运和水资源综合利用等效益

---

① 外媒关注三峡防洪作用 [N]. 中国三峡工程报，2010-7-27.

Australia for exchange and discussions. They offer the world-class Three Gorges knowledge and solutions by sharing experiences in navigation management, green development and emergency search and rescue.

During the construction period from 1993 to 2009 (17 years), China Three Gorges Corporation, the State Grid Corporation of China and Changjiang Institute of Survey, Planning, Design and Research (CISPDR) visited more than 40 countries, participated in more than 100 international trade meetings and international academic conferences, and conducted extensive exchanges with world-famous companies. As a result, they cultivated a group of highly qualified experts with a global vision in relevant specialties, some of whom have since been appointed as leaders of international academic organizations, thus enhancing the discourse power of China's engineering and technological industries in the international arena. China Three Gorges Corporation has also conducted extensive exchanges with international industrial associations, river basin authorities and international environmental groups, and has introduced advanced ideas and practical experience in the sustainable development of hydropower and environmental protection.

Section 2　The International Influence of the Three Gorges Project

I. The Three Gorges Project Has Been Well Received by the International Community

The Three Gorges Project is by far the largest water conservancy project in the world. It was given the International Federation of Consulting Engineers (FIDIC) Centenary Award, for in 2013 for its remarkable achievements in civil engineering. As a project of promoting the well-being of mankind, the Three Gorges Project has been well received by the international community for its enormous benefits in flood control, electricity generation, shipping and comprehensive utilization of water resources.

In July 2010, heavy flood waters swept across the Yangtze River basin in China. The role of the Three Gorges Project in this disaster came under the spotlight when foreign media covered the disaster. *China Three Gorges Projects News* offered comprehensive coverage of the comments by foreign media under the headline "Foreign media paying close attention to

世界超级工程——中国三峡工程建设开发的实践与经验
A Mega Project in the World——Practice and Experiences in the Construction and Development of the Three Gorges Project in China

也受到国际社会的高度关注。2009年，三峡工程作为世界上最大的水电站，被世界著名杂志《科学美国人》(Scientific American)评为"世界十大可再生能源工程"(The World's 10 Largest Renewable Energy Projects)。专门讨论严肃新闻话题及刊发观点文章的美国 Policymic 网于 2013 年 7 月 18 日刊文称："三峡水电站发电能力相当于 15 个核电站，是中国解决其能源危机的'绿色'举措的关键部分。"[①]Hydrolink 杂志（Hydrolink Magazine）在 2014 年第 2 期刊登三峡工程专刊（Themed Issue: The Three Gorges Project），从防洪抗旱、航运、发电、移民、环境、技术成就等各方面介绍三峡工程的综合效益，全面展现三峡工程在水资源综合利用上所发挥的重要作用。

为了让国外观众更好地了解和认识三峡工程，一些国外媒体相继来到三峡，拍摄和制作三峡工程纪录片。2009 年，美国国家地理频道 (NGC) 制作了系列纪录片《超大建筑狂想曲·大坝》（Big, Bigger, Biggest: Dam）。本片以三峡大坝为落脚点，利用生动的数字动画，详细展现了古今中外六座最具代表性的大坝所实现的六次重大科技创新，阐释了三峡工程集成了世界最先进的科技创新和工程技术，得以建造成为世界最大的水坝，谱写出新的超大建筑狂想曲。2015 年，英国广播公司 (BBC) 赴宜昌三峡实景拍摄了《奇迹工程：三峡大坝》(Impossible Engineering-Season 2 Episode 3-Three Gorges Dam) 专题纪录片。该片从能源需求与电力供应、工程规模、混凝土温度控制技术创新、双线五级船闸和升船机、滑跃式泄洪道、混流式水轮机、输配电工程等视角，通过古今中外工程类比的方式全面介绍了三峡工程的概况和功能。纪录片中说："长江上的三峡大坝是一个史诗般的工程项目，其规模之大系地球上绝无仅有。""它真的是一个工程奇迹。"

---

① 梁晨，丁廷立. 三峡大坝成世界大工程参照系 [N]. 环球时报，2013-9-2.

the Three Gorges Project's function in flood control". Bloomberg, in the US[①], reported that "China's Three Gorges Dam, as the largest of its kind in the world, contained the momentum of the biggest flood in 12 years, alleviating the flood damage suffered by areas in Central China". Canada Online reported that "the Three Gorges Project tamed flood peaks rarely seen in the last decade. It can be rated as a miracle of modern development since it can produce a large amount of badly needed electrical energy, ship passengers and cargoes and reduce water damage". Furthermore, *Lianhe Zaobao* in Singapore said that "the enormous positive role played by the Three Gorges Project in this year's flood control efforts may lighten people's criticism of it".

In addition to its remarkable function in flood prevention, the Three Gorges Project's benefits include power generation, shipping and multipurpose utilization of water resources, which have also attracted much attention from the international community. In 2009, the Three Gorges Project, as the biggest hydropower station in the world, was chosen as one of the World's 10 Largest Renewable Energy Projects by the world-famous journal *Scientific American*. PolicyMic, a website in the US devoted to the discussion of serious news topics and comment pieces published an article reporting that "The electricity generation capacity of the hydropower station of the Three Gorges Project is equivalent to 15 nuclear power stations, so it is a key part in China's green solution to its energy crisis."[②] Hydrolink Magazine dedicated its second issue of 2014 to The Three Gorges Project, explaining the comprehensive benefits of the Three Gorges Project, including flood control, combating drought, shipping, electricity generation, relocation of residents, environment and technological achievements. It gave a panoramic view of the important functions of the Three Gorges Project in the multipurpose utilization of water resources.

To give foreign audiences a better understanding of the Three Gorges Project, some foreign media came to China to shoot documentary films. In 2009, the National Geographic Channel in the US produced a documentary series *Big, Bigger, Biggest: Dam*. Centered on the Three Gorges Dam, the film explains in detail the major technological innovations realized by six iconic dams using vivid digital animation. It shows that the Three Gorges Project is a combination of the most advanced technology and engineering in the world, which has allowed the building of the largest dam in the world. In 2015, the BBC came to the Three Gorges in Yichang to shoot an episode of their documentary series, *Impossible Engineering –*

---

① Foreign media paying close attention to the Three Gorges Project's function in flood control[N]. China Three Gorges Projects News, July 27, 2010
② Liang Chen and Ding Tingli. The Three Gorges Dam becomes a reference of the major projects in the world. October 2, 2013

2019年7月23日，由《中国能源报》和中国大坝工程学会联合推出的微信公众号"大坝新闻"刊载了国际大坝委员会主席迈克尔·罗杰斯（Michael F. Rogers）撰写的《关于三峡工程安全状况的评述》一文。他在文中指出："我认为，大坝工程师把大坝安全提升到道德准则的高度。大坝安全对于所有人，尤其是大坝工程师，是非常严肃的议题。2011年以来，我先后五次到访中国，认识了许多中国同行，参加了多次学术交流会议，从中了解到，这一点在中国尤为突出。在我的职业生涯中，我密切关注三峡大坝的设计和建设，我可以确信地说，三峡大坝绝对是世界上质量最高、设计和建设最好的大坝之一。通过对诸多工程的了解，包括2012年考察三峡工程，我目睹了中国工程师为确保大坝具有尽可能高的安全度所做的巨大努力，见证了中国大坝工程界对大坝质量和安全的高度重视。以中国对大坝安全的重视程度来说，我对三峡大坝良好的运行维护和安全监测充满了信心。"

二、三峡工程提升了中国水电国际地位

三峡工程技术复杂，难题众多。中国水电建设者通过集成创新，提升了中国水利水电工程建设管理水平，助推了中国水电行业的腾飞。三峡工程的成功建设标志着中国由水电开发大国向水电开发强国转变。以三峡工程为代表的中国水电建设能力和技术标准得到了国际社会的普遍认可。

三峡集团在积极"走出去"的同时，积极推行中国水电的"三峡标准"，引领中国水电走向世界。通过积极参与权威国际行业组织的活动，参加行业文件的制定，积极与大自然保护协会（TNC）、世界自然基金会（WWF）等国际知名环保组织合作，在借鉴国际先进理念和经验的同时，也使中国水电建设的实践和经验更多地为国际社会所了解。

三峡集团积极参与了国际水电协会（IHA）组织的行业指导性文件《水电可持续性评估规范》中文版的修订及在中国水电项目上应用的推广工作；作为IHA核心成员，为IHA战略发展、工作方向、热点问题研讨等出谋划策，并以IHA为平台在国际上介绍中国水电开发相关政策、规划与进展。同时，还参与UNESCO/IHA的水库温室气体课题研究，为GHG项目的开展及国内外相关工作接轨提供支持。

titled 'The Three Gorges Dam' (season 2, episode 2). This episode explains the overview and functions of the Three Gorges Project through a comparison with other projects throughout history, from different angles including energy demand and the supply of electric power, the scale of the project, technological innovation in concrete temperature control, the double-way five-level ship lock and ship lifter, the ski-jump type flood spillway, the Francis hydro turbine, and power transmission and distribution. The documentary film concludes that "The Three Gorges Dam on the Yangtze River is an epic-like project, and it is second to none on the earth in terms of scale" and "It really is a miracle."

On July 23, 2019, the WeChat official account "Dam News", representing both China Energy News and the Chinese National Committee on Large Dams, posted an article titled "Commentary on the Safety of the Three Gorges Project," written by Michael F. Rogers, president of the International Commission on Large Dams. In it, he points out that "In my opinion, the dam engineers have raised the issue of dam safety to the level of ethical criteria. Dam safety is a very serious topic for all, especially the dam engineers. Since 2011, I have paid five visits to China, got to know a lot of Chinese peers and participated in several academic exchange meetings. To the best of my knowledge, safety is a particularly important issue in China. In my career, I have paid close attention to the design and construction of the Three Gorges Dam. I can say with assurance that the Three Gorges Dam is definitely one of the world's dams that has the highest quality with the best design and construction. Based on my understanding of many projects, including the field visit to the Three Gorges Project in 2012, I witnessed the enormous efforts the Chinese engineers made to ensure the dam is as safe as possible and bore witness to the great importance China's dam engineering circle attached to the quality and safety of the dam. In terms of the priority China gives to the safety of the dam, I am full of confidence in the operation maintenance and safety monitoring of the Three Gorges Dam."

II. The Three Gorges Project Has Promoted China's International Status in the Hydropower Sector

The Three Gorges Project features complicated technology and numerous difficulties. China's hydropower generation builders have promoted its level in the construction and management of hydropower projects and facilitated the takeoff of China's hydropower sector through integrated innovation. The success of the Three Gorges Project marks China's shift towards becoming a strong power in the development of hydropower in this field. The building ability and technological standards in hydropower stations represented by the Three Gorges Project are widely recognized by the international community.

2012年，三峡集团主导成立国际大坝委员会（ICOLD）"水电站与水库调度联合调度专委会"。ICOLD是国际相关专业领域最高的权威机构，该专委会是首次由中国主导并成立的专委会，是中国水利水电界提升国际水电及能源领域影响力的重要平台。三峡集团负责该组织的运作，专委会陆续吸收了巴西、加拿大、法国、德国等14国的委员，汇聚全球主要国家相关领域的专业人士开展交流与研究。2018年，ICOLD成立水库移民专委会，三峡集团担任主席单位。至此，由中国主导的三个专委会中，有两个由三峡集团牵头。经过两年来的发展，水库移民专委会已有来自英国、德国、巴西、哥伦比亚、墨西哥、日本、韩国、印度尼西亚、尼日利亚、阿根廷、南非、巴基斯坦等国的16名委员，在三峡集团的牵头下，分享水库移民经验、交流实际工作中的难点，推动水库移民工作健康发展。2019年，经ICOLD执委会同意，三峡集团专家新增为公众认知与教育专委会委员，以专委会为平台，宣介中国在大坝公众认知方面取得的工作成效。

In seeking to go global, China Three Gorges Corporation is increasing efforts to promote the Three Gorges standards for China's hydropower development and introduce Chinese practice in this field to the world. China Three Gorges Corporation actively participates in the activities held by authoritative international industry organizations and in the formulation of industry documents. It also actively cooperates with well-known international environmental groups, including The Nature Conservancy (TNC) and the World Wildlife Fund (WWF). These pursuits enable the practices and experiences gained from the development of China's hydropower to be better understood, while China Three Gorges Corporation also draws on the advanced ideas and experience of foreign countries.

China Three Gorges Corporation has actively participated in the revision of the Chinese edition of the *Hydropower Sustainability Assessment Protocol*, an industry guidance document organized by the International Hydropower Association (IHA), and the promotion of its application in China's hydropower projects. As a core member of IHA, China Three Gorges Corporation gives advice and suggestions about strategic development, work direction and key topics, and explains relevant policies, planning and progress in the development of China's hydropower sector from the platform of the IHA. At the same time, it participates in the research of the subject of reservoir greenhouse gas initiated by UNESCO/IHA, providing support for the development of the greenhouse gas GHG project and for the connection of relevant work at home and abroad.

In 2012, China Three Gorges Corporation led the establishment of the Hydropower Station and Reservoir Joint Operation Special Committee under the International Commission on Large Dams (ICOLD). ICOLD is the highest authority organization in the world in its field. It is a professional committee led and established by China, that also serves as an important platform for China's water conservancy and hydropower industry to enhance its global influence in hydropower and energy. China Three Gorges Corporation is responsible for the operation of this organization, and it has welcomed 14 countries as members including Brazil, Canada, France and Germany. Its aim is to gather professionals from relevant fields of the world's most important countries to conduct exchanges and research. In 2018, ICOLD set up the Reservoir Relocation Special Committee, with China Three Gorges Corporation as the chairperson unit. So far, China Three Gorges Corporation has led two of the three special committees run by China. Through the development over the past two years, the Reservoir Relocation Special Committee has 16 members including the UK, Germany, Brazil, Colombia, Mexico, Japan, South Korea, Indonesia, Nigeria, Argentina, South Africa and Pakistan. Led by China Three Gorges Corporation, the special committee shares experience in reservoir relocation, and exchange suggestions and solutions for the difficulties met in this area, driving the healthy development of reservoir relocation. In 2019, an expert from China Three

三峡集团与国际上有影响力的环保组织如大自然保护协会（TNC）、世界自然基金会（WWF）等积极开展交流合作。2008年，三峡集团与TNC签署《关于开展金沙江下游生态系统保护项目的合作备忘录》，并按计划开展了一系列金沙江下游淡水生态系统保护工作，对于改善"长江上游珍稀、特有鱼类国家级自然保护区"内多种珍稀特有鱼类栖息地环境、保护水生生态多样性产生了积极影响。2013年8月，双方在美国华盛顿续签合作备忘录，共同开展兼顾坝下生态需求的大坝生态调度研究和改善流域关键淡水生物及其栖息地现状的专题研究，双方实践"可持续开发理念"的合作开始国际化，将致力于促进当地生物多样性和生态系统保护，建立保护性开发理念示范项目。2018年10月，与大自然保护协会续签2018年至2023年合作备忘录，在长江大保护、流域生态调度、长江淡水生态系统、海外项目环境与社会方面的可持续发展、清洁能源开发、典型区域生态修复、学术研究与宣传等七个领域重点开展合作。同时，三峡集团也与WWF积极开展生态放流、中华鲟放流活动、"地球一小时"等方面的合作。

此外，三峡集团广泛开展与联合国发展署（UNDP）、联合国环境规划署（UNEP）、联合国教科文组织（UNESCO）、国际能源署（IEA），以及国际标准电工委员会（IEC/TC）等机构的交流合作。另外，以三峡工程为起点，与德国企业开展的合作，使三峡集团成为中德经济顾问委员会的理事单位，中国水电在中德经贸合作中的地位得到切实提升。

Gorges Corporation was accepted as a new member of the Public Awareness and Education Special Committee, approved by the executive committee of ICOLD. With the help of this platform, this expert has the opportunity to publicize the progress made by China in the public awareness of large dams.

China Three Gorges Corporation is in active exchanges and cooperation with influential international environmental groups, including TNC and the WWF. In 2008, it signed the *Cooperation Memorandum Regarding Launching the Ecosystem Protection Project in the Lower Reaches of Jinsha River* with TNC, and adopted a series of protective measures for the freshwater ecosystem in the lower reaches of Jinsha River according to this plan. This has produced a positive effect for habitat improvement for many rare fish in the national-level nature reserve for the rare fish in the upper reaches of the Yangtze River. In August 2013, both sides renewed the cooperation memorandum in Washington for making joint efforts to carry out dam ecological operation research, focusing on the dam's downstream ecology and the improvement of key freshwater wildlife and habitat in the basin. The cooperation between both sides, practicing the philosophy of sustainable development, is going global, and the participants will continue to be committed to promoting local biodiversity, bolstering the protection of the ecosystem, and setting up demonstration projects embracing the philosophy of protective development. In October 2018, China Three Gorges Corporation renewed the cooperation memorandum with TNC until 2023, focusing on cooperation in seven fields: comprehensive protection of the Yangtze River, basin ecological operation, the Yangtze River's freshwater, overseas projects' sustainable development in the environment and society, the development of clean energy, the ecological restoration of key regions, and academic research and publicity. Furthermore, China Three Gorges Corporation is actively cooperating with the WWF in ecological release, the specific release of Chinese sturgeon, and Earth Hour.

In addition, China Three Gorges Corporation carries out extensive exchanges and cooperation with organizations such as the United Nations Development Programme (UNDP), the United Nations Environment Programme (UNEP), the United Nations Educational, Scientific and Cultural Organization (UNESCO), the International Energy Administration (IEA) and the International Electrotechnical Commission (IEC/TC). In addition, the cooperation with German enterprises starting from the Three Gorges Project makes China Three Gorges Corporation has become a director unit of the China-Germany Economic Advisory Committee. The status of China's hydropower sector in China-Germany cooperation, in terms of both economy and trade, has been raised substantially.

三、三峡工程形成了"三峡品牌"国家名片

三峡工程建设催生了富有国际竞争力的中国水电产业。三峡集团充分利用三峡品牌的全球影响力和在水电行业的领先优势,努力开拓海外清洁能源市场,大力开展海外投资和国际承包业务,促进国际产能合作。

目前,三峡集团在40多个国家和地区开展清洁能源项目投资建设和国际工程承包。建设、运行和管理了老挝南立1-2水电站(10万千瓦)、南椰2水电站(18万千瓦),尼泊尔上马蒂水电站(2.5万千瓦),巴基斯坦卡洛特水电站(72万千瓦),巴西朱比亚(155万千瓦)、伊利亚(344万千瓦)水电站,秘鲁圣加旺Ⅲ水电站(20.9万千瓦),几内亚凯乐塔水电站(23.5万千瓦)、苏阿皮蒂水电站(45万千瓦)等一大批对当地国计民生具有重要意义的水电站,充分体现了三峡集团用清洁能源服务人类社会发展、建设美丽家园的发展理念。

2012年,三峡集团成为葡萄牙电力公司的第一大股东;2015年,三峡集团卡洛特项目写入中国与巴基斯坦政府间联合声明,这是中国水电行业的第一次;2015年,几内亚国家货币20 000法郎面额钞票的背景图案,正是三峡集团中水电对外公司承建的凯乐塔水利枢纽工程整体形象;2016年1月,三峡集团获得了巴西伊利亚、朱比亚两座水电站特许经营权,一跃成为巴西第二大私营发电企业;2020年,三峡集团完成秘鲁第一大配电公司路德斯公司股权交割,是当年中资企业最大的海外并购,近三年全球最大的电力资产并购。以三峡集团为代表的中国水电企业已成为世界清洁能源市场最重要的力量之一。

III. The Three Gorges Brand Has Become an Ambassador for China

The building of the Three Gorges Project expedited the emergence of China's hydropower industry and its international competitiveness. China Three Gorges Corporation makes the most of the project's global influence as a brand and its cutting-edge position in the hydropower industry to open up the overseas clean energy market. It is increasing efforts to make overseas investment and develop international project contracting business to boost the capacity of cooperations in international production.

At present, China Three Gorges Corporation is investing in the construction of clean energy projects and contracting to build international projects in more than 40 countries and regions. It is responsible for the building, operation and management of a large number of hydropower stations which are of great significance for the national economy and people's livelihoods, including the Nam Lik 1-2 Project (100,000 kW) and Nam Ngiep 2 Hydropower Project in Laos, Upper Madi Hydropower Station (25,000 kW) in Nepal, Karot Hydropower Project (720,000 kW) in Pakistan, Jupiá Hydropower Station (1,550,000 kW) and Ilha Hydropower Station (3,440,000 kW) in Brazil, San Gabán III Hydropower Station (209,000 kW) in Peru, and Kaleta Hydropower Station (235,000 kW) and Suapiti Hydropower Station (450,000 kW) in Guinea. This has fully demonstrated the development philosophy of using clean energy to serve the development of human society and building a beautiful society that China Three Gorges Corporation embraces.

In 2012, China Three Gorges Corporation became the biggest shareholder of Portuguese electric utilities company Energias de Portugal (EDP) and in 2015, the corporation's Karot Project was written into the intergovernmental joint statement of China and Pakistan, a first for China's hydropower sector. In the same year the background pattern of the banknote of Guinea's 20,000 francs banknote was changed to an image of the Kaleta Hydropower Station, which was built by the China International Water and Electric Corporation (CWE), a subsidiary of China Three Gorges Corporation. In January 2016, China Three Gorges Corporation obtained the franchise for the two hydropower stations in Brazil, Jupia and Ilha, making it the second largest private power generation firm in Brazil. Then, in 2020, China Three Gorges Corporation closed the acquisition of equity interests in Luz Del Sur (LDS), the biggest power distribution company in Peru, representing the biggest overseas merger of any Chinese enterprise in that year and the biggest merger of electricity assets in the world in the past three years. Chinese hydropower enterprises represented by China Three Gorges Corporation have already become one of the most important players in the international clean energy market.

本章小结：

三峡工程辉煌灿烂的建设历程是一部开放包容、卓有成效的国际交流合作史，是一部凝聚人类共同智慧、推进全球水电事业发展的奋斗史。20多年来，国际交流合作始终伴随着三峡工程，不仅为工程建设提供了有效的技术、信息和管理支持，宣传展示了中国水电产业突飞猛进的发展成就，也为国际水电同行贡献了三峡经验、三峡智慧和三峡标准。以三峡工程为起点，中国水电行业更广泛地参与到世界水电技术创新和全球能源治理体系中，以非凡的发展成就和强劲的实力引领着世界水电产业的发展，用清洁能源更好造福人类美好未来。

参考文献：

[1] 本书编委会. 百问三峡 [M]. 北京：科学普及出版社，2012.

[2]《中国三峡建设年鉴》编纂委员会. 中国三峡建设年鉴（2018）[J]. 宜昌：中国三峡建设年鉴社，2019.

[3]《中国三峡建设年鉴》编纂委员会. 中国三峡建设年鉴（2020）[J]. 宜昌：中国三峡建设年鉴社，2020.

Chapter 10 International Cooperation and the Global Influence of the Three Gorges Project

Chapter Summary:

The successful construction of the Three Gorges Project is a historic example of international exchange and cooperation, characterized by openness, inclusivity and effectiveness. It is also an example of sharing intelligence internationally and pushing forward the development of the hydropower cause across the globe. Over the past 20 years, the Three Gorges Project has been supported by international exchanges and cooperation. These have not only provided effective support in technology, information and management for the construction of the project but have also demonstrated the amazing achievements made by China's hydropower industry, but also contributed the Three Gorges experiences, wisdom and standards to the foreign peers in this field. Starting from the Three Gorges Project, China's hydropower industry began to participate in the worldwide technological innovation in hydropower and the global system of energy governance. China is leading the development of the global hydropower industry depending on its remarkable achievements and powerful strength. Clean energy will help mankind create a bright future.

References:

[1] Editorial Board. *One Hundred Questions Regarding the Three Gorges Project* [M]. Beijing: Popular Science Press, 2012

[2] Editorial Board of China Three Gorges Construction Yearbook. *China Three Gorges Construction Yearbook (2018)* [J] Yichang: China Three Gorges Construction Yearbook Press, 2019.

[3] Editorial Board of China Three Gorges Construction Yearbook. *China Three Gorges Construction Yearbook (2020)* [J] Yichang: China Three Gorges Construction Yearbook Press, 2020.

图书在版编目（CIP）数据

世界超级工程：中国三峡工程建设开发的实践与经验 = A Mega Project in the World— Practice and Experiences in the Construction and Development of the Three Gorges Project in China：汉文、英文 / 本书编委会编著 . —北京：中国三峡出版社，2023.1

ISBN 978–7–80223–795–7

Ⅰ.①世… Ⅱ.①本… Ⅲ.①三峡水利工程—汉、英 Ⅳ.① TV632

中国版本图书馆 CIP 数据核字（2022）第 203461 号

责任编辑：任景辉

中国三峡出版社出版发行
（北京市通州区新华北街156号　101100）
电话：（010）57082645　57082577
http://media.ctg.com.cn

天津画中画印刷有限公司印刷　新华书店经销
2023 年 1 月第 1 版　2023 年 1 月第 1 次印刷
开本：787 毫米 ×1092 毫米　1/16　印张：24.25
字数：385千字
ISBN 978-7-80223-795-7　定价：160.00元